# Advanced Engineering Mathematics with MATLAB

# Advanced Engineering Mathematics with MATLAB

Contributors
_____

**Viliam Fedák, Tibor Balogh et al.**

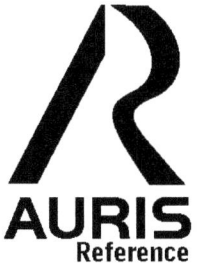

www.aurisreference.com

# Advanced Engineering Mathematics with MATLAB

Contributors: Viliam Fedák, Tibor Balogh et al.

**Published by Auris Reference Limited**
**www.aurisreference.com**

United Kingdom

**Advanced Engineering Mathematics with MATLAB**

ISBN: 978-1-78154-822-6

British Library Cataloguing in Publication Data
A CIP record for this book is available from the British Library

Printed in the United Kingdom

Exclusively distributed by CBS Publishers & Distributors Pvt. Ltd.

Sales & Distribution Rights only for India, Pakistan, Bangladesh, Sri Lanka, Nepal and Bhutan.This book is not to be sold outside these territories.

# Contents

# List of Abbreviations

| | |
|---|---|
| ANNs | Artificial neural networks |
| BLDC ` | Brushless Direct Current |
| BOC | Beginning of cycle |
| BP | Back-propagation |
| C/A | Course/Acquisition |
| CBOC | Composite Binary-Offset Carrier |
| CGL | Chebyshev-Gauss-Lobatto |
| DCMs | Direction cosine matrices |
| DOE | Design of experiments |
| DQM | Differential quadrature method |
| EOC | End of Cycle |
| EOM | Equations of motion |
| FDM | Finite difference method |
| FEM | Finite element method |
| FGM | Functionally graded material |
| FHCE | Fourier heat conduction equation |
| FMNF | First-mode natural frequency |
| GA | Genetic algorithm |
| GUI | Graphic User Interface |
| IN/GB | Interval-Newton Generalized-Bisection |
| INS | Inertial navigation system |
| MCR | Matlab Compiler Runtime |
| OFDM | Orthogonal Frequency Division Multiplexing |
| PSD | Power Spectral Density |
| RK4 | Runge-Kutta fourth order method |
| ROC | Receiver Operation Curves |
| SDP | Sequential Decoding Process |
| SINS | Strapdown inertial navigation system |
| SOC | Symmetrical Optimum Criterion |
| SOR | Successive over-relaxation |
| SRK | Soave-Redlich-Kwong |
| SSC | Spectral Separation Coefficient |
| VSI | Voltage-Source Inverter |

# List of Contributors

**Fedák Viliam**
Technical University of Košice, Slovakia

**Tibor Balogh[2] and**
Magneti Marelli, Electronic Systems Division, Industrial Park Kechnec, Slovakia

**Pavel Záskalický**
Technical University of Košice, Slovakia

**Avinash Ramsaroop**
Department of Mechanical Engineering, Durban University of Technology, Durban, South Africa

**Krishnan Kanny**
Department of Mechanical Engineering, Durban University of Technology, Durban, South Africa

**Daniele Borio**
EC Joint Research Centre, Institute for the Protection and Security of the Citizen, Italy

**Eduardo Cano**
EC Joint Research Centre, Institute for the Protection and Security of the Citizen, Italy

**Ke Wang**
School of Mechanical Engineering, University of Yangzhou, Yangzhou, China

**Jiping Zhou**
School of Mechanical Engineering, University of Yangzhou, Yangzhou, China

**Hasan Ozturk**
Department of Mechanical Engineering, Dokuz Eylul University 35397, Buca, Izmir, Turkey, hasan.

**Zeki Kiral**
Department of Mechanical Engineering, Dokuz Eylul University 35397, Buca, Izmir, Turkey, hasan.

**Binnur Goren Kiral**
Department of Mechanical Engineering, Dokuz Eylul University 35397, Buca, Izmir, Turkey, hasan.

**Mohamed Salah**
Department of Engineering Mathematics and Physics, Faculty of Engineering, Zagazig University, Zagazig, Egypt

**R. M. Amer**
Department of Engineering Mathematics and Physics, Faculty of Engineering, Zagazig University, Zagazig, Egypt

**M. S. Matbuly**
Department of Engineering Mathematics and Physics, Faculty of Engineering, Zagazig University, Zagazig, Egypt

**Monika Žecová**
Institute of Control and Informatization of Production Processes, Faculty of BERG, Technical University of Košice, 042 00 Košice, Slovakia

**Ján Terpák**
Institute of Control and Informatization of Production Processes, Faculty of BERG, Technical University of Košice, 042 00 Košice, Slovakia

**Chi-Wei Lin**
Department of Industrial Engineering and Systems Management, Feng Chia University, Taichung, Taiwan

**Mohamed Salah**
Department of Engineering Mathematics and Physics, Faculty of Engineering, Zagazig University, Zagazig, Egypt

**R. M. Amer**
Department of Engineering Mathematics and Physics, Faculty of Engineering, Zagazig University, Zagazig, Egypt

**M. S. Matbuly**
Department of Engineering Mathematics and Physics, Faculty of Engineering, Zagazig University, Zagazig, Egypt

**Jianbin Hao**
School of Geology Engineering and Geomatics, Chang'an University, Xi'an, Shaanxi 710054, China

**Banqiao Wang**
School of Geology Engineering and Geomatics, Chang'an University, Xi'an, Shaanxi 710054, China

**Wen Zhang**
College of Opto-Electronic Science and Technology, National University of Defense Technology, Changsha 410073, China

**Mounir Ghogho**
School of Electronic and Electrical Engineering, University of Leeds, Leeds LS2 9JT, UK
International University of Rabat, Rabat 11 100, Morocco

**Baolun Yuan**
College of Opto-Electronic Science and Technology, National University of Defense Technology, Changsha 410073, China

**Benito A. Stradi-Granados**
Department of Materials Science and Engineering, Institute of Technology of Costa Rica, Cartago 07050, Costa Rica

# Preface

Engineering mathematics is a branch of applied mathematics concerning mathematical methods and techniques that are typically used in engineering and industry. The text *Advanced Engineering Mathematics with MATLAB* integrates technology into the conventional topics of engineering mathematics. It employs MATLAB to reinforce concepts and solve problems that require heavy computation. The aim of first chapter is to present methodology and results in development of unified series of virtual models for electrical machines and drive systems using MATLAB GUI. Second chapter deals with the generation of MATLAB script files that assists the user in the design of a composite laminate to operate within safe conditions. The goal of third chapter is to provide a general overview of semi-analytic techniques for the simulation of communications systems. Fourth chapter carries out a kinematical analysis of a high speed plate carrying manipulator which is designed by using the module function of Robotics Toolbox in the MATLAB. Fifth chapter deals with the dynamic behavior of a cracked beam subjected to a concentrated force traveling at a constant velocity. In sixth chapter, two different numerical schemes, namely the Runge-Kutta fourth order method and the implicit Euler method with perturbation method of the second degree, are applied to solve the nonlinear thermal wave in one and two dimensions using the differential quadrature method. Seventh chapter deals with the fractional heat conduction models and their use for determining thermal diffusivity. The purpose of eighth chapter is to demonstrate the application of genetic algorithm (GA) on determining the optimal locations of bearings of motorized spindle shafts. In ninth chapter, an efficient technique of differential quadrature method and perturbation method is employed to analyze reaction-diffusion problems. In tenth chapter, combining artificial neural networks (ANNs) with the orthogonal trial design, we perform the sensitivity analysis of factors affecting the maximum displacement of excavation side. The objectives of eleventh chapter are to establish a mathematical model and to develop a comprehensive MATLAB implementation for strapdown inertial navigation system (SINS). Last chapter presents the solution of nonlinear problems using interval arithmetic in INTLAB by using the interval-Newton generalized-bisection method (IN/GB).

# Chapter 1

## DYNAMIC SIMULATION OF ELECTRICAL MACHINES AND DRIVE SYSTEMS USING MATLAB GUI

Fedák Viliam[1], Tibor Balogh[2] and Pavel Záskalický[3]

[1]Technical University of Košice, Slovakia
[2]Magneti Marelli, Electronic Systems Division, Industrial Park Kechnec, Slovakia
[3]Technical University of Košice, Slovakia

## INTRODUCTION

Since the first appearance, the fields of electrical machine and drive systems have been continuously enriched by introduction of many important topics. Progress in power electronics, microcontrollers, new materials and advances in numerical modeling have led to development of new types of electrical machines and in field of electrical drives to realization of complex control algorithms. Their verification is usually done by simulation during system design, thus the effort is concentrated to development of simulation models.

MATLAB offers almost infinite possibilities for easy development of system models. MATLAB (GUI) Graphic User Interface in connection with Simulink and specialized toolboxes present a suitable and easy programmable tool for development of purpose-oriented virtual model of any dynamical system. Easy and comfortable change of parameters by control elements in MATLAB GUIDE (GUI Development Environment), such as push- and radio-buttons, text boxes, and easy visualization of results, enable to develop virtual models without deep knowledge of their substance nor without a tedious programming and debugging the models.

Well-elaborated models of electrical drives and machines available on-line were developed by (Riaz, n.d.). (Saadat, 2012) presented application of MATLAB GUI for electrical engineering subjects, available online and

MATLAB GUI was utilized in (Petropol-Serb et al, 2007) for development of virtual model of induction machine.

Our contribution aims to present methodology and results in development of unified series of virtual models for electrical machines and drive systems using MATLAB GUI. The organization of the contribution is as follows: after brief description of tasks at virtual models design in GUI MATLAB (chapter 2) in the third chapter we describe development of few typical GUI oriented models of (more complex) electrical machines starting from a simple outline of mathematic model, following by simulation model. Emphasis is put on development of the virtual model itself and description of its features. The fourth chapter deals with CAD of drive controllers using GUI MATLAB. Finally, in the fifth chapter we share some experiences from development of the GUIs and their utilization for training of students. In conclusion we also present ideas for our future work.

# DESIGN METHODOLOGY FOR VIRTUAL MODELS OF ELECTRICAL MACHINES AND DRIVES

## Tasks in Design of the GUI Screen

The GUI providing human-computer interaction presents one of the most important parts when working with the system model. User interacts with the computer easily, intuitively, without need for derivation, design, development, composition, and debugging the simulation model; without necessity to learn its operation, and finally, he gets required information in transparent, well-arranged form. In the fact, such GUI presents a functional virtual model, where the user sets system parameters, chooses mode of operation and required outputs to observe results. Design of GUI starts with careful planning of the following tasks:

1. Derivation of system mathematical model

2. Getting, debugging, and verification of simulation model

3. Programming GUI

4. Determination of input parameters changes (editing boxes, sliders)

5. Determination of outputs in graphical and text form

6. Design of the screen (or a set of interconnected screens)

7. Choice of calculation modes and algorithm of their control

8. Final refining and verification of functionality of the designed GUI screen

## Principles of Ergonomics of the GUI Screen

When designing a functional GUI screen for the technical systems, designer must understand principles of good interface and screen design. Generally, the rules are described in (Galitz, 2007). We have adapted and extended them for design of virtual model – of a GUI MATLAB screen. The most important principles, when designing the placement of objects on the GUI screen, are:

- Legibility: saying that information should be distinguishable.

- Facitily: how easy is the designed GUI screen intuitively usable?

- Readability: how information is identifiable and interpretable.

- Attractivity: to attract and call attention to different screen elements (placement of control elements and outputs, using colors …).

- Guiding the eye: by placement and grouping command objects by visual lines/boxes.

Further, designer should deal with user considerations, as follows:

- Visually pleasing (user friendly) composition of the screen.

- Organizing screen elements (balance, symmetry, alignment, proportion, grouping).

- Screen navigation and flow.

- Choice of implicitly pre-setting system parameters and their range (so that virtual models can be generally used in larger range of parameter changes).

- Changing system parameters by sliders or by numerical values in editing boxes.

- Finally, designer has to maintain ergonomic of the screen where the control elements and outputs should be organized in a legible way.

## VIRTUAL MODELS FOR ANALYSIS OF DYNAMICAL PROPERTIES OF ELECTRICAL MACHINES

In background of every GUI MATLAB there is working a simulation model of the system derived from its mathematical model. The same procedure is applied at development of GUI for electrical machines and drives. Let's show the GUI MATLAB development procedure on few electrical machines – the AC induction machine (asynchronous motor) and the brushless DC motor.

## AC Drive with 3-Phase Asynchronous Motor

The AC drive consists of an AC machine supplied by a converter. The variables of AC machine (an asynchronous motor in our case) like electrical quantities (supply voltages and currents), magnetic variables (magnetic fluxes), and mechanical variables (motor torque and rotor angular speed) are usually to be investigated in:

- Various reference frames (rotating coordinate systems). In case of asynchronous motor two basic reference frames are considered:

- $\{\alpha,\beta\}$ reference frame associated with stator, whose angular speed $\omega_k=0$

- $\{x,y\}$ reference frame rotating by synchronous angular speed $\omega_k=\omega_1$

- Various modes of supply:

- harmonic (sinusoidal voltage)

- non-harmonic (stepped voltage), PWM

## Asynchronous Motor Model

For dynamic properties investigation of asynchronous motor (influence of non-harmonic supply to properties of the AC drive, etc.) a dynamical model of AC machine is used. The AC machine is described by set differential equations. For their derivation some generally accepted simplifications are used (not listed here) concerning physical properties, construction of the machine, electromagnetic circuit, and supply source.

In order to simplify mathematical model of the squirrel cage motor, the multiphase rotor is replaced by an equivalent three-phase one and its parameters are re-calculated to the stator. Equations describing behavior of the machine are transformed from three- to two-phase system what yields to decreased number of differential equations. The quantities in equations are transformed into reference systems. To derive dynamic model of asynchronous motor, the three-phase system is to be transformed into the two-phase one. In the fact, this transformation presents a replacement of the three-phase motor by equivalent two-phase one. The stator current space vector having real and imaginary components is defined by the equation:

$$\bar{i} = \frac{2}{3}(i_a + ai_b + a^2 i_c)$$

(1)

where

$$a = e^{j120^\circ} = -\frac{1}{2} + j\frac{\sqrt{3}}{2}, \quad a^2 = e^{j240^\circ} = -\frac{1}{2} - j\frac{\sqrt{3}}{2}$$

(2)

Basic equations of the AC machine with complex variables (denoted by a line over the symbol of the variable) in the reference frame rotating by general angular speed $\omega_k$ are:

$$\bar{u}_1 = R_1 \bar{i}_1 + \frac{d\bar{\Psi}_1}{dt} + j\omega_k \bar{\Psi}_1$$

(3)

$$\bar{u}_2 = R_2 \bar{i}_2 + \frac{d\bar{\Psi}_2}{dt} + j(\omega_k - \omega)\bar{\Psi}_2$$

(4)

$$\frac{J}{p}\frac{d\omega}{dt} = \frac{3p}{2}\operatorname{Im}(\bar{\Psi}_{1k}^c \bar{i}_{1k}) - m_z$$

(5)

Where the nomenclature is as follows:

- u1, i1, R1, L 1– voltage, current, resistance and inductance of stator phase winding

- u2, i2, R2, L2 – voltage, current and resistance and inductance of rotor phase winding (re-calculated to stator quantities)

- Ψ1 Ψ2 – total magnetic flux of the stator and rotor (recalculated to stator side)

- ω– (rotor) mechanical angular speed

- mz – load torque

- $\omega_k$– angular speed of a general rotating reference frame $\omega_k = \omega_1$or 0

- $\omega_1, \omega$– magnetic field angular speed, rotor angular speed, where $\omega = \omega_1 - \omega$

- σ– leakage factor $\sigma = (L_1 L_2 - L_h^2) / L_1^2$

For manipulation between various reference frames in the motor model the transformation formulas are used as listed in Tab. 1. All rotor parameters and variables are re-calculated to the stator side.

After inserting real and imaginary components into the complex of variables (e.g. for stator voltage $\bar{u}_1 = u_{1x} + ju_{1y}$ in synchronously rotating reference frame {x, y}), we get the AC motor mathematical model whose equations are listed in Tab. 2 and a block diagram shown in Fig. 1 where $K_1 = 1/(\sigma L_1)$, $K_2 = 1/(\sigma L_2)$, $K = L_h/(\sigma L_1 L_2)$

**Table 1:** Transformation relations between three-phase system and two-phase reference frame and between {x, y} and {α, β} reference frames

| Transformation | Matrix notation | Block diagram |
|---|---|---|
| {a, b, c} → {α, β}<br>from 3-phase system {a, b, c} to 2-phase reference frame {α, β} fixed with the stator (Clark transform) | $\begin{bmatrix} i_\alpha \\ i_\beta \end{bmatrix} = \begin{bmatrix} 1 & 0 & 0 \\ 0 & \frac{1}{\sqrt{3}} & -\frac{1}{\sqrt{3}} \end{bmatrix} \begin{bmatrix} i_a \\ i_b \\ i_c \end{bmatrix}$ | |
| {α, β} → {a, b, c}<br>from 2-phase reference frame fixed with stator {α, β} into 3-phase system {a, b, c} (inverse Clark transform) | $\begin{bmatrix} i_a \\ i_b \\ i_c \end{bmatrix} = \begin{bmatrix} 1 & 0 \\ -\frac{1}{2} & \sqrt{\frac{3}{2}} \\ -\frac{1}{2} & -\sqrt{\frac{3}{2}} \end{bmatrix} \begin{bmatrix} i_\alpha \\ i_\beta \end{bmatrix}$ | |
| {x, y} → {α, β}<br>from synchronously rotating reference frame {x, y} into the stationary frame {α, β} (Park transform) | $\begin{bmatrix} i_x \\ i_y \end{bmatrix} = \begin{bmatrix} \cos\rho & \sin\rho \\ -\sin\rho & \cos\rho \end{bmatrix} \begin{bmatrix} i_\alpha \\ i_\beta \end{bmatrix}$<br>$\rho = \omega_1 t$ | |
| {α, β} → {x, y}<br>from stator reference frame {α, β} }into the synchronously rotating frame {x, y} (inverse Park transform) | $\begin{bmatrix} i_\alpha \\ i_\beta \end{bmatrix} = \begin{bmatrix} \cos\rho & -\sin\rho \\ \sin\rho & \cos\rho \end{bmatrix} \begin{bmatrix} i_x \\ i_y \end{bmatrix}$<br>$\rho = \omega_1 t$ | |

**Table 2:** Equations of windings of asynchronous motor model in {x,y} reference frame

| | Magnetic fluxes | Relation between fluxes and currents |
|---|---|---|
| Stator | $\frac{d\psi_{1x}}{dt} = u_{1x} - R_1 i_{1x} + \omega_k \psi\_{1y}$ | $i_{1x} = K_1 \psi_{1x} - K\psi_{2x}$ |
| | $\frac{d\psi_{1y}}{dt} = u_{1y} - R_1 i_{1y} - \omega_k \psi_{1x}$ | $i_{1y} = K_1 \psi_{1y} - K\psi_{2y}$ |
| Rotor | $\frac{d\psi_{2x}}{dt} = u_{2x} - R_2 i_{2x} - (\omega_k - \omega)\psi_{2y}$ | $i_{2x} = K_2 \psi_{2x} - K\psi_{1x}$ |
| | $\frac{d\psi_{2y}}{dt} = u_{2y} + R_2 i_{2y} - (\omega_k - \omega)\psi_{2x}$ | $i_{2y} = K_2 \psi_{2y} - K\psi_{1y}$ |

The corresponding Simulink model is drawn in Fig. 1. The model of squirrel cage motor (rotor voltages = 0) contains 4 inputs and 10 outputs (Tab. 3).

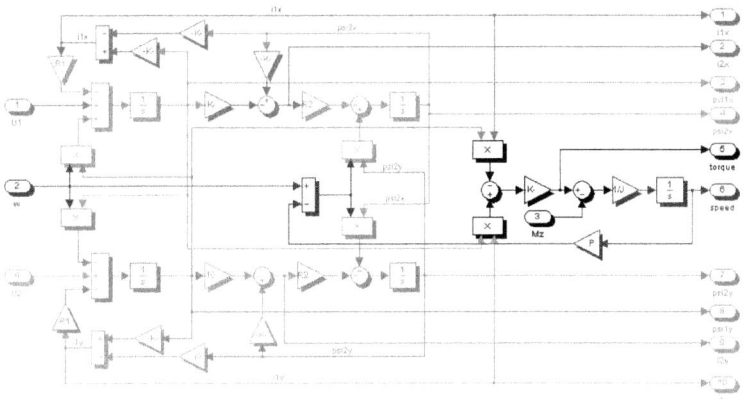

**Figure 1:** Simulink model of 3 phase squirrel cage asynchronous motor (the variables are denoted in the magnetic field reference frame $\{x, y\}$)

**Table 3:** Notation of inputs and outputs of the asynchronous motor model

| AM model inputs | AM model outputs |
|---|---|
| • $U_1$ input voltage (axis x or $\alpha$) | • current $\bar{i}$ (4 components) |
| • $U_2$ input voltage (axis y or $\beta$) | • magnetic fluxes $\bar{\Psi}$ (4 components) |
| • $M_z$ load torque | • motor torque $M$ |
| • $\omega_k$ reference frame angular speed | • rotor angular speed |

## Modeling of Supply Source

The asynchronous motor can be set into motion by various supply modes and control platforms:

- by direct connection to the supply network or to the frequency converter

- by frequency starting (with continuously increasing frequency of the supply voltage from the frequency converter)

Restrict our considerations to supply from indirect converter with the Voltage-Source Inverter (VSI). Based on the inverter control mode the output voltage can be:

- unmodulated (with 120 deg. switching in the power semiconductor devices)

- modulated by PWM

Developing inverter simulation schemes we have in mind two facts:

- the constant stator flux (i.e. fulfilling condition of constant ratio: U1/f1 = const.) should be preserved at all modes of motor control

- in range of very low frequency there should be kept an increased stator voltage (due to the voltage drop across the stator resistor) – so called V-curves (presenting a dependency of the supply voltage from the frequency). The V-curve can be modeled simply by a linear piecewise line.

The model of the motor supply source taking into consideration all described features is shown in Fig. 2(signals denoted as SL and op are control signals from the GUI buttons). It has 4 inputs: supply frequency and voltage magnitude, ramp frequency and voltage (to simulate frequency starting). The switches "step/ramp" are controlled by pushbuttons from the GUI control panel.

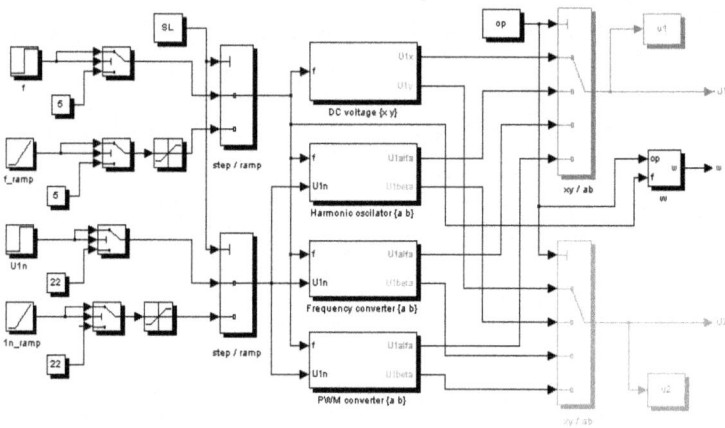

**Figure 2:** Simulink model of various modes of supply source (DC, harmonic, frequency converter and PWM)

## *Model of VSI converter (with constant output frequency)*

We start to model the inverter output voltage based on a similarity of output converter voltage with the perpendicular harmonic voltages (Fig. 3a). The VSI voltage vector changes its position 6 times per period, after every 60° (Fig. 3b).

Proper switching instants are realized by comparators and switches (Fig. 4). Harmonic oscillator creates a core of the inverter model. Generation of six switching states during period of the output voltage is adjusted by comparing values of the sin/cos signals with pre-set values of $\sin 60 = \overline{\varnothing} /2$ for the voltage $3\sqrt{3}$ and value of $\cos 60 = 1/2$ for the voltage $u1\alpha u1\alpha$. The amplitudes of output voltage are adjusted by constants with values 1; 0, 5 for u1βu1βand 0,866 =u1αu1α/2 for3√3.

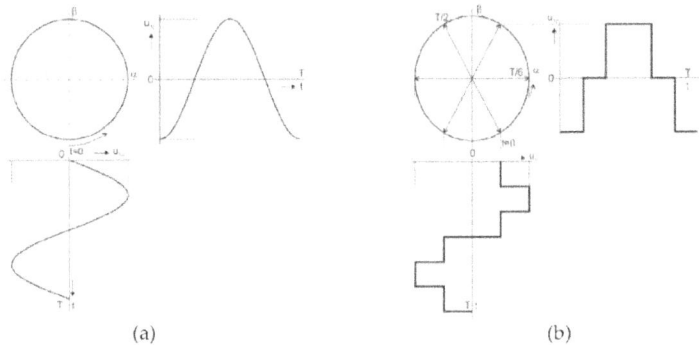

(a)    (b)

**Figure 3:** Simulink model of inverter

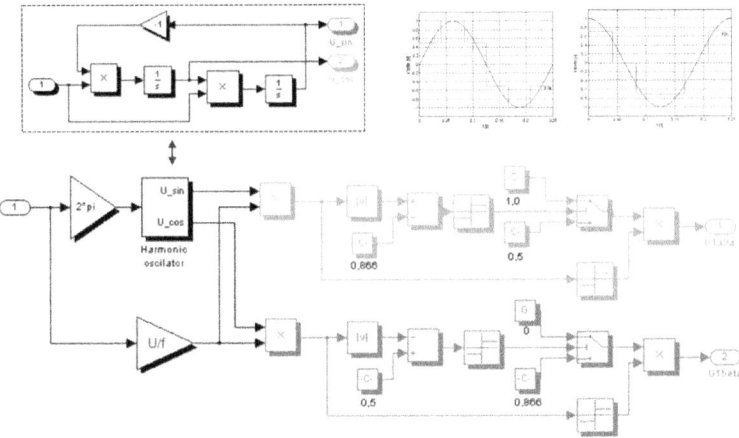

**Figure 4:** Simulation scheme realizing rectangle voltages $u_{1\alpha}, u_{1\beta}$ of the inverter

## Model of PWM source

The simplest way to generate a PWM signal uses the interceptive method. The three-phase PWM voltage is generated directly in two axes {α, β} as shown in Fig. 5. The courses of the inverter PWM voltages u1βu1β and u1αu1α are shown in Fig. 6. In frequency starting mode of the asynchronous motor, the frequency of supply voltage increases from zero to required final value. To get the stator flux constant, the voltage across the motor has to increase linearly with frequency (U/f = const.), except of very low frequency range (due to voltage drop across the stator resistor). For this purpose, the connection must be completed by a compensating circuit which increases the value of supply voltage keeping the ratio U/f = const (Fig. 7). Up to the frequency of approx. 5 Hz the input voltage is kept constant on 10 % of its nominal value.

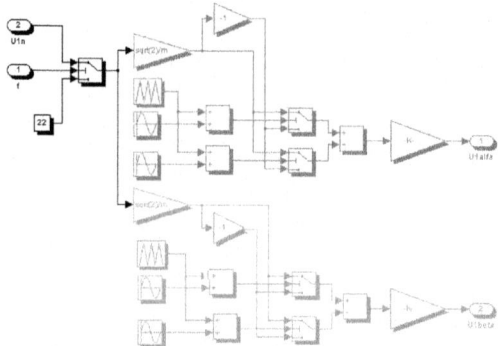

**Figure 5:** Model of voltages u1βu1β and u1αu1αfrom the inverter with PWM

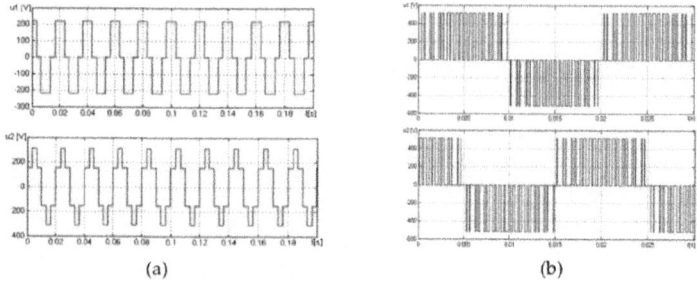

(a)                              (b)

**Figure 6:** Output voltages and u1βu1β and u1βu1β from the frequency converter: a) without and b) with PWM

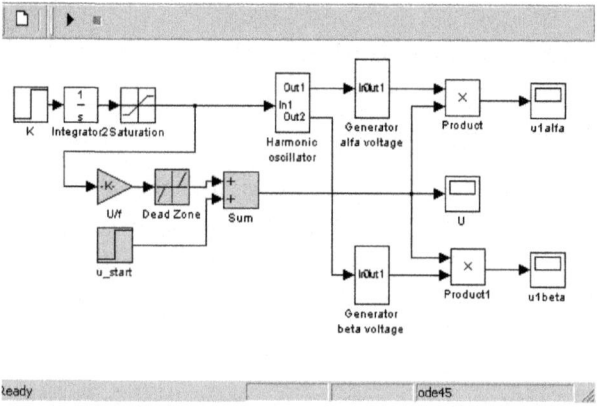

**Figure 7:** The model of converter realizing the frequency starting under consideration of the law of constant stator flux (U/f = const.)

The model supposes that amplitude of the DC link voltage is changed in the frequency converter. This solution is suitable for drives with low requirements to motor dynamics. The DC link contains a large capacitor what causes the DC link voltage cannot be changed step-by-step. The output inverter voltage can change faster if the PWM control is used. Output voltages of the inverter model with linear increasing frequency and voltage are shown in Fig. 8 (observe a non-zero amplitude of the voltage that at the starting what is consequence of described V-curve block).

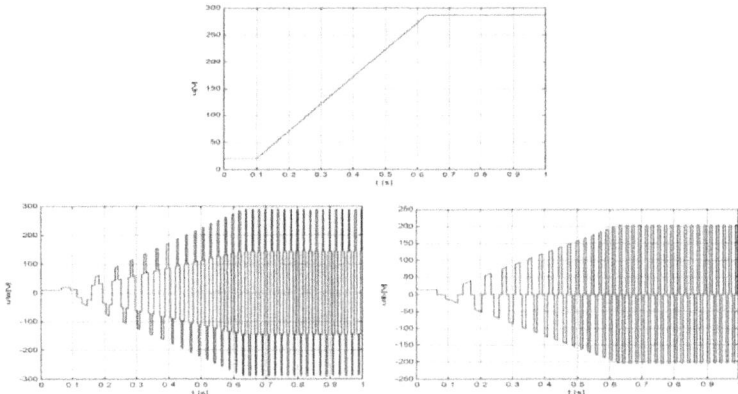

**Figure 8:** VSI output voltages u1αu1α and u1αu1α at increased frequency (the frequency time course is on the top figure)

## Model Verification

The AC induction motor model was simulated using following motor parameters: R1=1, 8 Ω; R2=1, 85 Ω; p=2; J=0, 05 kgm², K1=59, 35; K2=59, 35; K=56, 93.

Time courses of mechanical variables are shown in Fig. 10 (they are the same regardless the used reference frame). Motor dynamical characteristics ω =f (M) ω =fM at various modes of supply are compared in Fig. 11.

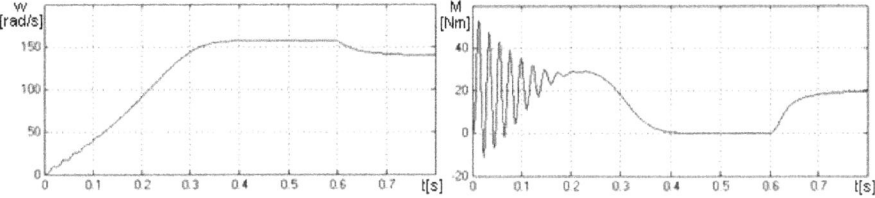

**Figure 9:** Time responses of asynchronous motor speed and torque at harmonic voltage supply at starting and loading the motor

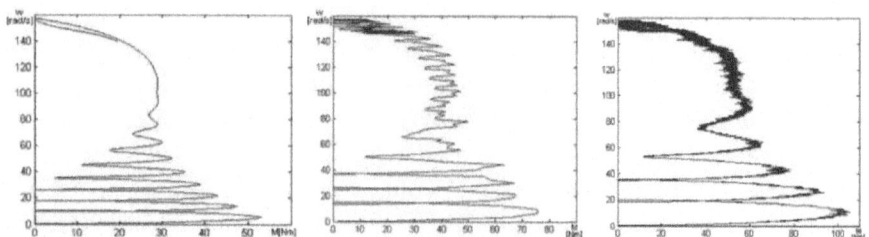

**Figure 10:** Dynamic characteristic of the asynchronous motor ω = f (M) supplied: a) by harmonic voltage, b) from frequency converter, c) from frequency converter with PWM

## GUI Design and Realisation

After debugging the motor model (Fig. 11), development of GUI continues with careful design of the program flowchart and design of GUI screen.

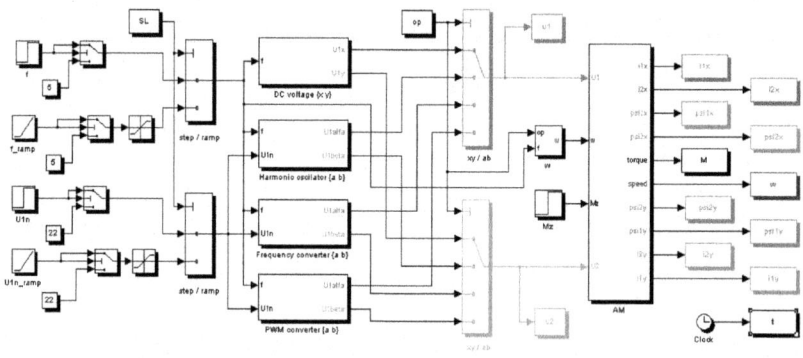

**Figure 11:** Arrangement of asynchronous motor subsystems in the Simulink GUI model

## Description of the GUI functionality

The GUI screen (Fig. 13) consists of several panels. Their functionality is as follows:

- Input data panel ("Motor parameters") in the bottom left part. The panel is used to system parameters entry. Their values can be changed by inserting numeric values into editing boxes. There is a possibility to return to original (default) parameters by pushing the button Default (in the pane

Mode).

- Choice of Coordinate reference frame system (the panel on the right top part) enables to display motor output variables:

- in the synchronously rotating reference frame

- in the reference frame associated with the stator

- at harmonic supply

- at nonharmonics supply from the VSI

- at nonharmonics supply from the VSI with PWM

- Output graphs. Output variables are displayed in four graphs:

- supply voltage time courses and in two coordinates

- mechanical variables - motor torque and speed

- stator currents or magnetic fluxes

- rotor currents or magnetic fluxes

- The graph to be displayed can be chosen by pushing radio button in the menu Graphs. Time courses are chosen by the button Time; dependency of one variable on other is chosen by the button Rectangular.

- Mode of starting the motor can be selected in the panel Motor supply:

- Direct connection to the supply – the button Step. The voltage $U_1$ (effective rms value) and frequency $f_1$ can be pre-set in the editing boxes.

- Frequency starting – the button Linear enables to pre-set the frequency time rise starting from zero.

- Using the buttons in the panel Mode we start Simulation, at pressing Default (original) parameters are set, and the Simulink scheme is shown by pushing the button Model.

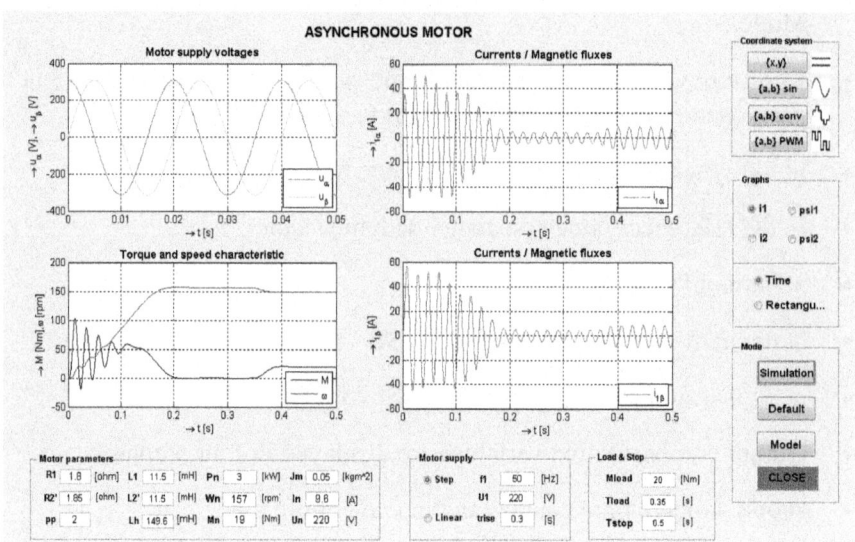

**Figure 12:** GUI screen of the AC drive with induction machine

Screen outputs

Samples of the screens displaying variables in the stator reference frameu1βu1β are shown in Fig. 13:

a.  time courses at supplying motor by frequency converter – button Time )

b.  chracteristics $\{\alpha,\beta\}$, M=f($\omega$), $i_{1\alpha}$=f($i_{1\beta}$)- button Rectangular )

c.  time courses$\psi_{1\alpha}$=f($\psi_{1\beta}$), $i_{1\alpha}$=f(t) at supplying from the PWM frequency converter

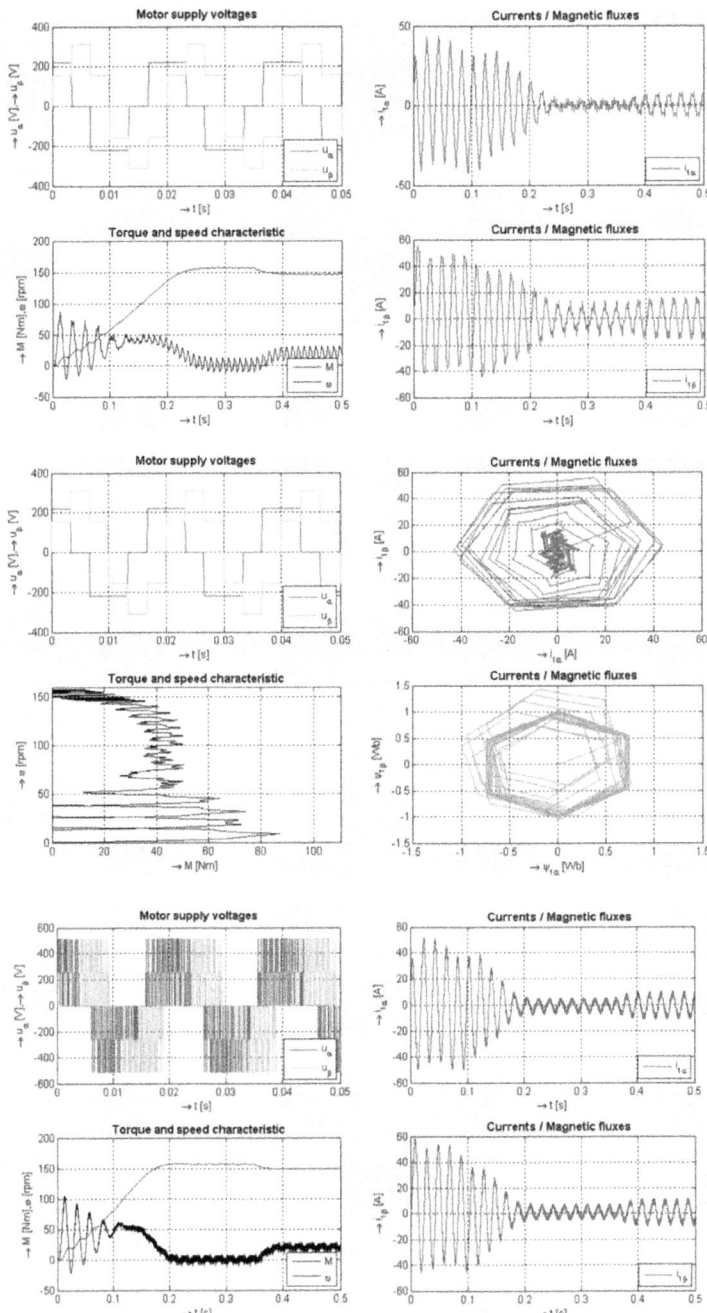

**Figure 13:** Examples of diplaying various graphs in the GUI for asynchronous motor

## BLDC Motor

The Brushless Direct Current (BLDC) motor is rapidly gaining popularity by its utilization in various industries. As the name implies, the BLDC motor do not use brushes for commutation; instead of this they are commutated electronically.

The BLDC motors have many advantages over brushed DC motors and induction motors. A few of these are: (1) Better speed versus torque characteristics; (2) High dynamic response; (3) High efficiency; (4) Long operating life; (5) Noiseless operation; (6) Higher speed ranges. In addition, the ratio of torque delivered to the size of the motor is higher, making it useful in applications where space and weight are critical factors (Indu, 2008).

The torque of the BLDC motor is mainly influenced by the waveform of back-EMF (the voltage induced into the stator winding due to rotor movement). Ideally, the BLDC motors have trapezoidal back-EMF waveforms and are fed with rectangular stator currents, which give theoretically constant torque. However, in practice, a torque ripple exists, mainly due to EMF waveform imperfections, current ripple, and phase current commutation. The current ripple follows up from PWM or hysteresis control. The EMF waveform imperfections result from variations in the shapes of slot, skew and magnet of BLDC motor, and are subject to design purposes. Hence, an error can occur between actual value and the simulation results. Several simulation models have been proposed for analysis of BLDC motor (Jeon, 2000).

## Construction and Operating Principle

The BLDC motor is also referred to as an electronically commuted motor. There are no brushes on the rotor and the commutation is performed electronically at certain rotor positions. In the DC commutator motor, the current polarity is reversed by the commutator and the brushes, but in the brushless DC motor, the polarity reversal is performed by semiconductor switches which are to be switched in synchronization with the rotor position. Besides of the higher reliability, the missing commutator brings another advantage. For the DC brushed motor the commutator presents also a limiting factor in the maximal speed. Therefore, the BLDC motor can be employed in applications requiring high speed (Jeon, 2000).

The BLDC motor is usually considered as a three-phase system and thus it has to be powered by a three-phase power supply. The rotor position must be known at certain angles, in order to align the applied voltage with the back-EMF. The alignment between the back-EMF and commutation events is very important.

A simple motor model of BLDC motor consisting of a three-phase power converter and a brushless DC motor is shown in Fig. 14.

## Mathematical Model of the BLDC Motor

Modeling of a BLDC motor can be developed in the similar manner as a three-phase synchronous machine. Since there is a permanent magnet mounted on the rotor, some dynamic characteristics are different. Similarly, the model of the armature winding for the BLDC motor is expressed as follows:

$$u_a = Ri_a + L\frac{di_a}{dt} + e_a \tag{6}$$

$$u_b = Ri_b + L\frac{di_b}{dt} + e_b \tag{7}$$

$$u_c = Ri_c + L\frac{di_c}{dt} + e_c \tag{8}$$

where L is armature self-inductance, R - armature resistance, ua, ub, uc - terminal phase voltages, ia, ib, ic - motor input currents, and ea, eb, ec - motor back-EMF.

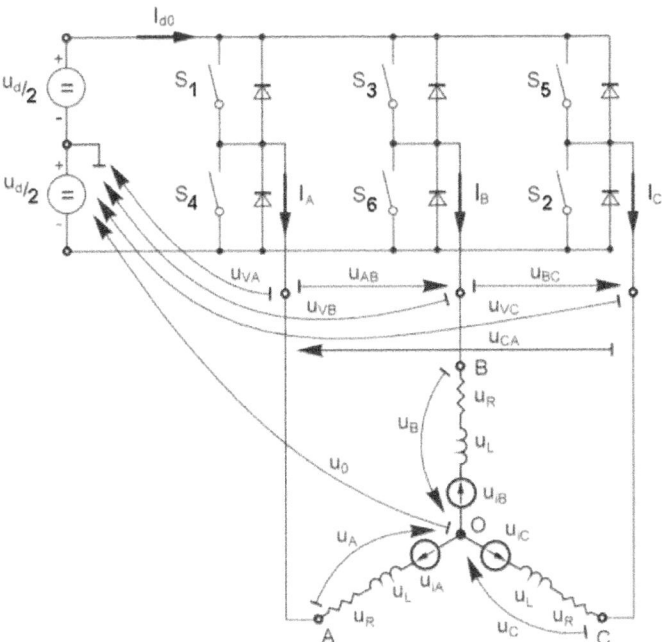

**Figure 14:** BLDC motor model

In the three-phase BLDC motor, the back-EMF is related to a function of rotor position and the back-EMF of each phase has 120° phase angle difference so the equation for each motor phase is as follows:

$$e_a = K_w f(\theta_e)\omega \tag{9}$$

$$e_b = K_w f(\theta_e - 2\pi/3)\omega \tag{10}$$

$$e_c = K_w f(\theta_e + 2\pi/3)\omega \tag{11}$$

where Kw is back EMF constant of one phase, θe - electrical rotor angle, ω - rotor speed. The electrical rotor angle θe is equal to the mechanical rotor angle θm multiplied by the number of poles p:

$$\theta_e = \frac{p}{2}\theta_m \tag{12}$$

Total torque output Te can be represented as summation of that of each phase:

$$T_e = \frac{e_A i_A + e_B i_B + e_C i_C}{\omega} \tag{13}$$

The equation of mechanical part is represents as follows:

$$T_e - T_l = J\frac{d\omega}{dt} + b\omega \tag{14}$$

where Tl is load torque, J - rotor inertia, b - friction constant.

Simulink Model of the BLDC Motor

Fig. 16 shows the block diagram of the BLDC motor SIMULINK model in the rotor reference frame.

Fig. 16 shows detail of the BLDC motor block. Fig. 17a shows Simulink diagram of trapezoidal back-EMF and in Fig. 17b there is Simulink model of sinusoidal back-EMF. The trapezoidal functions and the position signals are stored in lookup tables that change their output according to the value of the electrical angle (Indu, 2008).

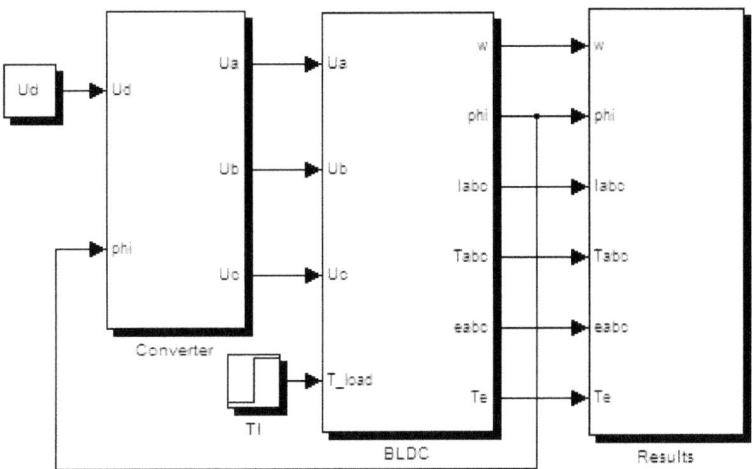

**Figure 15:** Simulink model of the BLDC motor

Unlike a brushed DC motor, the commutation of a BLDC motor is controlled electronically. To rotate the BLDC motor, the stator windings should be energized in sequences. In order to understand which winding will be energized following the energizing sequence, it is important to know the rotor position. It is sensed using Hall Effect sensors embedded into the stator. Most of the BLDC motors contain three Hall sensors embedded into the stator on the non-driving end of the motor. The number of electrical cycles to be repeated to complete a mechanical rotation is determined by rotor pole pairs. Number of electrical cycles/rotations equals to the rotor pole pairs. The commutation sequences are shown in Tab. 4.

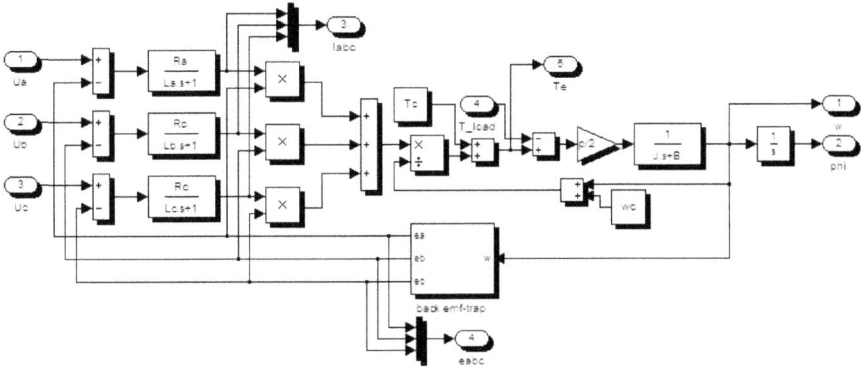

**Figure 16:** Detailed overview of the BLDC motor block

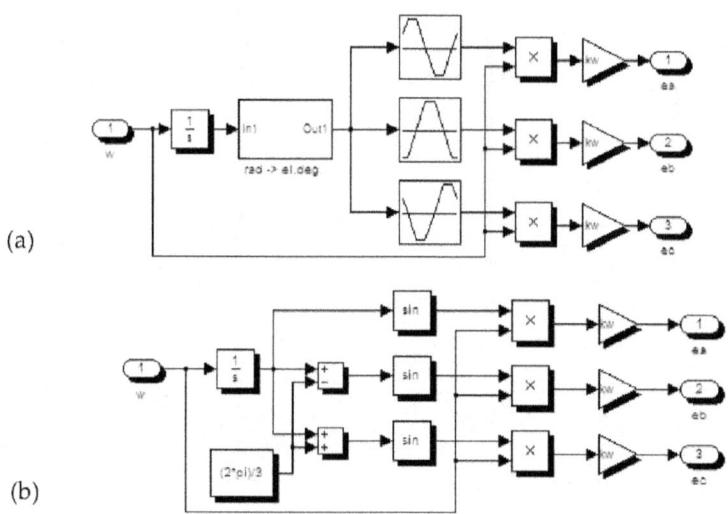

(a)

(b)

**Figure 17:** Trapezoidal (a) and simusoidal (b) model of the back-EMF

**Table 4:** Electrical degree, Hall sensor value and corresponding commuted phase in clockwise rotation of the rotor

| Electrical degree | Hall sensor value (ABC) | Phase | Switches |
|---|---|---|---|
| 0° - 60º | 101 | A-C | S1-S2 |
| 60° - 120º | 001 | B-C | S2-S3 |
| 120° - 180º | 011 | B-A | S3-S4 |
| 180° - 240º | 010 | C-A | S4-S5 |
| 240° - 300º | 110 | C-B | S5-S6 |
| 300° - 360º | 100 | A-B | S6-S1 |

Mathematical and Simulink Model of the Three-Phase Converter

The converter supplies the input voltage for three phases of the BLDC motor. Each phase leg comprises two power semiconductor devices. Fig. 18 shows the scheme of the considered three-phase converter.

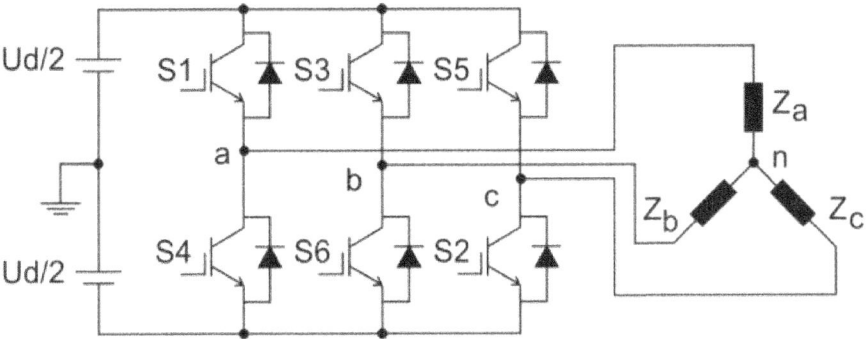

**Figure 18:** Modelled three-phase converter

Appropriate pairs of the switches (S1 to S6) are driven based on the Hall sensors input. Three phases are commutated in every 60° (el. degrees). The model of the converter is implemented using the equations:

$$U_{an} = S_1 \frac{Ud}{2} - S_4 \frac{Ud}{2} - U_f \qquad (15)$$

$$U_{bn} = S_3 \frac{Ud}{2} - S_6 \frac{Ud}{2} - U_f \qquad (16)$$

$$U_{cn} = S_5 \frac{Ud}{2} - S_2 \frac{Ud}{2} - U_f \qquad (17)$$

where $U_{an}$, $U_{bn}$, $U_{cn}$ are line-neural voltages, Ud – the DC link voltage, Uf – the forward diode voltage drop.

Fig. 19a shows the Simulink model of the three-phase converter block. In the simulation we assumed an ideal diode with neglected voltage drop Uf. The Commutation sequences block was developed based on the commutation sequence shown in Tab. 4. Converter voltage waveforms that are switched according to the commutation sequences in Tab. 4 are shown in Fig. 19b.

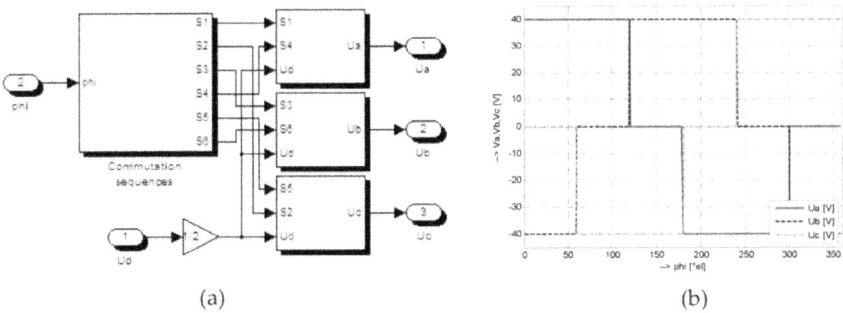

(a)                                        (b)

**Figure 19:** Detailed overview of the three-phase converter (a) and voltage source waveforms (b)

GUI of the BLDC Motor

The simulated BLDC motor is presented in a graphical user interface GUI (Fig. 20).

By the buttons in the panel Mode we start the Simulation, put Default (original) values and show the Simulink Model.

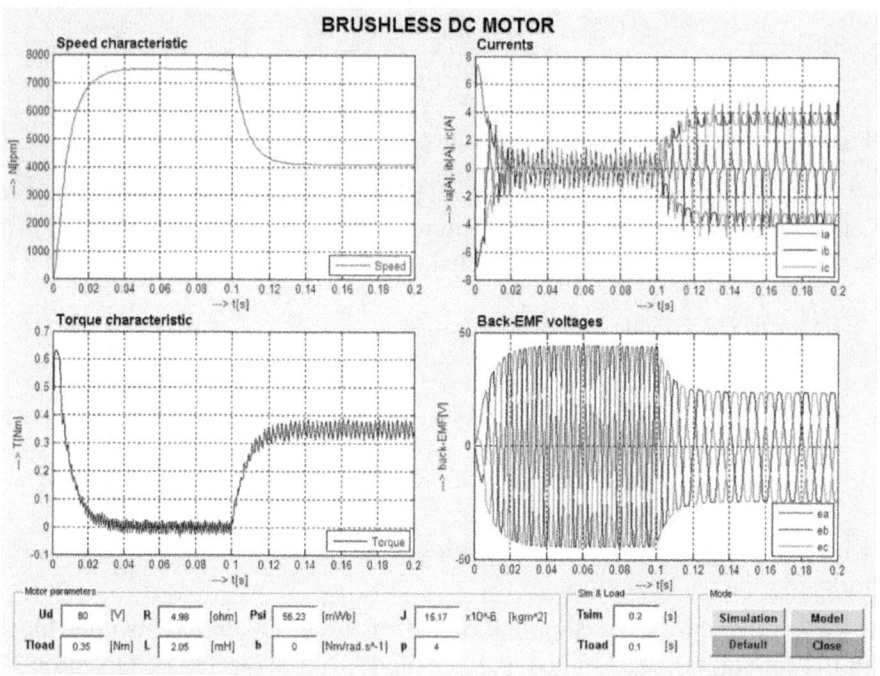

**Figure 20:** GUI for the BLDC motor

The default parameters of the BLDC motor for simulation are: Ud=80 V, Tl=0, 35 Nm, R=4, 98 $\Omega$, L=2, 05 mH, $\psi$=56, 23.10$^{-3}$ Wb, b=0 Nm/rad.s$^{-1}$, J=15, 17.10$^{-6}$ kgm$^2$, p=4.

# VIRTUAL MODELS APPLIED FOR SYNTHESIS OF DRIVE SYSTEMS

MATLAB GUI presents an extremely suitable tool for development of models to support CAD design of drive controllers, whose algorithms are known. Two simple cases are presented below: design of controllers in the frequency and time domains, other cases are mentioned in the subchapter 5.1.

## Cad Design of Controller Parameters for Dc Motor Drive in Frequency Domain

The DC drive controllers in the frequency domain are calculated mostly by using the following criteria:

1.  The current controller of the PI type is calculated on basis of the Optimum Modulus Criterion(OMC) from the drive system parameters:

$$F_{RI} = K_{RI} + \frac{1}{sT_{iI}} \qquad (18)$$

2.  After calculation of the current controller parameters and current control loop simplification, the speed controller is calculated based on the Symmetrical Optimum Criterion (SOC). It is again of PI type having the transfer function:

$$F_{R\omega} = K_{R\omega} + \frac{1}{sT_{i\omega}} \qquad (2)$$

Fig. 22 shows the principal block diagram of the system and in Fig. 23 there is view on the virtual GUI model of speed controlled DC drive.

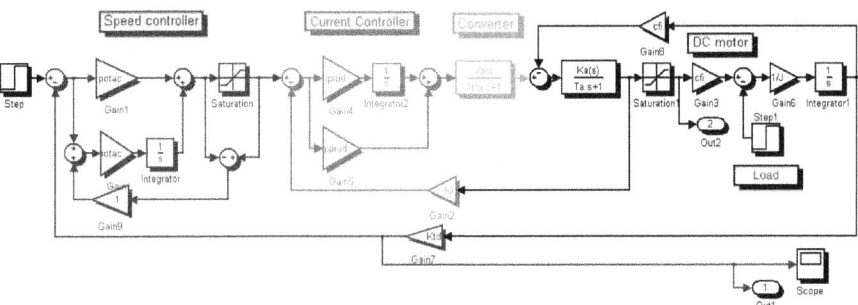

**Figure 21:** Control circuit of DC motor drive with current and speed controllers

Virtual model features

The user has a possibility to tune controller parameters in each design step according to displayed time response. The GUI screen consists of several panels:

*   Time response - the graph with time courses of the motor current and speed. Immediately after change of any system parameter (motor -, drive -, or controller parameters) by a slider or inserting a numeric value into editable box the simulation starts and new time responses are drawn (like in a real drive).

- Block diagram - displays the block diagram of the system

- System parameters are changed by sliders or inserting values into the boxes.

Before starting the model, implicit parameters are set up, but they can be changed later. After pushing the button Computed value the parameters of controllers are calculated from the actual values of parameters. Simultaneously a small window appears there with a question whether the calculated values of controller parameters are acceptable or not (if not, user can set up own parameters and can to tune them according to the time responses of the drive). To return to starting values, the user pushes the button Default (similar to the system restart).

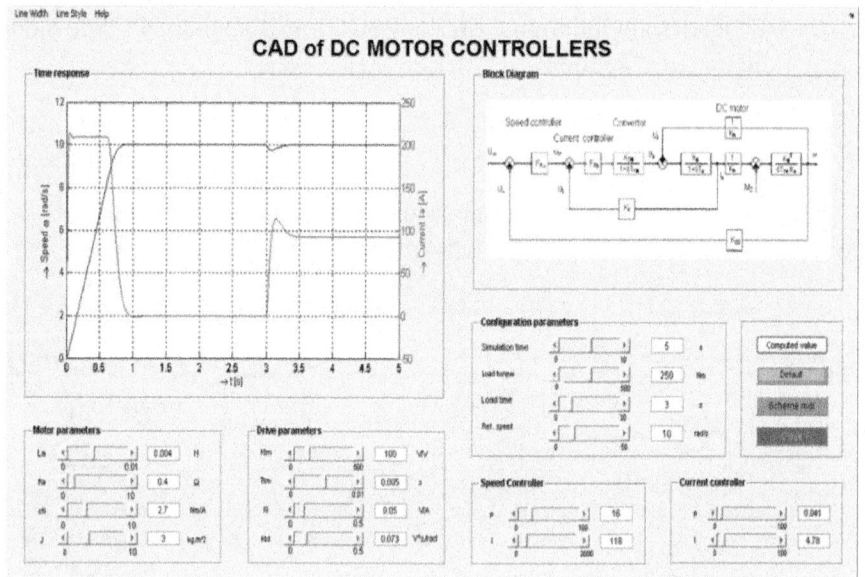

**Figure 22:** GUI screen for designing DC motor drive controllers in the frequency domain

## Cad Design of Controller Parameters for Dc Motor Drive in Time Domain

The computing algorithm is different from calculation of the controllers in the frequency domain and the task belongs to more complex one. The computation starts from the state-space model of the DC motor having two inputs in one output, in the form of state equations:

$$\dot{x} = A.x + b.u + e.z = \begin{bmatrix} 0 & \dfrac{K_m}{T_m.K_a} \\ -\dfrac{K_a}{K_m.T_a} & -\dfrac{1}{T_a} \end{bmatrix} x + \begin{bmatrix} 0 \\ K_T.K_a \\ T_a \end{bmatrix} u + \begin{bmatrix} -\dfrac{K_m^2}{T_m.K_a} \\ 0 \end{bmatrix} M_z$$

(20)

$$y = c^T x = \begin{bmatrix} 1 & 0 \end{bmatrix} x$$

(21)

where **A** is system matrix, **x** – state vector, **b** – input vector, $c^{T-}$ output (row) vector **e**-disturbance vector, u – input variable, y - output variable.

The final control structure with the feedback through the state controller vector $r^T$ is clear from the Simulink model Fig. 23. The integrator at the input serves to reject constant or slowly changing disturbances what is a common case.

The state control structure parameters: $y = c^T x = \begin{bmatrix} 1 & 0 \end{bmatrix} x$, $K_i$ and $r_1$ are designed by known pole placement method where for a prescribed position of poles the required polynomial is compared with the system polynomial and missing parameters of the controller are calculated from a set of linear algebraic equations.

The control structure in Simulink to simulate the system is shown in (Fig. 23).

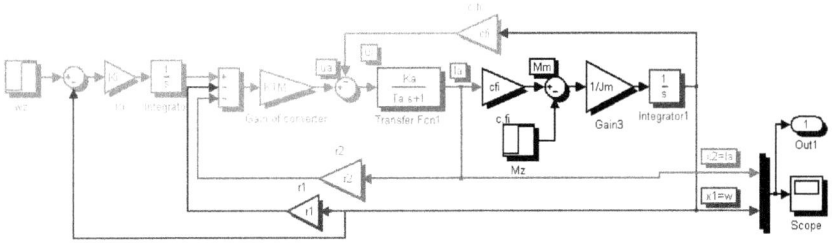

**Figure 23:** Simulink model of the state-space control of DC drive

Fig. 24 shows the GUI screen of the virtual model that enables to calculate state-space controller parameters and visualize time responses of the current and speed. It is a more complex GUI involving synthesis of the state-space controllers and giving the possibility to tune theoretically calculated parameters.

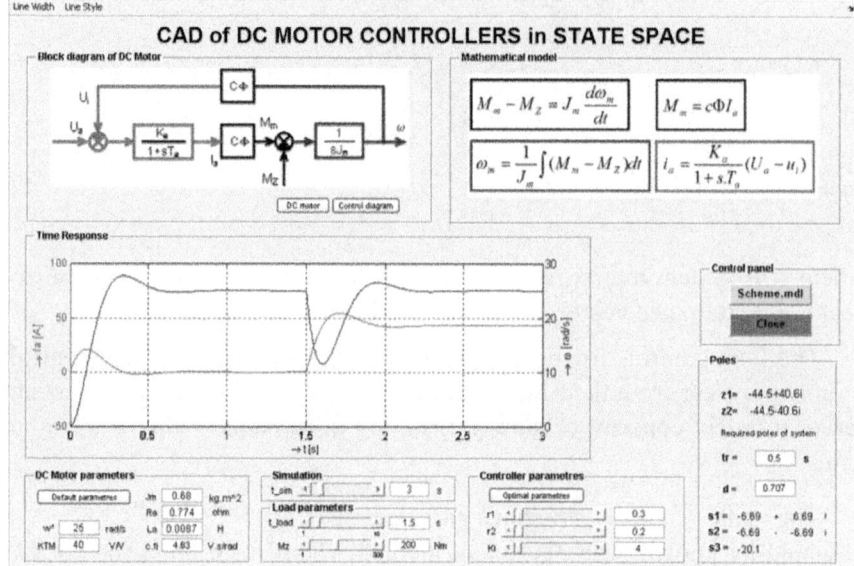

**Figure 24:** GUI screen for designing DC motor drive controllers in the state space domain

The panel Controller parameters serves to setting parameters of the state controller – by tuning or selecting the button Optimal parameters to calculate poles position placement.

Here:

- $r_1$ – feedback from state variable x1 (motor speed),

- $r_2$ – feedback from state variable x2 (motor current),

- $K_i$ – gain of the integrator (to reject steady-state disturbances).

The state controller parameters are calculated automatically on basis of required values of control time and damping (panel Poles, the item required poles of the system). In the upper part of the panel the real positions of poles are shown.

# EXPERIENCES WITH UTILIZATION OF VIRTUAL MODELS

## Utilization of Virtual Dynamical Models

Except of the presented GUI screens of virtual models a series of tens other models from fields of electrical engineering and mechatronic systems was developed to suit institutional needs. They cover topics from Electrical Machines, Power Electronics, Electrical Actuators and Drives, Servodrives, Mechatronic Systems, Control Theory, and others. We have also developed some more complex GUIs, reported e.g. by (Ismeal & Fedák, 2012), calculating artificial intelligence algorithms - to design PID controllers using fuzzy logic and genetic algorithms with various objective functions to evaluate the best PID controller.

The developed GUIs serve as virtual models to clarify phenomena and enhance features of the systems during lectures, and to prepare students for laboratory work. In order students to get more skills and practical experiences prior entering lab their work consists of two phases:

Design and simulation – for a given system motor or drive a student has:

- To derive mathematical model,

- To compose the block diagram,

- To design control law and controllers (in case of drives),

- To verify system behavior by simulation.

Verification and analysis – a student has:

- To verify the design using a virtual reference model,

- To perform system analysis ("to play himself" with the virtual model) in order to investigate system behavior at various values of system parameters and in various working points (small experiments round working point).

## Application of MATLAB Compiler in Virtual Model Development

A disadvantage of using GUI MATLAB consists in the fact, that the program can run only on a computer having installed the MATLAB program (and appropriate toolboxes containing instructions that are used in the main GUI program). This disadvantage can be suppressed by development of an executable (.exe) file from the original program. In this case the developer must install the Compiler Runtime program. The MATLAB Compiler creates a standalone executable file from the MATLAB code, which can then run in a runtime engine called Matlab Compiler Runtime (MCR). Once compiled, the standalone application, along with MCR, can be shared with other users for free. The only problem is that the original GUI MATLAB program cannot contain any Simulink model (mdl file). As the GUI MATLAB systems usually contain Simulink models, they have to be replaced by MATLAB programs solving mathematical model by various algorithms.

## CONCLUSION

The chapter describes principles and methodology of virtual models development in GUI MATLAB for few chosen electrical machines and controlled drives. The models perform analyses of real machines and drives in various working points and they enable easily to show system performance in various working points and to analyze influence of variable system parameters, modes of supply, and control parameters to system behavior. Presented virtual models have been of various complexity – the simpler ones enable virtual analysis of electrical machines and more complex virtual models also deal with algorithms for synthesis of drive controllers.

Strong advantage of developed virtual models consists in the fact user does not need to know the complexity of dynamical system whose simulation scheme is working in the background. He changes only system parameters, selects input signals (shape and amplitude of reference values, forcing and load signals), select mode of calculations and outputs (graphs displaying). The parameters of virtual models can be changed by a slider or by editing numerical values in editable boxes.

Based on the procedure a whole series of virtual models designed in GUI MATLAB has been developed at the authors' institution in recent years which are partly accessible through the website of the Virtual Laboratory for Control of Mechatronic Systems (KEM TU Kosice, 2010). The virtual dynamical models contribute to e-learning support at teaching and they also serve for preparation of students for laboratory experimentation. Their utilization makes more attractive lectures and considerably enhances explanation of systems properties. By simulation model students generally easier understand physical processes and they are better prepared to the laboratory work.

Shortcoming of developed models consists in the fact they run on computers having installed the MATLAB program. To overcome this, recently our work was concentrated to applications of the MATLAB Compiler to develop exececutable files. It should be noted that this application enables to run MATLAB operation without simulation (i.e. without a Simulink scheme), without 3D virtual reality views and without animation. The advantage on one side makes development of virtual models more difficult using more complex mathematical subroutines. Also some known problems with GUI MATLAB should be noted - cross platform appearance may not be the same and during the GUI development, often must be used tricks and somehow unfriendly techniques.

## ACKNOWLEDGEMENT

The financial support of the Slovak Research and Development Agency under the contract no. APVV-0138-10 is acknowledged. The work was also supported by Slovak Cultural and Educational Agency of the Ministry of Education of Slovak Republic under the contract KEGA 042TUKE-4/2012 "Teaching Innovation in Control of Mechatronic Systems".

## REFERENCES

1.  W. O. Galitz, 2007The Essential Guide to User Interface Design. An Introduction to GUI Design Principles and Techniques, Wiley Publishing, Inc., 978-0-47005-342-3Indianapolis, Indiana

2.  S. Hill, 2011How to make an Executable from MATLAB Code, Date of access: Mach 31, 2011, <http://www.ehow.com/how_8686360_make-executable-matlab-code.html>

3.  B. Indu, R. Ashly, M. Tom, 2008Dynamic Simulation of Brushless DC Drive Considering Phase Commutation and Backemf Waveform for Electromechanical Actuator. IEEE TENCON, Hyderabad 2008. 978-1-42442-408-5

4.  G. A. Ismeal, V. Fedák, 2012Overview of Control Algorithms for DC Motor Drive as a Basis for Development of GUI in MATLAB. XXVI. microCAD International Scientific Conference, Miskolc, 2012, 9 p., 978-9-63661-773-8

5.  Y. S. Jeon, H. S. Mok, G. H. Choe, D. K. Kim, J. S. Ryu, 2000A New Simulation Model of BLDC Motor with Real Back EMF waveforms. IEEE CNF on Computers in Power Electronics, COMPEL 2000, 217220July 2000

6.  KEM TU Kosice2010Virtual Laboratory of Mechatronic Systems Control., In: Date of access: March 31, 2012, Available from <http://andromeda.fei.tuke.sk/>in Slovak)

7.  Works. Math, (n.d., Deploytool, R2012a Documentation MATLAB Compiler       http://www.mathworks.com/help/toolbox/compiler/deploytool.html

8.  N. Parspour, R. Hanitsch, 1994Fuzzy Controlled Brushless DC Motor for Medical Applications. Industrial Electronic, Control and Instrumentation IECON, IEEE, Bologna, 1994, 0-78031-328-3

9.  G. D. Petropol-Serb, I. Petropol-Serb, A. Campeanu, A. Petrisor, 2007Using GUI of Matlab to create a virtual laboratory to study an induction machine. EUROCON, 2007. The International Conference on

Computer as a Tool, 978-1-42440-813-9Warsaw, September 9-12, 2007

10.    M. Riaz, (n.d., of. Simulation, Machine. Electric, Systems. Drive, M. A. T. L. A. B. Using, Simulink, University of Minnesota.www.ece.umn. edu/users/riaz/

11.    H. Saadat, 2012MATLAB Graphical User Interface for EE Students. Date of access: March 31, 2012, Available from <http://people.msoe. edu/~saadat/matlabgui.htm

# Chapter 2

# USING MATLAB TO DESIGN AND ANALYSE COMPOSITE LAMINATES

Avinash Ramsaroop, Krishnan Kanny

Department of Mechanical Engineering, Durban University of Technology, Durban, South Africa

## ABSTRACT

This work deals with the generation of MATLAB script files that assists the user in the design of a composite laminate to operate within safe conditions. The inputs of the program are the material properties, material limits and loading conditions. Equations based on Hooke's Law for two-dimensional composites were used to determine the global and local stresses and strains on each layer. Failure analysis of the structure was performed via the Tsai-Wu failure theory. The output of the program is the optimal number of fibre layers required for the composite laminate, as well as the orientation of each layer.

## INTRODUCTION

Composites are extremely versatile materials and may be tailored to suit any function. They have found uses from the aerospace industry to common everyday applications. However, one drawback of these materials is tedious design processes [1]. Therefore, in an attempt to reduce this time consuming phase, it was decided to develop a computer program that assisted the user in designing a composite structure. The program needed to perform the necessary calculations in the fraction of the time it would take if done using conventional techniques.

Conventional methods for designing composite structures involve the use of Hooke's law for two-dimensional unidirectional composites. Equations relating the stresses and strains in these materials have been developed and are available from various texts [2-4]. These equations, however, are limited to flat unidirectional laminates. The procedure to follow is quite laborious. The material properties, material limits, number of fibre layers, and the fibre orientation and thickness of each layer as well as the loading conditions need to be known. These quantities are then used in the governing equations and numerous matrix computations are required. The outputs from these equations are the global and local stresses and strains of each layer. The local stresses on each layer are then compared to the allowable material limits and, if they exceed these limits, the entire process needs to be repeated with new angles and, if necessary, more fibre layers.

If a composite structure has to be designed to optimally carry the applied loads, the manual calculations would take an extremely long time. Designers usually use standard fibre angles in their design to overcome the time factor because exact fibre orientation is difficult [5]. As a result composite structures are usually overdesigned as incorrect fibre angles may mean more fibre layers which in turn leads to increased material costs. Many researchers [6-9] have attempted to reduce material costs by reducing the laminate thickness but these were based on fixed fibre angles. Certain regions of a composite structure, such as holes and other high stress zones, may need more reinforcement than others. Only these areas would require more fibre layers. However, with the conventional approach to design, the entire structure would be laid up to compensate for the high stress concentrations which increases the material costs.

Therefore an approach was needed where the calculations can be performed easily and automatically, and where only the regions with high stresses have extra reinforcement. A logical solution was to write computer programs to perform the computations. MATLAB, a numerical computing package, was used as a basis for the programs as it is more than sufficient to handle the numerous matrix computations [10-14]. It may also be interfaced with programs written in other languages, such as C, C++, and FORTRAN [13-14].

Basically, two programs were generated and each program consisted of a main script file and various functions. The first program was written according to the conventional methods for designing composite structures. The purpose of this program was to determine whether its functions operated efficiently and performed calculations correctly as these would be used for the second program. Results from this program were compared to manually calculated examples. The second program used a different approach where the number of inputs were reduced to just the material properties, material limits and applied loading conditions. The outputs of this program are the number of layers and the fibre angle of each layer.

These programs are discussed in further detail after the equations governing the stress-strain relationships are examined.

## HOOKE'S LAW FOR TWO-DIMENSIONAL COMPOSITE STRUCTURES

**Figure 1** shows a schematic representation of a composite lamina. The direction along the fibre axis is designated 1. The direction transverse to the fibre axis but in the plane of the lamina is designated 2. The direction transverse to both the fibre axis and the plane of the lamina (out of page) is designated 3. This direction is not shown in the figure as it only becomes necessary in three-dimensional cases.

The 1-2 co-ordinate system can be considered to be local co-ordinates based on the fibre direction. However this system is inadequate as fibres can be placed at various angles with respect to each other and the structure. Therefore a new co-ordinate system needs to be defined that takes into account the angle the fibre makes with its surroundings. This new system is referred to as global co-ordinates (x-y system) and is related to the local co-ordinates (1-2 system) by the angle θ.

A composite material is not isotropic and therefore its stresses and strains cannot be related by the simple Hooke's Law ($\sigma = E\varepsilon$). This law has to be extended to two-dimensions and redefined for the local and global co-ordinate systems [2]. The result is Equations (1) and (2).

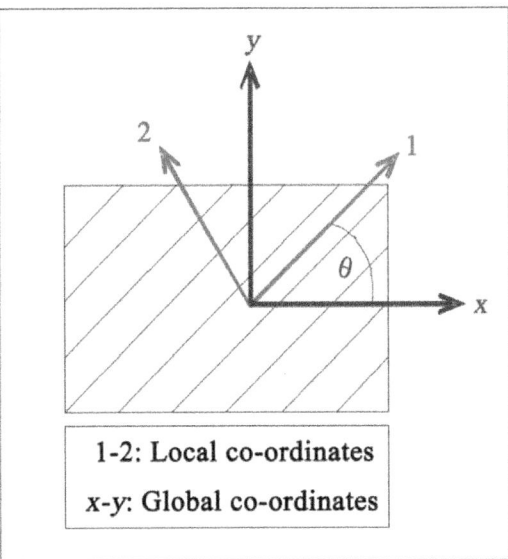

**Figure 1.** Global co-ordinate system in relation to local co-ordinate system.

$$\begin{bmatrix} \sigma_1 \\ \sigma_2 \\ \tau_{12} \end{bmatrix} = \begin{bmatrix} Q_{11} & Q_{12} & 0 \\ Q_{12} & Q_{22} & 0 \\ 0 & 0 & Q_{66} \end{bmatrix} \begin{bmatrix} \varepsilon_1 \\ \varepsilon_2 \\ \gamma_{12} \end{bmatrix}$$

(1)

$$\begin{bmatrix} \sigma_x \\ \sigma_y \\ \tau_{xy} \end{bmatrix} = \begin{bmatrix} \bar{Q}_{11} & \bar{Q}_{12} & \bar{Q}_{16} \\ \bar{Q}_{12} & \bar{Q}_{22} & \bar{Q}_{26} \\ \bar{Q}_{16} & \bar{Q}_{26} & \bar{Q}_{66} \end{bmatrix} \begin{bmatrix} \varepsilon_x \\ \varepsilon_y \\ \gamma_{xy} \end{bmatrix}$$

(2)

where $\sigma_{1,2}$ are the normal stresses in directions 1 and 2; $\tau_{12}$ is the shear stress in the 1-2 plane; $\varepsilon_{1,2}$ are the normal strains in directions 1 and 2; $\gamma_{12}$ is the shear strain in the 1-2 plane; [Q] is the reduced stiffness matrix; $\sigma_{x,y}$ are the normal stresses in directions x and y; $\tau_{xy}$ is the shear stress in the x-y plane; $\varepsilon_{x,y}$ are the normal strains in directions x and y; $\gamma_{xy}$ is the shear strain in the x-y plane; $[\bar{Q}]$ is the transformed reduced stiffness matrix.

The elements of the [Q] matrix in Equation (1) are dependent on the material constants and may be calculated using Equation (3).

$$Q_{11} = \frac{E_1}{1 - v_{12}v_{21}}, Q_{12} = \frac{v_{12}E_2}{1 - v_{12}v_{21}}, Q_{22} = \frac{E_2}{1 - v_{12}v_{21}}, Q_{66} = G_{12}$$

(3)

where $E_{1,2}$ are Young's modulus in directions 1 and 2; $G_{12}$ is the shear modulus in the 1-2 plane; $v_{12,21}$ are Poisson's ratios in the 1-2 and 2-1 planes.

The $[\bar{Q}]$ matrix in Equation (2) may be determined by Equation (4).

$$[\bar{Q}] = [T]^{-1}[Q][R][T][R]^{-1}$$

(4)

where [T] is the transformation matrix; [R] is the Reuter matrix. These matrices are given by:

$$T = \begin{bmatrix} c^2 & s^2 & 2sc \\ s^2 & c^2 & -2sc \\ -sc & sc & c^2 - s^2 \end{bmatrix} \quad and \quad R = \begin{bmatrix} 1 & 0 & 0 \\ 0 & 1 & 0 \\ 0 & 0 & 2 \end{bmatrix}$$

(5)

where $c = \cos\theta$; $s = \sin\theta$.

The local stresses and strains in Equation (1) are related to the global stresses and strains in Equation (2) by Equation (6).

$$\begin{bmatrix} \sigma_x \\ \sigma_y \\ \tau_{xy} \end{bmatrix} = [T]^{-1} \begin{bmatrix} \sigma_1 \\ \sigma_2 \\ \tau_{12} \end{bmatrix} \quad and \quad \begin{bmatrix} \varepsilon_1 \\ \varepsilon_2 \\ \gamma_{12} \end{bmatrix} = [R][T][R]^{-1} \begin{bmatrix} \varepsilon_x \\ \varepsilon_y \\ \gamma_{xy} \end{bmatrix}$$

(6)

Equations (1) to (6) are used to determine the stresses and strains for a single composite layer. Since composites are multi-layered entities, equations for this case must also be set up. The result is Equation (7).

$$\begin{bmatrix} N \\ M \end{bmatrix} = \begin{bmatrix} A & B \\ B & D \end{bmatrix} \begin{bmatrix} \varepsilon^0 \\ \kappa \end{bmatrix} \tag{7}$$

where N is the vector of resultant forces; M is the vector of resultant moments; $\varepsilon^0$ is the vector of the mid-plane strains; $\kappa$ is the vector of mid-plane curvatures. Vectors $\varepsilon$ and $\kappa$ are related to the global co-ordinates by Equation (8).

$$\begin{bmatrix} \varepsilon_x \\ \varepsilon_y \\ \gamma_{xy} \end{bmatrix} = \begin{bmatrix} \varepsilon_x^0 \\ \varepsilon_y^0 \\ \gamma_{xy}^0 \end{bmatrix} + z \begin{bmatrix} \kappa_x \\ \kappa_y \\ \kappa_{xy} \end{bmatrix} \tag{8}$$

where z is an arbitrary distance from the mid-plane.

The [A], [B], and [D] matrices in Equation (7) are known as the extensional, coupling, and bending stiffness matrices, respectively. The elements of these matrices may be determined from Equations (9) to (11).

$$A_{ij} = \sum_{k=1}^{n} \left[ \left( \overline{Q_{ij}} \right) \right]_k \left( h_k - h_{k-1} \right) \quad i = 1,2,3 \quad j = 1,2,3 \tag{9}$$

$$B_{ij} = \frac{1}{2} \sum_{k=1}^{n} \left[ \left( \overline{Q_{ij}} \right) \right]_k \left( h_k^2 - h_{k-1}^2 \right) \quad i = 1,2,3 \quad j = 1,2,3 \tag{10}$$

$$D_{ij} = \frac{1}{3} \sum_{k=1}^{n} \left[ \left( \overline{Q_{ij}} \right) \right]_k \left( h_k^3 - h_{k-1}^3 \right) \quad i = 1,2,3 \quad j = 1,2,3 \tag{11}$$

where n is the number of layers; $\left[ \left( \overline{Q_{ij}} \right) \right]_k$ is the i-thj-th element of the $[\overline{Q}]$ matrix of the k-th layer; $h_k$ is the distance of the top or bottom of the k-th layer from the mid-plane of the composite. **Figure 2** illustrates how to determine the distance $h_k$ from the mid-plane.

The final aspect involved in the design of a composite structure is the failure analysis. There are various failure theories; however, the Tsai-Wu failure criterion is only one that closely correlates with experimental data. This failure theory is given by Equation (12).

$$H_1\sigma_1 + H_2\sigma_2 + H_6\tau_{12} + H_{11}\sigma_1^2 + $$
$$H_{22}\sigma_2^2 + H_{66}\tau_{12}^2 + 2H_{12}\sigma_1\sigma_2 < 1 \tag{12}$$

The parameters for the Tsai-Wu failure criterion are given by Equation (13).

$$H_1 = \frac{1}{\left(\sigma_1^T\right)_{ult}} - \frac{1}{\left(\sigma_1^C\right)_{ult}} \; ; H_2 = \frac{1}{\left(\sigma_2^T\right)_{ult}} - \frac{1}{\left(\sigma_2^C\right)_{ult}} \; ; H_6 = 0$$

$$H_{11} = \frac{1}{\left(\sigma_1^T\right)_{ult}\left(\sigma_1^C\right)_{ult}} \; ; H_{22} = \frac{1}{\left(\sigma_2^T\right)_{ult}\left(\sigma_2^C\right)_{ult}} \; ; H_{66} = \frac{1}{\left(\tau_{12}\right)_{ult}^2}$$

$$H_{12} = -\frac{1}{2}\sqrt{\frac{1}{\left(\sigma_1^T\right)_{ult}\left(\sigma_1^C\right)_{ult}\left(\sigma_2^T\right)_{ult}\left(\sigma_2^C\right)_{ult}}}$$

(13)

where $\left(\sigma_{1,2}^T\right)_{ult}$ are the ultimate tensile stresses in direction 1 and 2; $\left(\sigma_{1,2}^C\right)_{ult}$ are the ultimate compressive stresses in direction 1 and 2; $\left(\tau_{12}\right)_{ult}$ is the ultimate shear stress in the 1-2 plane.

In order to better facilitate the use of this failure theory, each stress component of Equation (12) was multiplied by a variable [2]. This variable is referred to as the strength ratio (SR) and combining this in Equation (12) resulted in Equation (14).

$$(H_1\sigma_1 + H_2\sigma_2 + H_6\tau_{12})SR +$$
$$(H_{11}\sigma_1^2 + H_{22}\sigma_2^2 + H_{66}\tau_{12}^2 + 2H_{12}\sigma_1\sigma_2)SR^2 < 1$$

(14)

The purpose of SR is to directly determine by what ratio the local stresses must be increased or decreased to avoid failure. This also directly relates to the applied forces. The criterion for SR is that it can only be positive. If SR is less than 1, then failure occurs because it means that the loading needs to decrease to avoid failure. A SR value of 1 implies that the composite structure is perfectly suited for the applied loading conditions. A value of greater than 1 means that the structure is more than capable of carrying the applied loading and that the loading may also be increased. For example, a SR value of 1.5 means that the loading may be increased up to 50% without failure occurring.

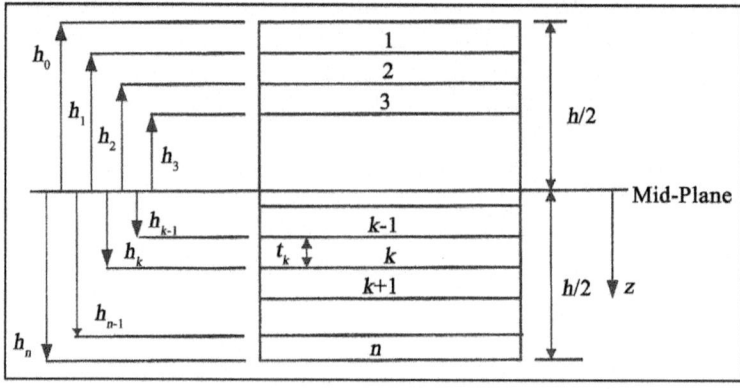

**Figure 2**: Locations of layers in a composite structure [2].

# PROGRAMMING

## Conventional Approach Program

This approach follows the commonly found methods laid out in the various texts [2-4]. As stated above, the material constants, material limits, number of fibre layers, fibre angle of each layer, and the thickness of each layer as well as the loading conditions need to be known. The following procedure illustrates the steps involved in this approach:

1) Calculate the [Q] matrix for each layer using Equation (3).

2) Using Equation (5), compute the [T] matrix for each layer.

3) Determine $[\bar{Q}]$ matrix for each layer via Equation (4).

4) Find the location of the top and bottom surface of each layer, $h_k$ (k = 1 to n).

5) Use Equations (9), (10) and (11) to calculate the [A], [B], and [D] matrices.

6) Substitute these matrices as well as the applied loading into Equation (7) and solve the six simultaneous equations to determine the mid-plane strains ($\varepsilon^0$) and curvatures ($\kappa$).

7) Find the global strains and stresses in each layer using Equations (8) and (2), respectively.

8) Find the local strains and stresses for each layer by using Equation (6).

9) Determine the parameters for the Tsai-Wu failure criterion via Equation (13).

10) Using these parameters and the local stresses for each layer, find SR via Equation (14).

Once SR has been determined, the designer will know whether the structure will fail or not. If failure occurs, the fibre angles and possibly the number of layers have to be adjusted, and the above procedure needs to be repeated until no failure takes place. Manual calculations would take a long period, therefore a program was written to perform the necessary calculations.

The program prompts the user for the following inputs:

- the material properties ($E_1$, $E_2$, $G_{12}$ and $v_{12}$)
- the material limits ($(\sigma_1^T)_{ult}$, $(\sigma_1^C)_{ult}$, $(\sigma_2^T)_{ult}$, $(\sigma_2^C)_{ult}$, $(\tau_{12})_{ult}$)
- the number of layers (N)
- the fibre angle ($\theta$) of each layer

- the thickness (t) of each layer
- applied loading conditions

These input parameters were used to determine the [Q], [T], and $[\overline{Q}]$ matrices for each layer. All data concerning each layer ($E_1$, $E_2$, $G_{12}$, $v_{12}$, $\theta$, t, [Q], [T], and $[\overline{Q}]$) were stored in an array. The data in this array was used to calculate the [A], [B], and [D] matrices. Using these matrices and the applied loading conditions, the global and local stresses and strains were computed. The local stresses were compared to the material limits, via the Tsai-Wu failure criterion, to determine whether the composite will fail.

The computation of the various matrices and other relevant data was each carried out in a separate function. This means there was one function to determine the [Q] matrix, another to calculate the [T] matrix, and so on. A script file was written that controlled the use of each function. The purpose of the functions was to enable easier programming in future.

This program was tested against manually calculated examples in the various texts [2, 3] and the results were exactly comparable to the manual computations. In an example (Example 4.3) by Kaw [2], the stresses and strains in a graphite/epoxy composite laminate were examined. This laminate consisted of three layers, with fibre angles of 0°, 30° and –45°, and each layer had a thickness of 5 mm. The material properties, material limits and loading conditions used in this example are given in **Table 1**.

The resulting global strains, global stresses, local strains and local stresses at the top, middle and bottom of each layer are shown in Figures 3 to 6, respectively.

The values in Table 1 and the fibre angles of 0°, 30° and –45° were used as inputs in the Conventional Approach script file. The global and local stresses and strains were computed and are shown in Figures 7 to 10. The values in these tables and the ones in Figures 3 to 6 are exactly comparable. The conclusion from this is that the Conventional Approach program can accurately determine the stresses and strains in a flat unidirectional composite structure.

The Conventional Approach script file was further extended to include failure analysis via the Tsai-Wu failure theory. The resulting SR values for the top, middle and bottom of each layer is shown in Figure 11. The rows represent the layers and columns 1 to 3 represent the top, middle and bottom of each layer, respectively.

The SR values in Figure 11 indicate that the composite structure will fail first on layer 1 (SR = 0.6147) and then layer 2 (SR = 0.8870). It is clear from this that the input angles and number of layers were insufficient to carry the applied loads.

Although this program performs all the relevant calculations easily and automatically, and significantly shortens the design process, using the program to design a composite structure would require much guesswork. The designer would enter the necessary input values, allow the program to perform the various calculations, and then discover whether the composite structure would fail. If failure does occur, as in the above case, the designer may have to change the fibre angles and/or the thickness of each layer and/or the number of layers. The choice of values may be uncertain and values would have to be assumed. Furthermore, it would also be an uncertainty on which values to change. Much time may be spent in achieving a structure that would not fail. On the other hand, if failure does not occur, the structure may prove to be overdesigned in which case the manufacturing costs of the structure may increase as unnecessary material will be used. To overcome these problems, a new script file was generated.

**Table:** Material properties, limits and loading conditions for graphite/epoxy composite example in Kaw [2].

| Material Properties | | | | | | | | | Forces | Moments |
|---|---|---|---|---|---|---|---|---|---|---|
| $E_1$ (GPa) | $E_2$ (GPa) | $G_{12}$ (GPa) | $v_{12}$ | $(\sigma_1^T)_{ult}$ (MPa) | $(\sigma_1^C)_{ult}$ (MPa) | $(\sigma_2^T)_{ult}$ (MPa) | $(\sigma_2^C)_{ult}$ (MPa) | $(\tau_{12})_{ult}$ (MPa) | $N$ | $M$ |
| | | | | | | | | | 1000 | 0 |
| 181 | 10.3 | 7.17 | 0.28 | 1500 | 1500 | 40 | 246 | 68 | 1000 | 0 |
| | | | | | | | | | 0 | 0 |

| Table 4.1 Global Strains (m/m) in Example 4.3 | | | | |
|---|---|---|---|---|
| Ply # | Position | $\varepsilon_x$ | $\varepsilon_y$ | $\gamma_{xy}$ |
| | Top | 8.944 ($10^{-8}$) | 5.955 ($10^{-6}$) | −3.836 ($10^{-6}$) |
| 1 (0°) | Middle | 1.637 ($10^{-7}$) | 5.134 ($10^{-6}$) | −2.811 ($10^{-6}$) |
| | Bottom | 2.380 ($10^{-7}$) | 4.313 ($10^{-6}$) | −1.785 ($10^{-6}$) |
| | Top | 2.380 ($10^{-7}$) | 4.313 ($10^{-6}$) | −1.785 ($10^{-6}$) |
| 2 (30°) | Middle | 3.123 ($10^{-7}$) | 3.492 ($10^{-6}$) | −7.598 ($10^{-7}$) |
| | Bottom | 3.866 ($10^{-7}$) | 2.670 ($10^{-6}$) | 2.655 ($10^{-7}$) |
| | Top | 3.866 ($10^{-7}$) | 2.670 ($10^{-6}$) | 2.655 ($10^{-7}$) |
| 3 (−45°) | Middle | 4.609 ($10^{-7}$) | 1.849 ($10^{-6}$) | 1.291 ($10^{-6}$) |
| | Bottom | 5.352 ($10^{-7}$) | 1.028 ($10^{-6}$) | 2.316 ($10^{-6}$) |

**Figure 3**: Table from example in Kaw [2] showing global strains.

| Table 4.2 Global Stresses (Pa) in Example 4.3 | | | | |
|---|---|---|---|---|
| Ply # | Position | $\sigma_x$ | $\sigma_y$ | $\tau_{xy}$ |
| | Top | 3.351 ($10^4$) | 6.188 ($10^4$) | -2.750 ($10^4$) |
| 1 ($0°$) | Middle | 4.464 ($10^4$) | 5.359 ($10^4$) | -2.015 ($10^4$) |
| | Bottom | 5.577 ($10^4$) | 4.531 ($10^4$) | -1.280 ($10^4$) |
| | Top | 6.930 ($10^4$) | 7.391 ($10^4$) | 3.381 ($10^4$) |
| 2 ($30°$) | Middle | 1.063 ($10^5$) | 7.747 ($10^4$) | 5.903 ($10^4$) |
| | Bottom | 1.434 ($10^5$) | 8.102 ($10^4$) | 8.426 ($10^4$) |
| | Top | 1.235 ($10^5$) | 1.563 ($10^5$) | -1.187 ($10^5$) |
| 3 ($-45°$) | Middle | 4.903 ($10^4$) | 6.894 ($10^4$) | -3.888 ($10^4$) |
| | Bottom | -2.547 ($10^4$) | -1.840 ($10^4$) | 4.091 ($10^4$) |

**Figure 4**: Table from example in Kaw [2] showing global stresses.

| Table 4.3    Local Strains (m/m) in Example 4.3 | | | | |
|---|---|---|---|---|
| Ply # | Position | $\varepsilon_1$ | $\varepsilon_2$ | $\gamma_{12}$ |
| | Top | 8.944 ($10^{-8}$) | 5.955 ($10^{-6}$) | -3.836 ($10^{-6}$) |
| 1 ($0°$) | Middle | 1.637 ($10^{-7}$) | 5.134 ($10^{-6}$) | -2.811 ($10^{-6}$) |
| | Bottom | 2.380 ($10^{-7}$) | 4.313 ($10^{-6}$) | -1.785 ($10^{-6}$) |
| | Top | 4.837 ($10^{-7}$) | 4.067 ($10^{-6}$) | 2.636 ($10^{-6}$) |
| 2 ($30°$) | Middle | 7.781 ($10^{-7}$) | 3.026 ($10^{-6}$) | 2.374 ($10^{-6}$) |
| | Bottom | 1.073 ($10^{-6}$) | 1.985 ($10^{-6}$) | 2.111 ($10^{-6}$) |
| | Top | 1.396 ($10^{-6}$) | 1.661 ($10^{-6}$) | -2.284 ($10^{-6}$) |
| 3 ($-45°$) | Middle | 5.096 ($10^{-7}$) | 1.800 ($10^{-6}$) | -1.388 ($10^{-6}$) |
| | Bottom | -3.766 ($10^{-7}$) | 1.940 ($10^{-6}$) | -4.928 ($10^{-7}$) |

**Figure 5**: Table from example in Kaw [2] showing local strains.

| Table 4.4 Local Stresses (Pa) in Example 4.3 | | | | |
|---|---|---|---|---|
| Ply # | Position | $\sigma_1$ | $\sigma_2$ | $\tau_{12}$ |
| | Top | 3.351 ($10^4$) | 6.188 ($10^4$) | -2.750 ($10^4$) |
| 1 ($0°$) | Middle | 4.464 ($10^4$) | 5.359 ($10^4$) | -2.015 ($10^4$) |
| | Bottom | 5.577 ($10^4$) | 4.531 ($10^4$) | -1.280 ($10^4$) |
| | Top | 9.973 ($10^4$) | 4.348 ($10^4$) | 1.890 ($10^4$) |
| 2 ($30°$) | Middle | 1.502 ($10^5$) | 3.356 ($10^4$) | 1.702 ($10^4$) |
| | Bottom | 2.007 ($10^5$) | 2.364 ($10^4$) | 1.513 ($10^4$) |
| | Top | 2.586 ($10^5$) | 2.123 ($10^4$) | -1.638 ($10^4$) |
| 3 ($-45°$) | Middle | 9.786 ($10^4$) | 2.010 ($10^4$) | -9.954 ($10^3$) |
| | Bottom | -6.285 ($10^4$) | 1.898 ($10^4$) | -3.533 ($10^3$) |

**Figure 6**: Table from example in Kaw [2] showing local stresses.

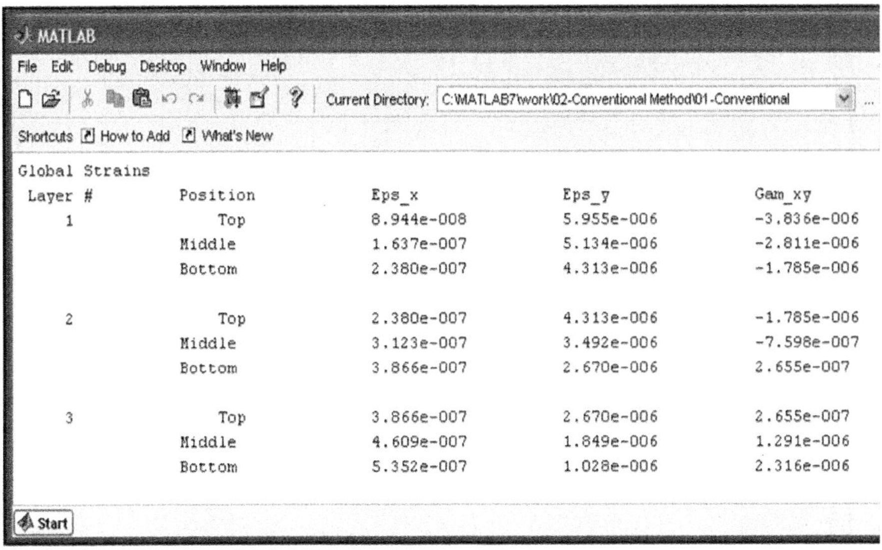


Global Strains

| Layer # | Position | Eps_x | Eps_y | Gam_xy |
|---|---|---|---|---|
| 1 | Top | 8.944e-008 | 5.955e-006 | -3.836e-006 |
| | Middle | 1.637e-007 | 5.134e-006 | -2.811e-006 |
| | Bottom | 2.380e-007 | 4.313e-006 | -1.785e-006 |
| 2 | Top | 2.380e-007 | 4.313e-006 | -1.785e-006 |
| | Middle | 3.123e-007 | 3.492e-006 | -7.598e-007 |
| | Bottom | 3.866e-007 | 2.670e-006 | 2.655e-007 |
| 3 | Top | 3.866e-007 | 2.670e-006 | 2.655e-007 |
| | Middle | 4.609e-007 | 1.849e-006 | 1.291e-006 |
| | Bottom | 5.352e-007 | 1.028e-006 | 2.316e-006 |

**Figure 7:** Global strains from Conventional Approach program with Table 1 as inputs.

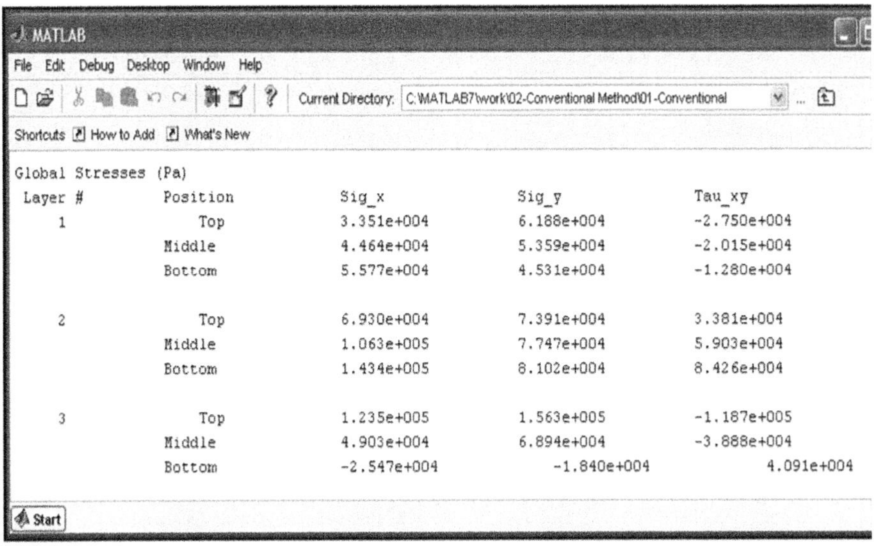


Global Stresses (Pa)

| Layer # | Position | Sig_x | Sig_y | Tau_xy |
|---|---|---|---|---|
| 1 | Top | 3.351e+004 | 6.188e+004 | -2.750e+004 |
| | Middle | 4.464e+004 | 5.359e+004 | -2.015e+004 |
| | Bottom | 5.577e+004 | 4.531e+004 | -1.280e+004 |
| 2 | Top | 6.930e+004 | 7.391e+004 | 3.381e+004 |
| | Middle | 1.063e+005 | 7.747e+004 | 5.903e+004 |
| | Bottom | 1.434e+005 | 8.102e+004 | 8.426e+004 |
| 3 | Top | 1.235e+005 | 1.563e+005 | -1.187e+005 |
| | Middle | 4.903e+004 | 6.894e+004 | -3.888e+004 |
| | Bottom | -2.547e+004 | -1.840e+004 | 4.091e+004 |

**Figure 8:** Global stresses from Conventional Approach program with Table 1 as inputs.

Local Strains

| Layer # | Position | Eps_1 | Eps_1 | Gam_12 |
|---|---|---|---|---|
| 1 | Top | 8.944e-008 | 5.955e-006 | -3.836e-006 |
| | Middle | 1.637e-007 | 5.134e-006 | -2.811e-006 |
| | Bottom | 2.380e-007 | 4.313e-006 | -1.785e-006 |
| 2 | Top | 4.837e-007 | 4.067e-006 | 2.636e-006 |
| | Middle | 7.781e-007 | 3.026e-006 | 2.374e-006 |
| | Bottom | 1.073e-006 | 1.985e-006 | 2.111e-006 |
| 3 | Top | 1.396e-006 | 1.661e-006 | -2.284e-006 |
| | Middle | 5.096e-007 | 1.800e-006 | -1.388e-006 |
| | Bottom | -3.766e-007 | 1.940e-006 | -4.928e-007 |

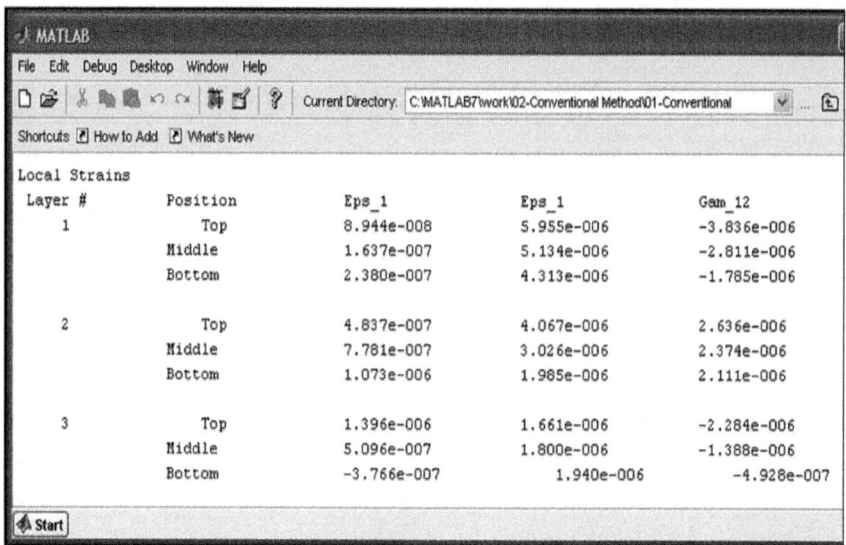

**Figure 9:** Local strains from Conventional Approach program with Table 1as inputs.

Local Stresses (Pa)

| Layer # | Position | Sig_1 | Sig_2 | Tau_12 |
|---|---|---|---|---|
| 1 | Top | 3.351e+004 | 6.188e+004 | -2.750e+004 |
| | Middle | 4.464e+004 | 5.359e+004 | -2.015e+004 |
| | Bottom | 5.577e+004 | 4.531e+004 | -1.280e+004 |
| 2 | Top | 9.973e+004 | 4.348e+004 | 1.890e+004 |
| | Middle | 1.502e+005 | 3.356e+004 | 1.702e+004 |
| | Bottom | 2.007e+005 | 2.364e+004 | 1.513e+004 |
| 3 | Top | 2.586e+005 | 2.123e+004 | -1.638e+004 |
| | Middle | 9.786e+004 | 2.010e+004 | -9.954e+003 |
| | Bottom | -6.285e+004 | 1.898e+004 | -3.533e+003 |

**Figure 10:** Local stresses from Conventional Approach program with Table 1 as inputs.

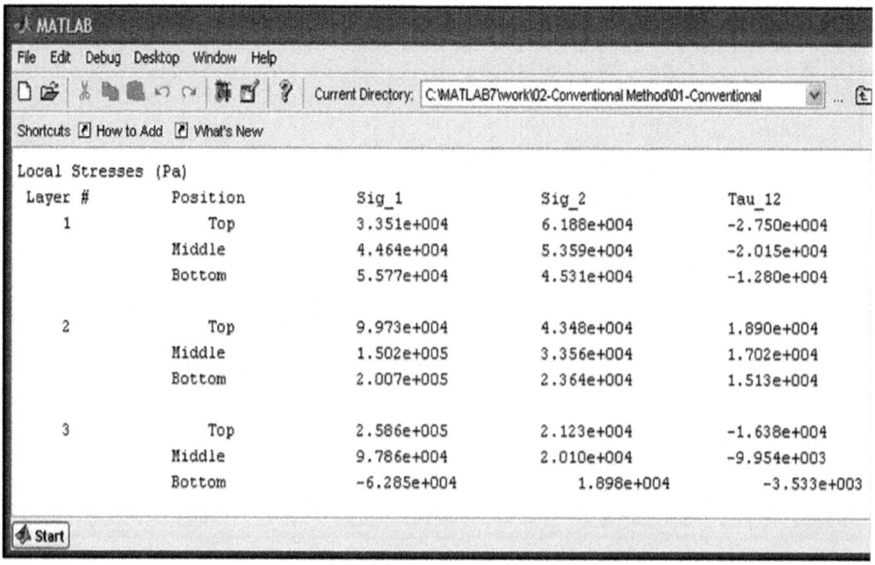

## Fibre-Angle-Output Program

Initially it was decided that this program should take the applied loading conditions, material properties and material limits as inputs. The program was to assume that the number of layers equals the number of input forces, and that the fibre angle of each layer equals the direction of each force. It would then perform all the necessary matrix calculations as in the Conventional Approach. The global and local stresses and strains would be determined as before and failure analysis with the Tsai-Wu criterion may be performed.

However, the assumptions made above cannot be used in a real situation as it may result in an overdesign of a composite structure as in the case of the conventional approach. Hence a program was written that does not make any assumptions but rather builds the laminate up one layer at a time. Basically this program varies the fibre angles to achieve a composite structure that is perfectly suited to carry the applied loading conditions.

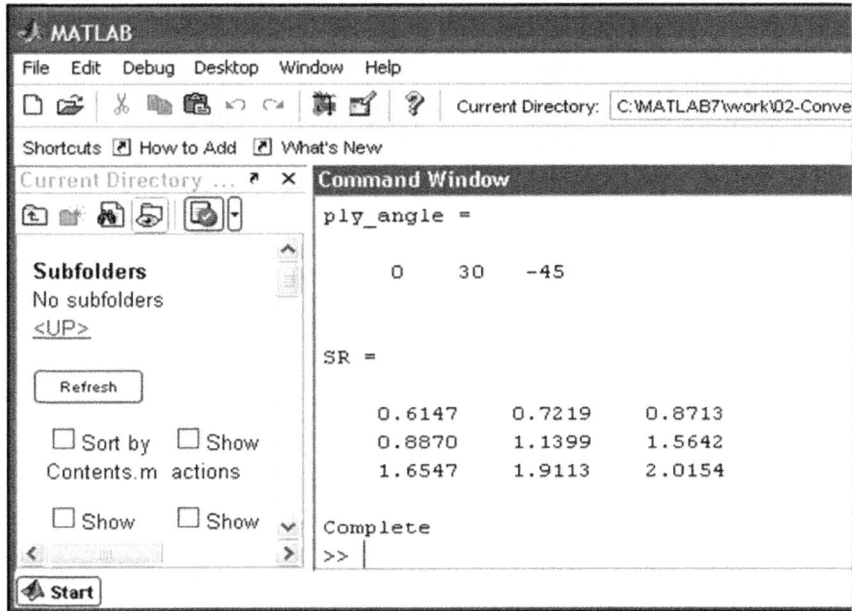

**Figure 11:** SR ratios from Conventional Approach program with Table 1 as inputs.

The Outputs of the program are the number of layers and the fibre angle of each layer.

The program asks the user to input the following:

- Material properties
- Material limits

- Loading conditions

The [Q] matrix and the Tsai-Wu parameters can be calculated immediately as these do not vary with fibre angle but rather with material properties and limits. The program begins with one layer at an angle of 0° and computes $h_k$, as well as the [T], [$\bar{Q}$], [A], [B] and [D] matrices. Thereafter the mid-plane strains and curvatures, and, the global and local stresses and strains are calculated. The Tsai-Wu failure theory is applied and a value for SR is obtained for each layer.

The program then analyzes the SR values and confirms whether it is in a certain range. The lower limit of this range is 1 as any value below this would mean failure. The upper limit in this range is 1.2 and this is to avoid overdesigning.

If the SR value for each layer is between 1 and 1.2, then the design of the composite structure is complete. On the other hand, if the SR value for any of the layers is out of this range, the program varies that particular fibre angle to obtain a SR value between 1 and 1.2. However, due to the angle change all the matrix calculations have to be redone. A new SR value is obtained for each layer and this is compared to the old values. If the new SR value for a particular layer is higher than the old SR value, then the angle of that layer is changed to the new one. For example, after the initial calculation a layer with fibre angle of 30° has a SR value of 0.8. This is not acceptable and the program changes the angle to obtain a better SR value. Say at 35° the SR value is 0.9, the program will then change the angle of the 30° layer to 35°. If at 35° the SR value was 0.7, then the program will retain the 30° angle.

The program continues to change angles and compare old and new SR values until the SR value of each layer is between 1 and 1.2, in which case the design is completed. However, if this range cannot be achieved, a new layer is added on. The program starts from the beginning and recalculates the various matrices. SR values are obtained and the comparison between new and old values resumes. New layers will be added on until the SR value for each layer is between 1 and 1.2. The program then outputs the number of layers and the fibre angle of each layer. **Figure 12**illustrates the procedure the script file follows to obtain the desired outputs.

The Fibre-Angle-Output program was subjected to several runs. The results, namely the number of layers and the fibre angles, needed to be verified. Therefore these were used as inputs in the Conventional Approach program. The Tsai-Wu failure theory was used and in each case there was no failure. As an example, consider the graphite/epoxy composite laminate examined earlier. The material properties, material limits and loading conditions for this laminate were given in Table 1. The SR values for three layers with fibre angles of 0°, 30° and –45° were shown in Figure 11. The values in Table 1 were input into the Fibre-Angle-Output program. The output of the program is shown in Figure 13.

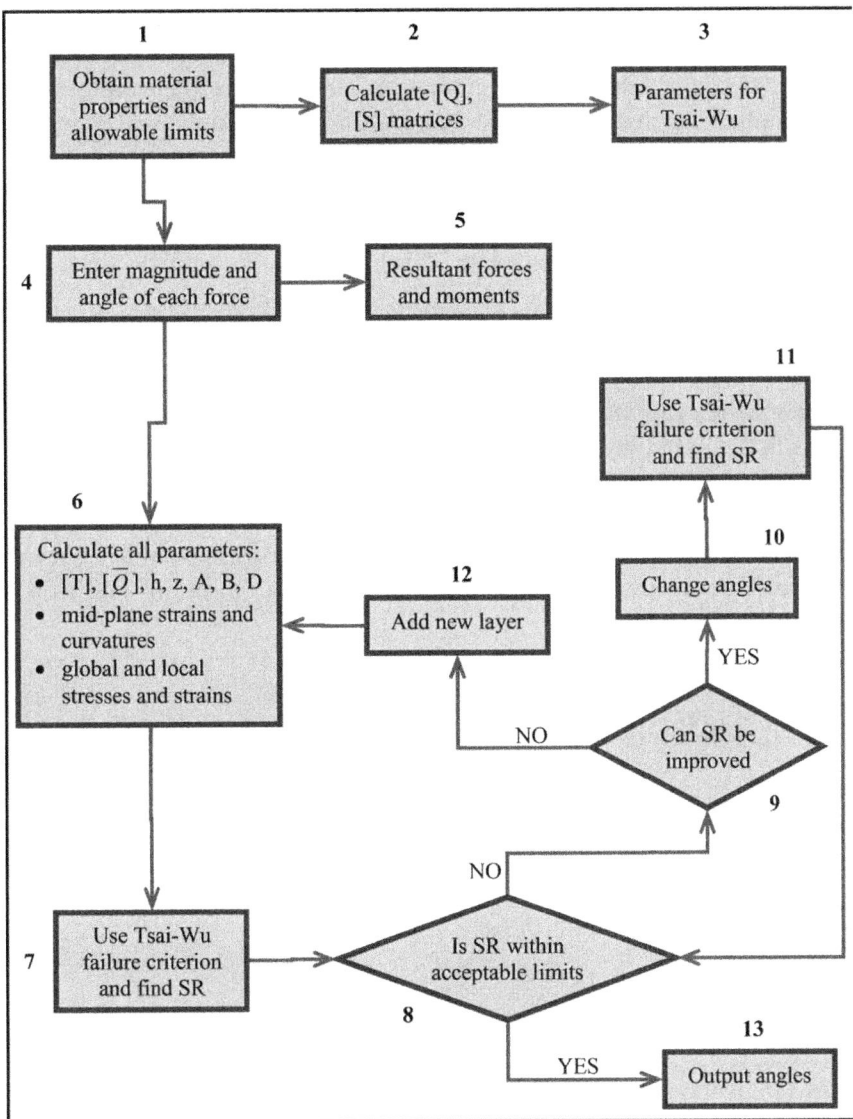

**Figure 12**: Flowchart describing the Fibre-Angle-Output program.

It can be seen from Figure 13 that the number of layers is the same as when the example was analyzed by the Conventional Approach program; however, the fibre angles are different. Here the required angles are 49°, −79° and 49° instead of the previous 0°, 30° and −45°. Further the SR values are all between 1 and 1.2 indicating no failure or overdesign.

The new fibre angles (49°, −79° and 49°) along with the material properties, material limits and loading conditions in Table 1 were then input into the Conventional Approach program. The resulting SR values were exactly the same as in Figure 13. Further, the local stresses of the new laminate, shown in Figure 14, was compared with the local stresses of the earlier laminate, shown in Figure 10. It can be seen that the local stresses in Figure 14 are constant through the thickness of each layer. This shows that the stresses were evenly distributed and not varied through the thickness of each layer.

Previously the SR value on the first layer was less than 1 (Figure 11) and lower than the other two layers indicating earliest failure. However with the new layup, the SR value has increased to a value greater than one. This means that the first layer is now capable of carrying a higher load and this is evident by the values for the local stresses. The previous value at the top of the first layer was 33.5 MPa (Figure 10), and the new value is 75.7 MPa (Figure 14). This clearly demonstrates the higher load bearing capability of the new laminate. Similar trends may be seen from the other positions and from the other two layers.

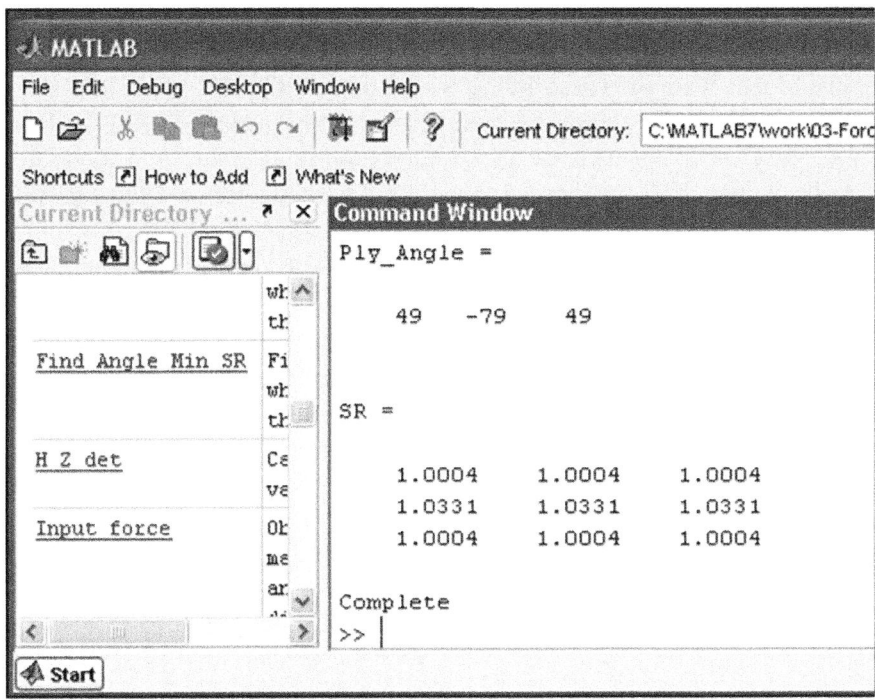

**Figure 13:** Output of Fibre-Angle-Output program with Table 1 as input.

| Local Stresses (Pa) | | | | |
|---|---|---|---|---|
| Layer # | Position | Sig_1 | Sig_2 | Tau_12 |
| 1 | Top | 7.575e+004 | 3.662e+004 | -2.281e+004 |
| | Middle | 7.575e+004 | 3.662e+004 | -2.281e+004 |
| | Bottom | 7.575e+004 | 3.662e+004 | -2.281e+004 |
| 2 | Top | 1.414e+005 | 3.389e+004 | 2.692e+004 |
| | Middle | 1.414e+005 | 3.389e+004 | 2.692e+004 |
| | Bottom | 1.414e+005 | 3.389e+004 | 2.692e+004 |
| 3 | Top | 7.575e+004 | 3.662e+004 | -2.281e+004 |
| | Middle | 7.575e+004 | 3.662e+004 | -2.281e+004 |
| | Bottom | 7.575e+004 | 3.662e+004 | -2.281e+004 |

**Figure 14**: Local stresses for laminate with angles of 49°, –79°, and 49°.

It was then decided to increase the loading conditions of the 49°, −79° and 49° laminate to determine whether an increase in load will result in failure. The applied loads were increased by 10% and the resulting SR values are shown in **Figure 15**. It can clearly be seen that all the SR values for all the layers were less than 1. An increase in loading resulted in failure and therefore it may be deduced that the Fibre-Angle-Output program optimally designed the composite laminate for the loading conditions in **Table 1**. This further implies a saving in material cost as there would be no extra material used because of overdesigning.

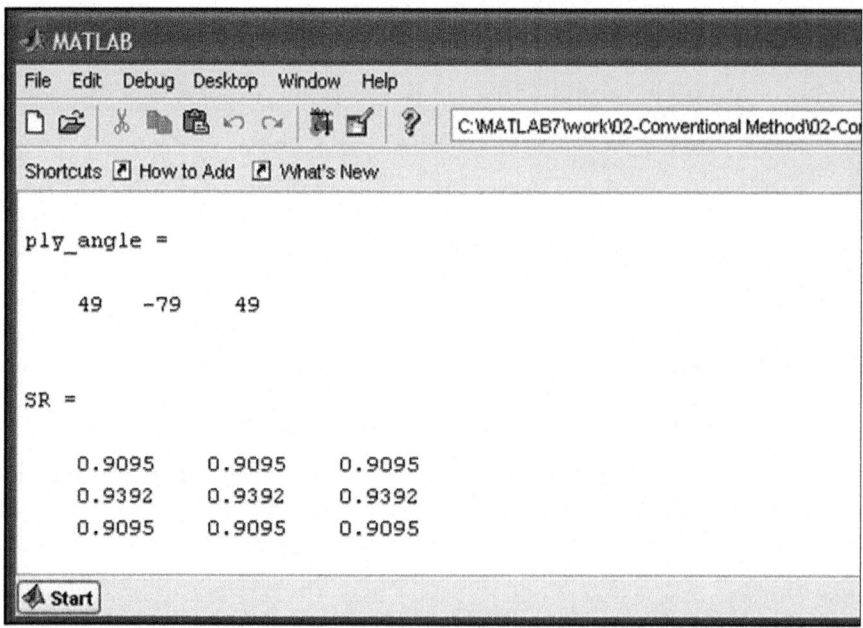

**Figure 15**: SR ratios for 49°, −79° and 49° laminate with 10% increase in loading.

## Finite Element Analysis

For further verification of the computational results, it was decided to perform finite element analyses on the laminates examined in the Conventional Approach program and the Fibre-Angle-Output program. The package used for the analysis was MSC Patran/Nastran 2007.

Two finite element models were setup, and both models represented a flat composite plate with material properties, material limits and loading conditions from **Table 1**. The first model had the same fibre layup that was used in the example for the Conventional Approach program, that is, 0°, 30°, and –45°. The analysis was run and the resulting fringe plots of the stress components of each layer were plotted. These plots are shown in Figure 16.

The stress plots from the finite element package corresponds to the local stresses experienced by a laminate, in particular the stresses occurring at the middle of each layer. Comparing the stress values in Figure 16 with the stresses incurred at the middle of each layer in either Figure 6 or Figure 10, it can be seen that the values are exactly the same. This implies that the model was correctly setup and that it may be used to evaluate local stresses of laminates designed with the Fibre-Angle Output program.

The second model was setup and was similar to the first model in every aspect except for the fibre layup. This model had the new fibre layup (49°, –79°, and 49°) that was determined by the Fibre-Angle-Output program. The fringe plots of the stress components from this analysis are shown in Figure 17. Here it is also evident that the values in the plots exactly correspond to the stress values at the middle of each layer in Figure 14. This verifies the above results from the Fibre-Angle Output program and supports the conclusion that the program optimally designed the composite laminate, with the fibre layup of 49°, –79° and 49°, for the loading conditions in Table 1.

## CONCLUSIONS

A MATLAB script file was generated that uses the conventional approach in the design of composite laminates. The inputs are the material properties, material limits, number of fibre layers, and the fibre orientation and thickness of each layer as well as the loading conditions. These values are then used in the governing equations, based on Hooke's law for two-dimensional unidirectional laminates, to determine the global and local stresses and strains. The local stresses are compared to allowable limits via the Tsai-Wu failure theory. The results from this program were compared to manually calculated examples in the various texts [2,3] and it was found that they were exactly comparable.

The above program required much guesswork in choosing inputs and was therefore inadequate to design composite laminates. Hence it was adapted to take material properties, material limits and the loading conditions as inputs. The necessary calculations were performed and the optimum number of fibre layers required as well as the fibre angle of each layer was determined.

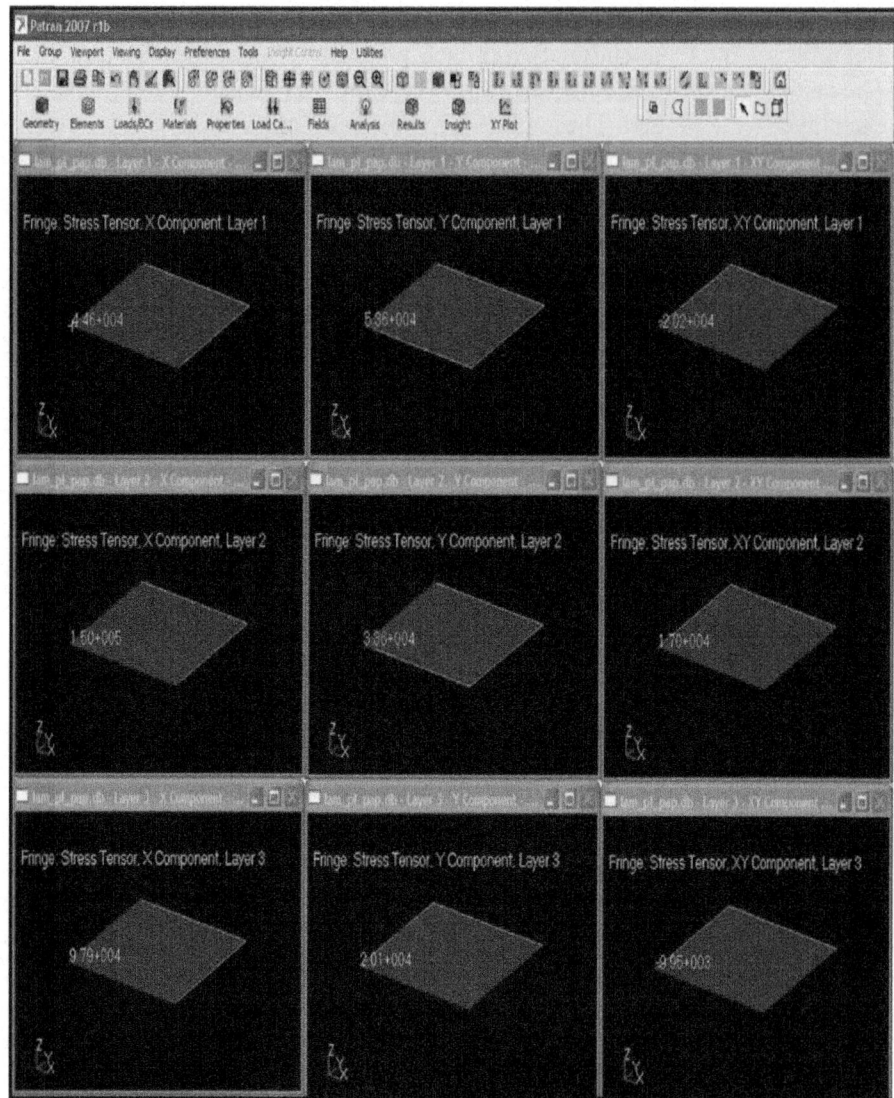

**Figure 16**: Fringe plots from Patran showing stress components for 0°, 30°, and −45° laminate.

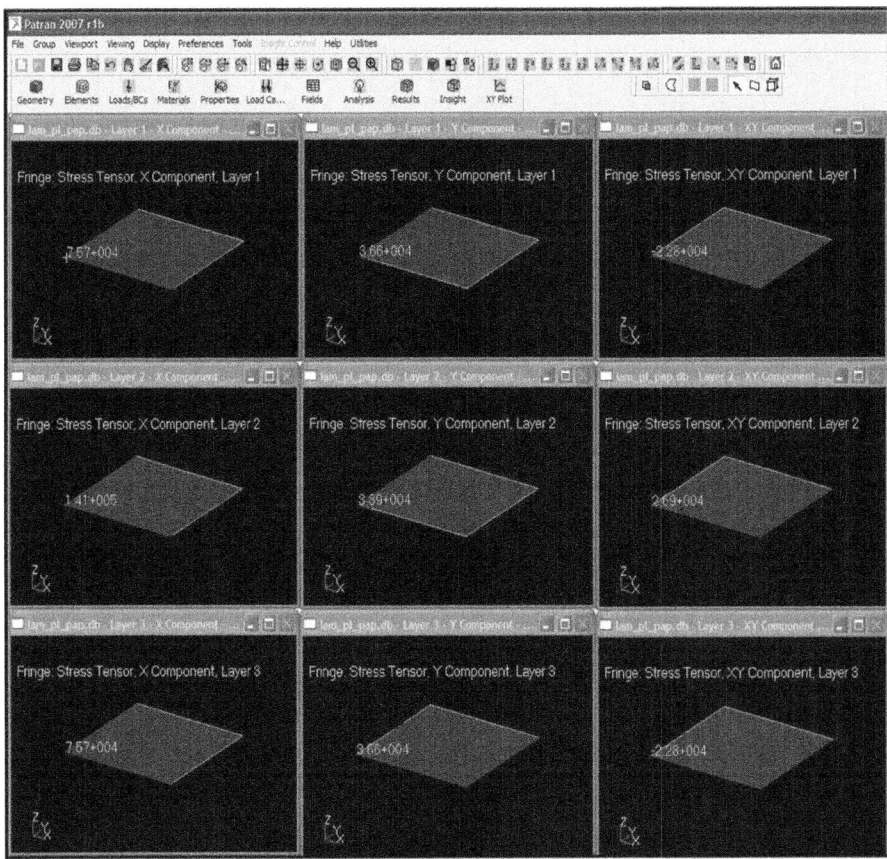

**Figure 17**: Fringe plots from Patran showing stress components for 49°, –79°, and 49° laminate.

These outputs were input into the Conventional Approach program in order to verify the results. In each case the resulting composite structure did not fail. The example used showed that the Fibre-Angle-Output program optimally designed the composite laminate for the given loading conditions. Finite element analyses were conducted to verify this conclusion.

The use of these programs will greatly reduce the design time of composite structures as the numerous computations are completed in a fraction of the time that it would take if done manually. The designer merely has to run the Fibre-Angle-Output program, and view the number of layers required and the fibre angle for each layer. The program will also aid in reducing material costs. Presently the program is limited to flat unidirectional structures but work is underway to extend it to three dimensions. There is no limit on the type of composite material as far as the matrix and fibre reinforcement is concerned.

## REFERENCES

1.   S. K. Mazumdar, "Composite Manufacturing: Materials, Product, and Process Engineering," CRC Press LLC, Boca Raton, 2002.

2.   A. K. Kaw, "Mechanics of Composite Materials," CRC Press LLC, Boca Raton, 1997.

3.   C. T. Herakovich, "Mechanics of Fibrous Materials," John Wiley & Sons Inc., Hoboken, 1998.

4.   R. M. Christensen, "Mechanics of Composite Materials," Dover Publications, New York, 2005.

5.   M. Akbulut and F. O. Sonmez, "Optimum Design of Composite Laminates for Minimum Thickness," Computers and Structures, Vol. 86, No. 21-22, 2008, pp. 1974-1982.

6.   K. Sivakumar, N. G. R. Iyengar and K. Deb, "Optimum Design of Laminated Composite Plates with Cutouts Using a Genetic Algorithm," Composite Structures, Vol. 42, No. 3, 1998, pp. 265-279.

7.   S. A. Barakat and G. A. Abu-Farsakh, "The Use of an Energy-Based Criterion to Determine Optimum Configurations of Fibrous Composites," Composite Science and Technology, Vol. 59, No. 12, 1999, pp. 1891-1899.

8.   P. Kere, M. Lyly and J. Koski, "Using Multicriterion Optimization for Strength Design of Composite Laminates," Composite Structures, Vol. 62, No. 3-4, 2003, pp. 329-333.

9.   C. H. Park, W. I. Lee, W. S. Han and A. Vautrin, "Weight Minimization of Composite Laminated Plates with Multiple Constraints," Composite Science and Technology, Vol. 63, No. 7, 2003, pp. 1015-1026.

10.  MediaWiki, "MATLAB," 2008. http://en.wikipedia.org/ wiki/MATLAB

11.  R. V. Dukkipati, "MATLAB for Mechanical Engineers," New Age International (P) Limited, New Delhi, 2008.

12.  G. Z. Voyiadjis and P. I. Kattan, "Mechanics of Composite Materials with MATLAB," Springer Netherlands, Dordrecht, 2005.

13.  A. Gilat, "MATLAB: An Introduction with Applications," John Wiley & Sons Inc., Hoboken, 2008.

14.  B. D. Hahn and D. T. Valentine, "Essential MATLAB for Engineers and Scientists," Butterworth-Heinemann, Rome, 2007.

# Chapter 3

# SEMI-ANALYTIC TECHNIQUES FOR FAST MATLAB SIMULATIONS

Daniele Borio[1] and Eduardo Cano[1]

[1]EC Joint Research Centre, Institute for the Protection and Security of the Citizen, Italy

## INTRODUCTION

Advances in electronics and telecommunications are leading to complex systems able to efficiently use the available resources. Fast electronics, complex modulation schemes and correction codes enable transmissions on channels with unfavorable characteristics, coexistence between different services in the same frequency bands and high transmission rates. However, the complexity of such communications systems often prevents analytical characterizations. For example, figures of merit such as the BER are difficult to determine analytically for transmission schemes involving correction codes and communication channels with ISI and fading. In such cases, the system is characterized through Monte Carlo simulations [1, 2]. The Monte Carlo framework involves simulations of the whole system under analysis. For example, when considering a communications system, the whole transmission-reception chain is simulated. A large number of sequences are sent through the simulated system and the message recovered by the simulated receiver is compared to the original transmitted sequence. This comparison allows one to determine the average number of transmission errors and compute the BER.

Monte Carlo simulations can be applied to almost any system although their implementation and computation requirements can be significantly high. In addition to this, precision problems can arise when the quantity to be estimated is significantly low. For example, BER of the order of $10^{-8}-10^{-9}$ require at least $10^9-10^{10}$ simulation runs. Conversely, analytical models have a limited applicability and usually adopt approximations (i.e., model linearization) that can yield a poor description of the system under analysis.

In order to overcome the limitation of Monte Carlo and analytical techniques, semi-analytic approaches have been previously implemented [2, 1]. In a semi-analytic framework, the knowledge of the system under analysis is exploited to reduce the computational load and complexity that full Monte Carlo simulations would require. In this way, the strengths of both analytical and Monte Carlo methods are effectively combined. Semi-analytic techniques are a powerful tool for the analysis of complex systems.

The characteristics and relationships among the three aforementioned methods are shown in Figure 1 : semi-analytic approaches represent a good compromise in terms of applicability and complexity, combining the strengths of Monte Carlo and analytical approaches.

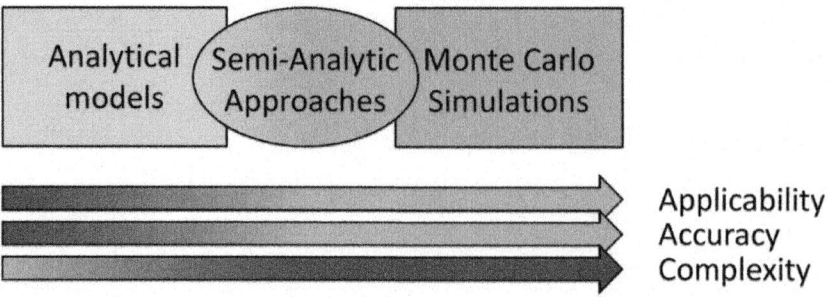

**Figure 1:** Different approaches available for the analysis of complex systems. Semi-analytic approaches represent a compromise in terms of applicability and complexity (computational and implementation) between analytical models and Monte Carlo simulations.

The main goal of this chapter is to provide a general overview of semi-analytic techniques for the simulation of communications systems. Specific emphasis is given to their implementation in MT and two examples from the communications context are analyzed in detail.

Despite their potential, semi-analytic techniques have received limited attention from the communications community. Reference books on simulation of communications systems such as [1], [2] and [3] dedicate only a few pages to this kind of techniques. The focus is usually on the computation of the BER, which represents one of the first applications of semi-analytic techniques in communication system analysis [4, 5]. In this case, the model depicted in Figure 2 is adopted. The communications channel is modeled as a non-linear device, which distorts the signal component alone, with the addition of a noise term that is assumed to be Gaussian. This model is quite general and can be used to represent several communications channels.

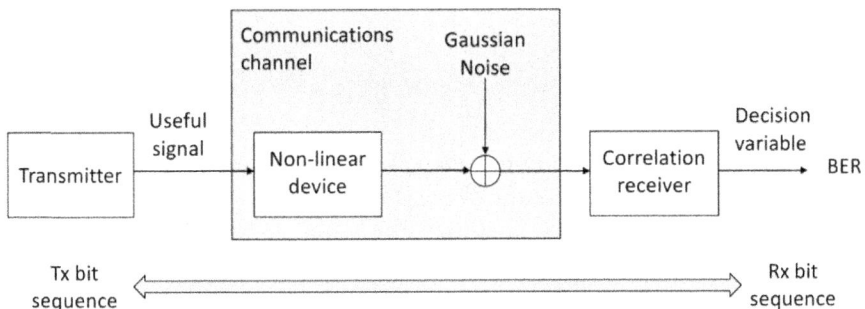

**Figure 2:** Model adopted for the evaluation of the BER using a semi-analytic approach. The communications channel is modeled as a non-linear device, which affects only the signal component, and the addition of a noise term supposed to be Gaussian.

A classic example is the model of a transmission chain where a TWTA is used to amplify the useful signal before transmission. The TWTA is highly non-linear and can lead to signal distortions. Since the signal is injected into the TWTA before transmission, the noise component entering the amplifier is negligible. In this case, the model depicted in Figure 2 is appropriate for describing the transmission chain including a non-linear amplifier.

The TWTA is a memory-less device and can be characterized using *AM-AM conversion* and *AM-PM conversion* (AM = Amplitude Modulation, PM = Phase Modulation) curves [3]. When a base-band signal model is used, the amplifier input and output are complex quantities; moreover, the response of the device usually depends only on the amplitude (instantaneous power) of the input signal. AM-AM and AM-PM conversion curves define the relationship between the input/output signal amplitudes and phases as a function of the input amplitude. Using these conversion curves, it is possible to simulate the behavior of the TWTA and other non-linear devices.

In the semi-analytic framework, the additivity of the noise component is exploited to compute the BER. More specifically, only the signal transmission chain is simulated and for each possible signal symbol, the Eb is computed. Since the noise properties are known, the BER for the $_i$th symbol, $s_i$, is given by

$$BER_i = \frac{1}{2}\mathrm{erfc}\left(\sqrt{\frac{E_{b,i}}{N_0}}\right),$$

(1)

Where Eb,i has been obtained by simulating the transmission chain (including the non-linear device) in the absence of noise and transmitting the symbol $s_i$. The parameter $N_0$ is the noise power spectral density and it is a known value of the system. Finally, the BER of the system is obtained from

$$BER = E\,[BER_i] = \frac{1}{2} \sum_{i=0}^{N_s-1} p_i \text{erfc}\left(\sqrt{\frac{E_{b,i}}{N_0}}\right),$$

(2)

Where $p_i$ is the probability that the symbol $s_i$ will be transmitted and Ns is the number of symbols of the signal constellation.

This simple example clearly illustrates the principles of semi-analytic techniques: the analytical knowledge of the system is exploited to reduce the computation load and complexity that full Monte Carlo simulations would require. In this case, only the transmission of the signal component is simulated.

In the literature, several generalizations of the aforementioned BER computation technique have been proposed. For example, [5] considered the case where the noise term at the input of the non-linear amplifier is not negligible. An equivalent model is proposed where the noise at the input is propagated after the non-linearity. [5] also considered the presence of a bandpass channel. More recently, [6] proposed a methodology for estimating the BER in the presence of ISI. All these examples show the potential and flexibility of the semi-analytic approach.

## Building Blocks

When considering the previous example it is possible to identify three building blocks that play different roles in the semi-analytic framework:

• simulation block

• estimation block

• Analytical model.

The simulation block corresponds to that part of the system that is actually simulated. In the previous example, this block corresponds to the signal generation, the non-linear amplifier and the correlation receiver simulation. These blocks were used to determine the decision variable employed for recovering the transmitted symbol. The analytical model exploits the properties of the system to determine the quantities of interest. In the previous example, the fact that the noise introduced by the communication channel is Gaussian was exploited to determine the BER as a function of the Eb of each transmitted symbol. The estimation block is used as the interface between the simulated and analytical parts of the system. In BER computation case, the simulation part allows one to generate the different decision variables, whereas the analytical model is expressed as a function of energy per bit. The estimation block is used to determine Eb from the simulated decision variables.

The three functional blocks can be connected according to different configurations leading to different types of semi-analytic approaches. In the next section, two of these configurations are briefly discussed. Examples for each type of semi-analytic system are given in Section 2 and Section 3.

## Main Configurations

When considering the BER example, it is possible to note that the simulation, estimation and analytical blocks are connected in series. The simulation block is used at first to compute the different decision variables. The estimation block determines the Eb associated with each variable and finally the analytical model is used to compute the BER from the EN0.

This type of configuration is defined here as *sequential* since there is no feedback between the different blocks and each element of the chain is run sequentially. The principle of this configuration is shown in the upper part of Figure 3.

Sequential approach

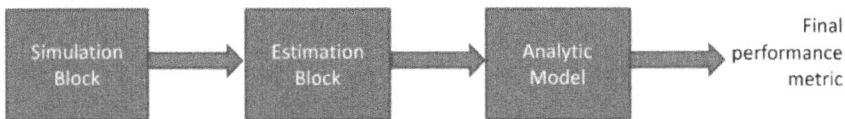

Closed-loop approach

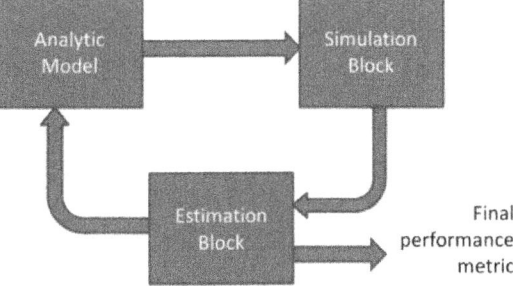

**Figure 3:** Basic blocks and main configurations adopted in semi-analytic approaches. An estimation block is used for determining key signal and system parameters and interfacing the simulation and analytic components of the system.

A second type of configuration has been recently considered for the analysis of tracking loops in DSSS and GNSS receivers. The most computationally demanding operation in a DSSS/GNSS receiver is despreading, i.e, the correlation of the incoming samples with local replicas of the code and carrier. This operation is performed by the ID blocks that rely on simple operations that can be analytically modeled. For this reason, semi-analytic models exploiting the knowledge of the ID blocks and simulating only the non-linear parts of the system have been developed [7, 8, 9, 10]. This resulted in efficient analysis tools, which require reduced processing time with the applicability of Monte Carlo simulations.

Different techniques for modeling the output of the ID have been suggested. [7] modeled the correlator outputs evaluated by the ID blocks as linear combinations of independent Gaussian random variables. Correlation among the different correlators was obtained by using, for the generation of different ID outputs, a subset of the same random variables. This technique becomes complex as the number of correlators increases. Another attempt made in [8] assumed that the correlator outputs were independent which, in general, not a realistic condition is. Finally, [9, 10] suggested the use of a technique based on the Cholesky decomposition detailed in [11]. This approach allows one to easily generate an arbitrary number of correlated Gaussian random variables. In this way, [9, 10] were able to simulate advanced tracking loops for new GNSS signals.

Regardless of the type of approach used for modeling the correlator outputs, the aforementioned semi-analytic configuration can be represented as in the bottom part of Figure 3. In this case, an analytical model is used to generate quantities that will be propagated by simulation. In the tracking loop case, an analytical model is used to generate the correlator outputs that are then processed through simulations. The non-linear parts of the system are fully simulated and quantities such as the loop discriminator and filter outputs are computed. Finally, an estimation block is used to interface the simulation and analytical components of the semi-analytic scheme. A new estimate of the signal parameters is obtained and used to generate new correlator outputs. The estimation block is also used to compute performance metrics such as tracking jitter or MTLL [12].

The three functional blocks described in Section 1.1 are connected in a loop and thus this type of configuration is named *closed-loop approach*. The technique developed by [9, 10] and the closed-loop approach will be detailed in Section 3.

# SEQUENTIAL APPROACH: THE INTER-SYSTEM INTERFERENCE CASE

As shown in Figure 3, the sequential approach, at first, requires an initial simulation block to generate the random processes and system functions that cannot be analytically described. Subsequently, the simulated processes are employed by the estimation unit to obtain key parameters required by the analytical model to compute metrics of interest. In the analytical model, the estimated parameters are plugged into mathematical expressions to obtain the desired final metrics. In the sequential architecture, the gain in computational load mainly depends on the simplifications allowed by the analytical model. The use of such a model allows one to simulate only a part of the system and eventually avoid computationally demanding error counting processes.

The computational complexity of Monte Carlo simulations increases significantly when two or more communication systems coexist within the same environment. In this case, it is necessary to account for the interaction of the different systems and determine potential inter-system interference. In addition, computational requirements of Monte Carlo simulations increase dramatically with the number of random elements included in each block of the communication chain. These requirements can be considerably reduced by adopting semi-analytic techniques in which the evaluated metrics are analytically expressed as a function of parameters estimated through simulations. This principle is the core idea behind the sequential semi-analytic approach.

In order to better illustrate the principles of the semi-analytic sequential approach, an inter-system interference scenario is considered in this section. The case of a satellite navigation receiver affected by interference generated by a communications system is considered. The primary system (i.e., the victim system) considered here is a GPS L1 receiver affected by an interference signal caused by third order harmonics of a DVB-T signal. A comprehensive description of the sequential approach applied to an inter-system interference case is provided in the following.

The reception of GNSS signals is challenging due to low signal power, possible severe channel conditions and the presence of RF interference. The presence of RF interference can be particularly troublesome and the performance of a GNSS receiver can vary significantly depending on the type of interference. For this reason, significant research efforts have been devoted to the characterization of the receiver performance in the presence of different types of interference [13, 14]. Furthermore, the impact of interference originated by specific communication technologies, such as UWB transmissions [15], DME signals [16] and DVB-T harmonics [17, 18], has been thoroughly investigated. It is noted that the interference impact strongly depends on the strategy adopted by a GNSS receiver for processing the useful signals. Moreover, different impacts are expected depending on the receiver operating mode. The first task of a GNSS receiver is to determine the presence of a specific GNSS signal. This task is accomplished by the acquisition block that implements a statistical test for the detection of useful signals. After acquisition, the useful GNSS signals are passed to the tracking stage that refines the estimates of different signal parameters. Since acquisition and tracking implement different processing strategies, RF interference will affect these two receiver blocks differently. In the following, the acquisition stage is considered. A semi-analytic approach for the analysis of GNSS tracking loops is discussed in Section 3.

The acquisition of a GNSS signal can be formulated as a classical detection problem [19], where the signal of interest is buried in noise. The outcome of the acquisition process is twofold. First, a decision relative to the signal presence is provided. If the signal is present, a rough estimate of signal parameters (defined in the following) is also obtained. The received useful GNSS signal, which is impaired by AWGN and interference, is processed by the acquisition block yielding a decision variable. If the decision variable is greater than a decision threshold the signal presence is declared. This decision variable is calculated by using the digital samples provided by the receiver front-end. The signal model and the acquisition process are briefly summarized in the following sections.

## The GNSS Signal

The signal at the input of a GNSS receiver, in a one-path AWGN channel and in the presence of RF interference, can be modeled as

$$r(t) = \sum_{l=0}^{L-1} y_l(t) + i(t) + \eta(t),$$

(3)

where $y_l(t)$ is the signal transmitted by the lth GNSS L1 satellite, L is the total number of satellites in view, $i(t)$ is the received interference signal and $\eta(t)$ is the noise term

Each useful signal, $y_l(t)$, can be expressed as

$$y_l(t) = \sqrt{2C_l} d_l (t - \tau_{0,l}) c_l (t - \tau_{0,l}) \cos \left( 2\pi (f_{RF} + f_{d0,l}) t + \varphi_{0,l} \right),$$

(4)

where

- $C_l$ is the power of the lth useful signal;

- $d_l(\cdot)$ is the navigation message;

- $c_l(\cdot)$ is the lth pseudo-random sequence extracted from a family of quasi-orthogonal codes and used for spreading the signal spectrum;

- $\tau_{0,l}$, $f_{d0,l}$ and $\varphi_{0,l}$ are the delay, Doppler frequency and phase introduced by the communication channel, and

- $f_{RF}$ is the centre frequency of the GNSS signal.

It is noted that GNSS signals adopt a DSSS modulation. The pseudo-random sequences, $c_l(t)$, allow the simultaneous transmission of several signals at the same time and in the same band. Moreover, $c_l(t)$ sequences are characterized by sharp correlation functions that allow the precise measurement of the signal travel time. The travel time is then converted into distances that allows a GNSS receiver to determine its position.

The pseudo-random sequence, $c_l(t)$, is composed of several terms including a primary spreading sequence and a subcarrier:

$$c_l(t) = \sum_{i=-\infty}^{+\infty} c_{l,(i \bmod N_c)} s_b(t - iT_h).$$

(5)

The signal $s_b(t - iT_h)$ in (5) represents the subcarrier of duration $T_h$, which determines the spectral characteristics of the transmitted GNSS signal. The GPS L1 Coarse/Acquisition (C/A) component is Bi-Phase Shift Keying (BPSK) modulated, whereas the Galileo E1 signal adopts a Composite Binary-Offset Carrier (CBOC) scheme. The sequence $c_{l,i}$, of length $N_c$, defines the primary spreading code of the lth GNSS signal. In the following, only the BPSK case is considered. The results can be easily extended to different subcarriers.

A GNSS receiver is able to process the L useful signals independently since the spreading codes are quasi-orthogonal. Therefore, expression (3) can be simplified to

$$r(t) = y(t) + i(t) + \eta(t),$$

(6)

where the index l has been dropped for ease of notation.

The received signal in (6) is filtered and down-converted by the receiver front-end. Filtering is of particular importance since it determines which portion of the interfering signal, i (t), will effectively enter the receiver. After down-conversion and filtering, the input signal is sampled and quantized. In this analysis, the impact of quantization and sampling is neglected. After these operations, (6) becomes:

$$r_{BB}[n] = y_{BB}(nT_s) + i_{BB}(nT_s) + \eta_{BB}(nT_s) = y_{BB}[n] + i_{BB}[n] + \eta_{BB}[n], \tag{6}$$

where the notation x[n] is used to denote discrete time sequences sampled at the frequency $f_s = \frac{1}{T_s}$. In addition, the index "BB" is used to denote a filtered signal down-converted to baseband. Furthermore, the signal $y_{BB}[n]$ in (7) can be written as

$$y_{BB}[n] = \sqrt{C}d(nT_s - \tau_0)c(nT_s - \tau_0)\exp\{j2\pi f_0 nT_s + \varphi_0\}. \tag{7}$$

The noise term, $\eta_{BB}[n]$, is AWGN with variance $\sigma^2_{BB}$. This variance depends on the filtering, down-conversion and sampling strategy applied by the receiver front-end and can be expressed as $\sigma^2_{BB} = N_0 B_{RX'}$, where $B_{RX}$ is the front-end bandwidth and $N_0$ is the Power Spectral Density (PSD) of the input noise $\eta(t)$. The ratio between the carrier power, C, and the noise PSD, $N_0$, defines the Carrier-to-Noise density power ratio (C/$N_0$), one of the main signal quality indicators used in GNSS.

## The DVB-T Interfering Signal

The interference term in (6), i (t), originates from DVB-T emissions. The DVB-T system is the European standard for the broadcasting of digital terrestrial television signals and has been adopted in many countries, mainly in Europe, Asia and Australia. The standard employs an Orthogonal Frequency Division Multiplexing (OFDM)-based modulation scheme operating in the VHF III (174 − 230 MHz), UHF IV (470 − 582 MHz) and UHF V (582 − 862 MHz) bands [20]. It is noticeable that none of these bands fall within the bands allocated for GNSS signals. However, the second harmonics of UHF IV and third order harmonics of UHF V could coincide with the GPS L1 band, and, thus cause harmful interference. The case of the third order harmonics of a DVB-T signal is considered here. The DVB-T transmitted signal can be represented as

$$i_{DVB-T}(t) = \frac{1}{\sqrt{M}} \sum_{p=0}^{N_d-1} \sum_{h=0}^{M-1} I_{p,h} \exp\left(\frac{j2\pi ht}{M}\right), \tag{9}$$

where M is the modulation order, h is the subcarrier index, p is the symbol index, $N_d$ represents the total number of transmitted symbols and $I_{p,h}$ models the hth constellation point of the pth symbol. Here, the term "subcarrier" should not be confused with the subcarrier used to modulate the GNSS signals in (5). In the OFDM context, several components are transmitted in parallel on different overlapping frequency bands. The term subcarrier denotes each individual transmitted component. In GNSS, the subcarrier is an additional component that modulates the transmitted signal and plays a role analogous to the carrier used for the signal up-conversion.

Third order harmonics are the consequence of the malfunctioning of the transmitter electronics. In particular, the presence of these harmonics are due to the non-linearities of an amplifier. The output of an amplifier can be modeled using a polynomial expansion of the amplifier input/output function:

$$p(t) = \sum_{n=1}^{\infty} a_n i_{DVB-T}^n(t),$$

$$(10)$$

where an are the polynomial coefficients of the Taylor series expansion of the amplifier input/output function. This type of model is an alternative to the AM-AM and AM-PM conversion functions discussed in Section 1 for the TWTA case.

The terms of order n > 1 in (10) model the amplifier non-linearities and the ratios an/a1 are expected to be small for n > 1. Since only the third harmonics will fall into the GPS L1 band, the interference signal at the antenna of a GNSS receiver is given by

$$i(t) = a_3 i_{DVB-T}^3(t).$$

$$(11)$$

The signal i (t) is filtered and down-converted by the receiver front-end and signal $i_F$ (t), the filtered version of i(t), will affect receiver operations.

Finally, the interference term in (7) can be modeled as $i_{BB}[n] = i_F(nT_s)$

## The Acquisition Process

After signal conditioning, the sequence $r_{BB}[n]$ is correlated with local replicas of the useful signal code and carrier as shown in Figure 4. Since the code delay, $\tau_0$, and the Doppler frequency, $f_0$, of the useful signal in (8) are unknown to the receiver, several delays and frequencies are tested by the acquisition block. In addition to this, several correlators, computed using subsequent portions of the input signal $r_{BB}[n]$, can be computed in order to produce a decision variable less affected by noise and interference. In this way, the output of the kth complex correlator can be expressed as

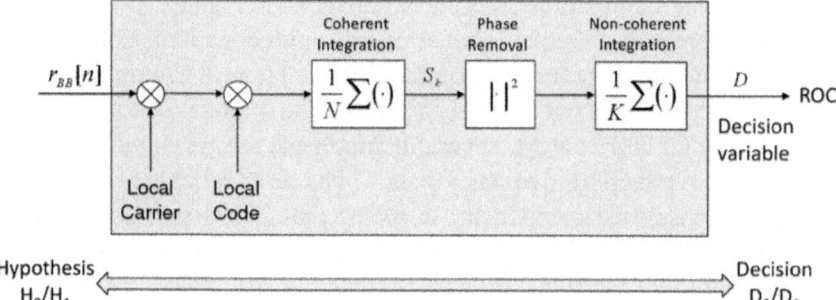

**Figure 4:** Schematic representation of the operations performed by the acquisition block of a GNSS receiver.

$$S_k = \frac{1}{N} \sum_{n=kN}^{(k+1)N-1} r_{BB}[n]c\,(nT_s - \tau)\exp\{-j2\pi f_d nT_s - j\varphi\},$$
(12)

where $\tau$, $f_d$ and $\varphi$ are the code delay, Doppler frequency and carrier phase tested by the receiver. The parameter N is the number of samples used for computing a single correlation output and $T_c = NT_s$ defines the coherent integration time. It is noted that the computation of correlation outputs is essential for the proper functioning of a GNSS receiver and they are both used in acquisition and tracking modes [21]. To further improve the acquisition performance, non-coherent integration can be implemented as illustrated in Figure 4. More specifically, the impact of the navigation message, d (·), is removed through squaring, $|Sk|^2$, and the final decision variable is computed as

$$D = \frac{1}{K} \sum_{k=0}^{K-1} |S_k|^2,$$
(13)

where K is the total number of correlation samples that are non-coherently integrated. It should be noted that for K = 1 only coherent integration is used. In order to determine the signal presence, the receiver compares D with a decision threshold, $\beta$. If D is greater than $\beta$ then the useful signal is declared present.

It is noted that, as in any binary test, two hypotheses are possible:

- $H_0$: the signal is not present or it is not correctly aligned with the local code and carrier replica.

- $H_1$: the signal is present and the local code and carrier replica are aligned.

The null hypothesis, $H_0$ assumes that the correlator outputs, $S_k$, are made of noise alone. Since the pseudo-random sequences, c(·), are selected to have good autocorrelation properties, if the code delay and Doppler frequencies tested by the receiver do not match the parameters of the input signal, $y_{BB}[n]$, then the useful signal component is almost completely filtered out at the correlator output. Thus, also in this case, the $H_0$ hypothesis is verified

Furthermore, $H_1$ is the alternative hypothesis and assumes that the signal is present and the local code and carrier replica are perfectly aligned. If $H_1$ is declared, then rough estimates of $\tau_0$ and $f_0$ are also obtained.

Depending on the result of the test, $D > \beta$, two decisions can be taken by the receiver

- $D_0$: the signal is declared not present

- $D_1$: the signal is declared present

**Table 1:** Confusion matrix describing the four events that can happen in the binary test performed by the acquisition process.

|        | $D_0$                                | $D_1$            |
|--------|--------------------------------------|------------------|
| $H_0$  | Signal absence correctly declared    | False alarm      |
| $H_1$  | Missed detection                     | Signal detection |

The off-diagonal events in Table 1 correspond to the different errors that the acquisition block can commit. The following probabilities are usually associated with the events in Table 1 :

$$P_d(\beta) = \mathrm{Prob}(D > \beta | H_1) \quad \text{Probability of detection} \tag{14}$$

and

$$P_{fa}(\beta) = \mathrm{Prob}(D > \beta | H_0) \quad \text{Probability of false alarm.} \tag{15}$$

The probabilities of missed detection and correct signal absence decision are obtained as $1 - P_d(\beta)$ and $1 - P_{fa}(\beta)$, respectively. The performance of the acquisition process is characterized in terms of Receiver Operation Curves (ROC) [22], which plots the detection probability as a function of the false alarm rate. ROC curves capture the behavior of a detector as a function of the different decision thresholds.

## The Semi-Analytic Approach

The goal of the sequential semi-analytic approach considered in this section is the evaluation of the ROC in the presence of DVB-T interference. The full Monte Carlo approach would consist of simulating the full transmission/reception scheme shown in Figure 5 and generating several realizations of D both under $H_0$ and $H_1$. Probabilities of detection and false alarm would then be determined through error counting techniques. This approach is computationally demanding and does not exploit the analytical knowledge of the system. More specifically, under the hypothesis that the correlator outputs $S_k$ are independent and identically distributed (i.i.d.) complex Gaussian random variables with independent real and imaginary parts, it is possible to show [23] that

$$P_{fa}(\beta) = \exp\left\{-\frac{\beta}{2\sigma_n^2}\right\} \sum_{i=0}^{K-1} \frac{1}{i!}\left(\frac{\beta}{2\sigma_n^2}\right)^i \tag{16}$$

and

$$P_{det}(\beta) = Q_K\left(\sqrt{K\lambda};\ \sqrt{\beta\sigma_n^2}\right), \tag{17}$$

Where $Q_K(a;b) = \int_b^{+\infty} x\left(\frac{x}{a}\right)^{K-1}\exp\left\{-\frac{x^2+a^2}{2}\right\} I_{K-1}(ax)dx$ the generalized Marcum Q-function of order K is the function $I_K(\cdot)$ is the modified Bessel function of first kind and order K. In (16) and (17), $\sigma_n^2$ is the variance of the real and imaginary part of $S_k$. The parameter $\lambda$ is given by and defines the SNR at the correlator outputs.

$$\lambda = \frac{|E[S_k]|^2}{\sigma_n^2} \tag{18}$$

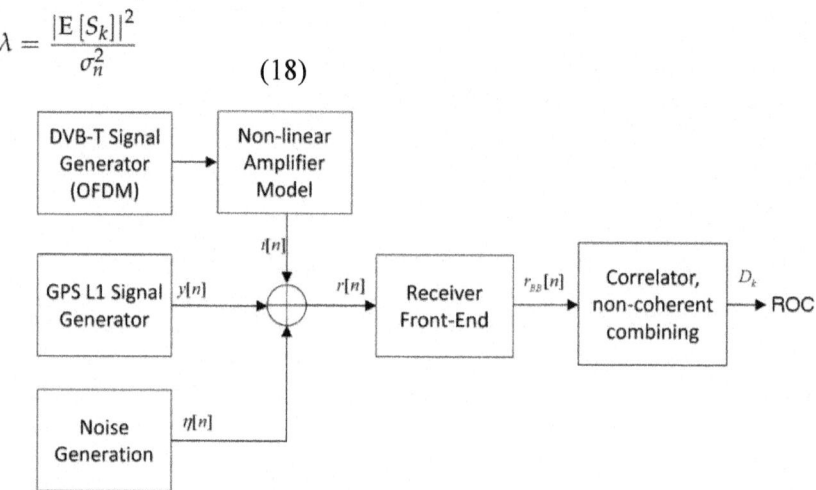

**Figure 5:** Schematic representation of the full Monte Carlo simulation system for the ROC evaluation of the acquisition of a GPS L1 signal in the presence of DVB-T third order harmonics.

The correlator output can be considered i.i.d. complex Gaussian random variables even in the presence of DVB-T interference. More specifically, the large number of terms in the sum performed in (12) allows one to invoke the central limit theorem and assume $S_k$ is Gaussian. Independence derives from the down-sampling performed by the correlators. Since only one correlator is produced every N samples, the statistical correlation between subsequent correlators is significantly reduced. The lack of correlation translates into independence for Gaussian random variables. Thus, models (16) and (17) can be used and the only parameters that need to be estimated are $\sigma_n^2$ and $\lambda$.

The analytical knowledge of the system can be further exploited to simplify the evaluation of $\sigma_n^2$ and $\lambda$. In particular, since $i_{BB}[n]$ and $\eta_{BB}[n]$ are modeled as zero mean random processes, they only contribute to the variance of the correlator outputs. Thus, neglecting residual errors due to delay and frequency partial misalignments, yields

$$|E[S_k]|^2 = C,$$
(19)

where C is the useful signal power and is one of the known parameters of the system. Thus, $\lambda$ can be derived from C and $\sigma_n^2$.

Finally, exploiting the linearity of the correlation process, it is possible to express $S_k$ as

$$S_k = S_{r,k} + S_{\eta,k} + S_{i,k},$$
(20)

which is a linear combination of a useful signal term, derived from $y_{BB}[n]$, a noise term, derived from $\eta_{BB}[n]$, and an interference term derived from $i_{BB}[n]$. The variance $\sigma_n^2$ can be obtained as

$$\sigma_n^2 = \frac{1}{2}\text{Var}\{S_k\} = \frac{1}{2}\text{Var}\{S_{\eta,k}\} + \frac{1}{2}\text{Var}\{S_{i,k}\}.$$
(21)

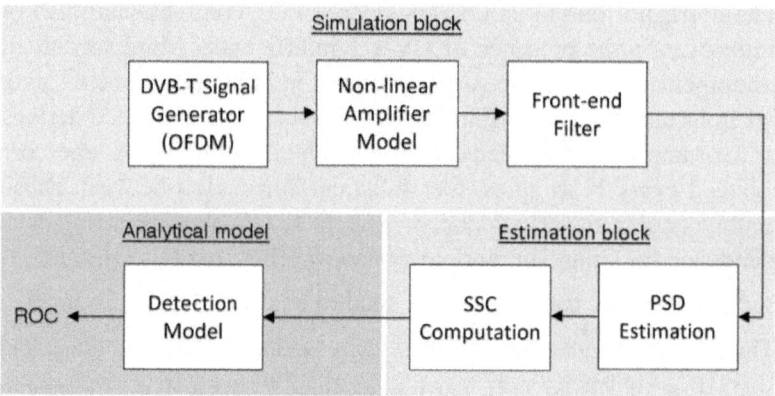

**Figure 6:** Schematic representation of the semi-analytic approach adopted for the evaluation of the ROC in the presence of DVB-T interference. The three functional elements of the semi-analytic approach are highlighted in different colors.

Using the results derived in [24], [13] and [23], it is possible to show

$$\frac{1}{2}\text{Var}\left\{S_{\eta,k}\right\} = \frac{N_0}{2T_c}$$
(22)

and

$$\frac{1}{2}\text{Var}\left\{S_{i,k}\right\} = \frac{C_i}{2T_c}k_a,$$
(23)

where $C_i$ is the interference power and $k_a$ is the Spectral Separation Coefficient (SSC) defined as [13, 24]

$$k_a = \int_{-B_{RX}/2}^{B_{RX}/2} G_i(f)G_c(f)df.$$
(24)

The function $G_i(f)$ in (24) is the normalized PSD of the DVB-T interference signal after front-end filtering. In addition, $G_i(f)$ is normalized such that

$$\int_{-B_{RX}/2}^{B_{RX}/2} G_i(f)df = 1.$$
(25)

The function $G_c(f)$ models the effect of the correlation on the interfering signal. Correlation can be modeled as an additional filtering stage and $G_c(f)$ can be shown to be well approximated by the PSD of the subcarrier used in the despreading process. Also, $G_c(f)$ is normalized to have a unit integral. It is noted that different subcarriers lead to different $G_c(f)$ and thus, $i_{BB}[n]$ will have different effects depending on the type of modulation considered.

The only unknown parameter in the previous equation is the SSC, which needs to be estimated using Monte Carlo simulations. Also, the interfering DVB-T signal is fully simulated. The resulting signal is filtered by the receiver front-end and the sequence $i_{BB}[n]$ is obtained. The samples of $i_{BB}[n]$ are used to estimate the normalized PSD, $G_i(f)$. This can be easily obtained using the MATLAB functions developed for spectral analysis. In this case, the pwelch function is used. The function $G_i(f)$ is used to compute the SSC, which is then used to determine the system ROC.

The developed semi-analytic approach is shown in Figure 6 where the simulation, estimation and analytic components are clearly highlighted.

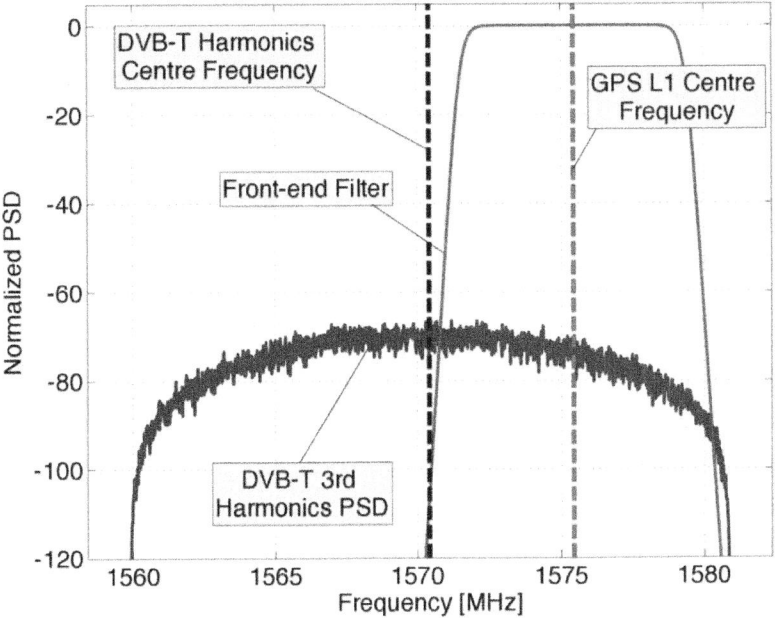

**Figure 7:** Representation of a normalized PSD realization of the third harmonic of the DVB-T signal and the frequency response of the GPS L1 front-end filter.

## Performance Comparison

A comparison between a full Monte Carlo simulation and a semi-analytic technique, implemented for the evaluation of the acquisition performance of a GPS L1 receiver impaired by third order harmonics of a DVB-T signal, is presented in this section. Initially, the DVB-T interfering signal in time domain is programmed in MT by following the DVB-T standard and the non-linear amplifier model, as illustrated in Figure 5. Note that the simulation of the interfering signal is required for both Monte Carlo and semi-analytic techniques. Subsequently, the estimated PSD of the interfering signal, needed for the estimation of the SSC in the semi-analytic method, is obtained by applying the pwelch function of MT. A realization of the normalized PSD of the interfering signal is depicted in Figure 7. The centre frequency of the interfering signal is set to $f_I = f_{RF} + \Delta f$, where $\Delta f$ is the frequency shift of the interference signal with respect to the centre frequency of the GPS L1 signal. The impact of selecting different values of

$\Delta f$ on the acquisition performance of a GPS L1 receiver is analyzed in [25]. Furthermore, the frequency response of the GPS L1 front-end filter is also plotted in Figure 7. In this case, the selected filter bandwidth is 8 MHz.

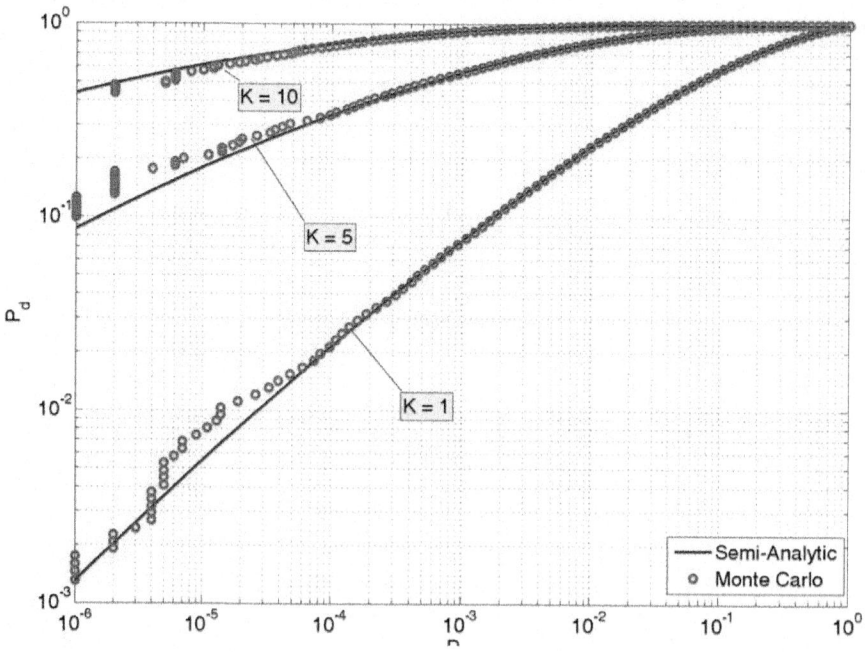

**Figure 8:** Comparison between ROC curves obtained using semi-analytic and Monte Carlo simulations. The semi-analytic framework considered provides increased precision and requires a lower computational complexity.

Sample results comparing ROC curves obtained using semi-analytic and Monte Carlo simulations are shown in Figure 8. The parameters used for the analysis are reported in Table 2. From Figure 8, it can be observed that the Monte Carlo and semi-analytic approaches provide similar results and the curves obtained using the two methods overlap. However, the semi-analytic approach provides increased precision, particularly when small values need to be estimated, and a significant reduction in terms of computational complexity. Full Monte Carlo simulations require the implementation of the full transmission/reception chain and the evaluation of the ROC with a computational complexity significantly higher than that of the semi-analytic approach described above.

**Table 2:** Parameters used for the evaluation of the ROC curves shown in Figure 8.

| Parameter | Value |
|---|---|
| $C/N_0$ | 35 dB-Hz |
| Coherent integration time, $T_c$ | 1 ms |
| Interference to signal power ratio, $\frac{C_i}{C}$ | 30 dB |
| Centre frequency difference, $\Delta f$ | 0 Hz |
| Receiver bandwidth, $B_{RX}$ | 8 MHz |
| Number of Monte Carlo Simulation runs | $10^6$ |

Additional results relative to the impact of DVB-T interference on GNSS can be found in [25].

# CLOSED-LOOP APPROACH: DIGITAL TRACKING LOOPS

As anticipated in Section 1, a second configuration, called *closed-loop approach*, has been recently proposed for the simulation of digital tracking loops in DSSS/GNSS receivers. The SATL Sim toolbox is a set of MT functions implementing the semi-analytic closed-loop approach for the analysis of digital tracking loops. The SATL Sim toolbox has been developed by [9, 10] and can be downloaded from the following websites:

- http://www.ngs.noaa.gov/gps-toolbox/SATLSim.htm

- http://plan.geomatics.ucalgary.ca/publications.php.

In the following, the closed-loop approach for the simulation of digital tracking loops is considered and the MT code developed in the SATL Sim toolbox is briefly analyzed.

A description of the correlator model used for reducing the computational complexity of the system is at first provided. The samples given by (7) at the input of a GNSS receiver are processed by the different functional blocks with different objectives. The acquisition process described in Section 2 is the first stage of a GNSS receiver and has the goal of determining the signal presence and provide a rough estimate of its parameters. These parameters include the code delay $\tau_0$ and Doppler frequency $f_0$.

If the signal is successfully acquired then different tracking loops are used to refine the estimate of the signal parameters. A DLL is usually used to provide accurate estimates of the code delay, $\tau_0$, and track delay variations due to the relative motion between receiver and satellite. The Doppler frequency, $f_0$, is recovered using either a FLL or a PLL. If a PLL is used then the carrier phase, $\varphi 0$, is also estimated. The code delay and carrier phase allow the receiver to determine its position whereas the Doppler frequency can be used for computing the user velocity.

As indicated in Section 2, a subcarrier can be used for shaping the spectrum of the transmitted GNSS signal and improving its robustness against multipath. The presence of a subcarrier makes code tracking more complex since the correlation function of the transmitted signal may have multiple peaks. More specifically, fine delay estimation is obtained by maximizing the correlation between input signal and local code: the correlation function is maximized only when the delay of the locally generated code matches the delay of the input signal. The presence of several peaks in the correlation function may cause the DLL to converge to a local maximum causing biases in the delay estimation. For this reason, several solutions have been proposed to avoid lock on secondary correlation peaks [26, 27]. An effective solution is represented by the SLL proposed by [27]. In this case, the subcarrier is seen as a periodic waveform that further modulates the transmitted signal. The delays of code and subcarrier are decoupled and estimated separately. In this way, the ambiguous one-dimensional signal correlation is projected in an unambiguous bi-dimensional function. In the following, the joint simulation of DLL and SLL is considered.

In a GNSS tracking loop, the incoming signal is correlated with several locally generated code and carrier replicas and different correlator outputs are produced. This process is analogous to the correlation operations described in Section 2 and is performed by the ID blocks.

Each correlator output is a function of the input signal and the parameters previously estimated by the tracking loop. The correlator outputs are passed to the non-linear discriminator that produces a first estimate of the tracking error that the loop is trying to minimize. The tracking error is filtered and passed to the NCO that is used for generating new local signal replicas.

Efficient tracking loop simulations can be obtained by substituting the ID blocks with their analytical model. More specifically, a correlator output can be modeled as:

$$\sqrt{C}\frac{\sin\left(\pi\Delta f_d T_c\right)}{\pi\Delta f_d T_c}R_l\left(\Delta\tau_d,\Delta\tau_s\right)\exp\left\{j\Delta\varphi\right\}+\eta_c, \tag{26}$$

where

- $\Delta f_d$ and $\Delta$ are the residual frequency and phase errors;

- $\Delta\tau_d$ and $\Delta\tau s$ are the code and subcarrier delay errors. The delay $\Delta\tau s$ is present only when a SLL is used to correctly align the signal subcarrier [27];

- $T_c = NT_s$ is the coherent integration time where N is the number of samples used to compute a single correlator;

- $Rl\left(\Delta\tau_d,\Delta\tau_s\right)$ is the correlation function between incoming and locally generated code and is a function of both code and subcarrier delay errors. When the SLL is not used, $Rl\left(\Delta\tau_d,\Delta\tau_s\right)$ is replaced by the standard code correlation function;

- $\eta_c$ is a zero-mean noise term whose variance depends on the input noise power, front-end filtering and the correlation process operated by the I&D blocks. More details on the properties of $\eta c$ can be found in [9].

From (26), it is possible to reconstruct the correlator outputs given the estimation errors generated by the tracking loops. Thus, the correlation process does not need to be simulated and only the estimation errors are determined using a Monte Carlo approach. Based on this principle, the simulation scheme shown in Figure 9 can be adopted for the fast simulation of digital tracking loops.

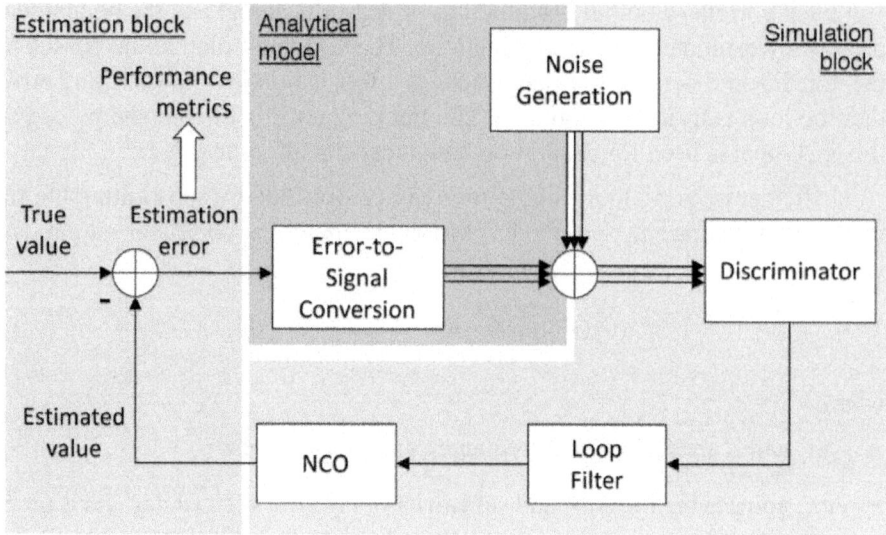

**Figure 9:** Semi-analytic scheme adopted for the simulation of GNSS tracking loops. Each element of the scheme proposed for the analysis of tracking loops has been implemented in a different function of the SATL Sim toolbox.

The functional elements in Figure 9 have been grouped to form the simulation block, the analytical model and the estimation part. The analytical model is used to convert the signal parameter errors/ $\Delta f_d$, $\Delta \varphi$, $\Delta \tau_d$ and $\Delta \tau_s$, into the signal components of the correlator outputs. At the same time, the analytical model is used to determine the variance and correlation of the different noise terms used to simulate $\eta_c$. Since the noise components are simulated using parameters determined by the analytical model, the "noise generation" block is shared between the analytical and simulation parts. The remaining parts of the loop, including the non-linear discriminator, loop filter and NCO, are fully simulated. Finally, the estimation block determines the residual signal parameter errors by comparing true values (determined by the simulation scenario) and estimates produced by the NCO.

By modifying these functional blocks, it is possible to simulate different tracking loops. In the simulation scheme implemented in the SATL Sim toolbox, a new estimate of the tracking parameters (Doppler frequency, carrier phase and code and subcarrier delays) is generated by an NCO model. This model accounts for the integration process performed by a real NCO and different update equations can be used [28]. A commonly used model is the rate-only feedback NCO [28], characterized by the following update equation:

$$\hat{\varphi}_k = \hat{\varphi}_{k-1} + \frac{T_c}{2} \left( \delta\hat{\varphi}_{k-1} + \delta\hat{\varphi}_{k-2} \right),$$

(27)

where $\hat{\varphi}_k$ denotes the k-th estimate of the tracking parameter under consideration and $\delta\hat{\varphi}_k$ is its estimated rate of change. The rate $\delta\hat{\varphi}_k$ is generally provided by the loop filter. It is noted that when several parameters are considered, equation (27) is used to update each term independently. The new parameter estimate is compared to the true value and a new estimation error is computed. This error is then used for the generation of the signal component at the output of the ID block using equation (26). The noise term, generated separately, is then added to the signal component. When several correlators are required, the correlation among the different noise components has to be accounted for. This is simulated using the approach described in [9].

The operations required to convert the correlator outputs into a new estimate of the parameter rate, $\delta\hat{\varphi}_k$, are fully simulated and correspond to the functional blocks that can be found in a real tracking loop. For instance, the correlator outputs are used to update the nonlinear discriminator and the loop filter. It is noted that a similar simulation scheme can be used for analyzing Kalman filter based tracking. In this case, the correlator outputs are fed to a Kalman filter that is used to produce new estimates of the tracking parameters.

## Code Structure

The structure of the code developed in the SATL Sim toolbox is provided in Figure 10. In this case, the code is used to estimate the tracking jitter of the loop as a function of different parameters, such as the Early-minus-Late spacing and the input CN. In particular, the non-linear discriminator may use several correlators to compute the cost function that the loop is trying to minimize. A DLL usually requires at least two correlators, named Early and Late correlators, computed for the delays

$$\hat{\tau} \pm \frac{1}{2}d_s$$

(28)

Where $\hat{\tau}$ the best code delay estimate and ds is the Early-minus-Late spacing. Early and Late correlators are computed symmetrically with respect to the best delay estimate and the non-linear discriminator computes a cost function proportional to the misalignment between these two correlators. Since the code correlation function is symmetric, the output of the discriminator is minimized when $\hat{\tau}$ corresponds to the delay of the input signal. The SLL works using similar principles. The performance of DLL and SLL depends on the Early-minus-Late spacing that is a simulation parameter. The CN is used to determine the correlator amplitude and the variance of the noise component $\eta_c$

**Initialization:**

- Initial parameters: sampling frequency, integration time...
- Loop filter design (FilterDesign.m)
- Input (true) parameters generation

**Main Simulation Loop**

**Loop on the Early-minus-Late spacing:**

   **Loop on the $C/N_0$ values:**

      - Noise generation (GenerateNoiseVector.m)

      **Loop on the simulation runs:**

         1. NCO update (UpdateNCO.m)
         2. Evaluation of the estimation error
         3. Error-to-Signal conversion (GenerateSignalCorrelation.m)
         4. Signal and Noise combining
         5. Discriminator update (UpdateDiscriminator.m)
         6. Loop Filter update (UpdateFilter.m)

      **End Loop on the simulation runs**

      - Evaluate tracking jitter

   **End Loop on the $C/N_0$ values**

   - Plot tracking results

**End Loop on the Early-minus-Late spacing**

**Figure 10:** Structure of the SATLSim toolbox and list of the different MT functions.

The parameters required for initializing the simulation procedure are accessible through the function Init Settings. These parameters include the sampling frequency, the loop bandwidth and the coherent integration time that are used to design the loop filters through the function Filter Design. In the code provided, standard formulae from [21] are used. However, Filter Design can be modified in order to adopt a different approach, such as the controlled-root formulation proposed by [28]. During the initialization phase, the true input parameters are also generated. The simulation core consists of three nested loops, on the Early-minus-Late spacing, for different CN values and for the number of simulation runs. The loop on Early-minus-Late spacing can be absent if, for example, only a PLL is considered. For each Early-minus-Late spacing and for a fixed CN, a noise vector containing the noise components of the correlator outputs is generated. The vector length is equal to the number of simulation runs and all the noise components are generated at once for efficiency reasons.

All intermediate results, such as the discriminator and loop filter outputs, are stored in auxiliary vectors and are used at the end of the loop on the simulation runs to evaluate quantities of interest such as the tracking jitter.

In the code provided, theoretical formulae for the computation of the tracking jitter are also implemented and used as a comparison term for the simulation results.

## Standard Pll (Pll.M)

The simulation of a standard PLL requires the generation of the Prompt correlator alone (GenerateSignalCorrelation). The Prompt correlator is the output of the ID block computed with respect to the best delay estimate provided by the loop [21]. For this reason, the noise generation (GenerateNoiseVector) simply consists of simulating a one dimensional complex Gaussian white sequence with independent and identically distributed real and imaginary parts with variance [9]

$$\sigma_n^2 = \frac{1}{C/N_0 T_c}.$$

(29)

When simulating a standard PLL alone, perfect code synchronization is assumed and (26) simplifies to

$$\sqrt{C} \frac{\sin{(\pi \Delta f_d T_c)}}{\pi \Delta f_d T_c} \exp{\{j\Delta\varphi\}} + \eta_c,$$

(30)

where $\Delta f_d$ is obtained by comparing the true Doppler frequency against the loop filter output. $\Delta\varphi$ is the phase error obtained as the difference between the true phase and the phase estimate produced by the NCO.

In SATLSim, the function UpdateDiscriminator implements a standard Costas discriminator. Different phase discriminators, as indicated in [21], can be easily implemented by changing this function.

## Double Estimator (Doubleestimator.M)

In the DE case, i.e. when DLL and SLL are jointly used, the function GenerateNoiseVector, responsible for the generation of the correlator noise, produces a 5×Nsim matrix, where $N_{sim}$ is the number of simulation runs. The five rows of this matrix correspond to the five correlators required by the DE that are characterized by the following correlation matrix

$$C_n = \begin{bmatrix} 1 & R_l\left(\frac{d_s}{2},\frac{d_{sc}}{2}\right) & R_l\left(\frac{d_s}{2},0\right) & R_l\left(\frac{d_s}{2},\frac{d_{sc}}{2}\right) & R_l\left(d_s,0\right) \\ R_l\left(\frac{d_s}{2},\frac{d_{sc}}{2}\right) & 1 & R_l\left(0,\frac{d_{sc}}{2}\right) & R_l\left(0,d_{sc}\right) & R_l\left(\frac{d_s}{2},\frac{d_{sc}}{2}\right) \\ R_l\left(\frac{d_s}{2},0\right) & R_l\left(0,\frac{d_{sc}}{2}\right) & 1 & R_l\left(0,\frac{d_{sc}}{2}\right) & R_l\left(\frac{d_s}{2},0\right) \\ R_l\left(\frac{d_s}{2},\frac{d_{sc}}{2}\right) & R_l\left(0,d_{sc}\right) & R_l\left(0,\frac{d_{sc}}{2}\right) & 1 & R_l\left(\frac{d_s}{2},\frac{d_{sc}}{2}\right) \\ R_l\left(d_s,0\right) & R_l\left(\frac{d_s}{2},\frac{d_{sc}}{2}\right) & R_l\left(\frac{d_s}{2},0\right) & R_l\left(\frac{d_s}{2},\frac{d_{sc}}{2}\right) & 1 \end{bmatrix},$$

(31)

where $d_{sc}$ is the subcarrier Early-minus-Late spacing.

The NCO update (Update NCO) is performed on both code and subcarrier loops and the estimated errors, $\Delta_{\tau d}$

And $\Delta_{\tau s}$, are used to compute new correlator signal components (Generate Signal Correlation). Two nonlinear discriminators (Update Discriminator) and loop filters (Update Filter) are run in parallel to determine the rate of change of both code and subcarrier delay.

The DE provides an example of how several tracking lo ops, operating in parallel, can be easily coupled in order to provide more realistic simulations accounting for the interaction of different tracking algorithms [9].

## Sample Results

In this section, sample results obtained using the SATL Sim toolbox are shown for the DE case. Results for the analysis of the PLL can be found in [10]. Specific focus is devoted to the analysis of the tracking jitter, which is one of the main metrics used for the analysis of digital tracking loops. The tracking jitter quantifies the amount of noise transferred by the tracking loop to the final parameter estimate [29]. The tracking jitter is the standard deviation of the final parameter estimate normalized by the discriminator gain. The non-linear discriminator is usually a memoryless device characterized by an input/output function relating the parameter estimation error to the discriminator output. The discriminator gain is the slope of this function in the neighborhood of zero (hypothesis of small estimation error).

Tracking jitter results obtained using non-coherent discriminators [21] for both DLL and SLL are shown in Figure 11. The figure is divided into three parts: [a)]

1. Tracking jitter of the DLL alone

2. Tracking jitter of the SLL alone

3. Jitter of the combined delay estimate.

This is due to the fact that the DE jointly uses a DLL, for estimating the code delay, and a SLL, for determining the subcarrier delay. Subcarrier and code delay are then combined to obtain the final estimate of the travel time of the transmitted signal [27]. Thus, three different jitters are evaluated for the different estimates produced by the system. Tracking jitter has been expressed in meters by multiplying the standard deviation of the delay estimates by the speed of light.

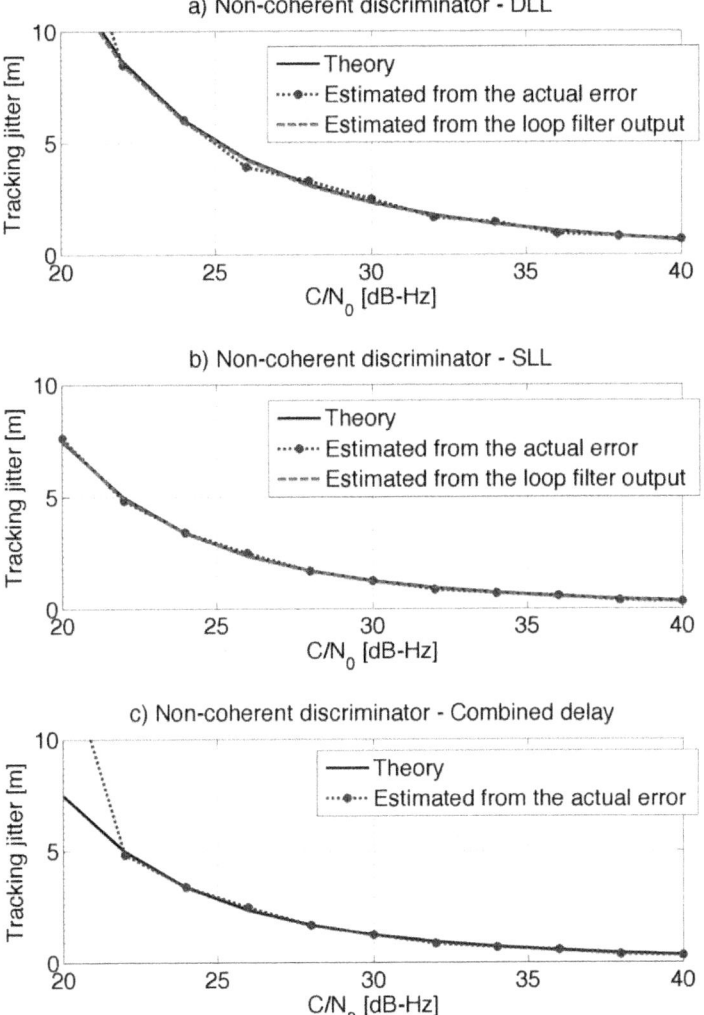

**Figure 11:** Tracking jitter obtained using the SATLSim toolbox. a) Tracking jitter of the DLL alone. b) Tracking jitter of the SLL alone. c) Jitter of the combined delay estimate.

In addition to this, the curves are shown in Figure 11 a) and Figure 11 b). More specifically, three different methodologies have been employed for determining the tracking jitter. The theoretical curve corresponds to approximate formulas obtained by linearizing the input/output function of the non-linear discriminator. These formulas are valid only for small tracking errors or equivalently for high CN. The jitter obtained from the actual error has been obtained by evaluating the variance of the code phase error. It is noted that in a real tracking loop the code phase error is not directly accessible since the true code phase is unknown. Thus, the tracking error can be evaluated by measuring the error at the loop filter output, which is an observable point, and propagating its variance through the loop. The tracking jitter obtained by propagating the variance at this measurable point corresponds to the curve denoted by "Estimated from the loop filter output". The relationship between the variances of the discriminator output and the true tracking error can easily be evaluated when the loop is working in its linear region. The measured curve was introduced to further validate the theoretical model and test the correctness of the simulation methodology. This latest curve is not available for the combined delay estimate.

The parameters used for the evaluation of the tracking jitter, shown in Figure 11, are provided in Table 3.

**Table 3:** Parameters used for the evaluation of the tracking jitter shown in Figure 11.

| Parameter | Value |
|---|---|
| Sampling Frequency | $f_s = 8$ MHz |
| Integration Time | $T_c = 4$ ms |
| DLL Early-minus-Late spacing | 0.1955 $\mu$s (0.2 chips) |
| SLL Early-minus-Late spacing | 0.1955 $\mu$s (0.2 chips) |
| DLL Loop Order | 1 |
| SLL Loop Order | 1 |
| DLL Loop Bandwidth | 0.5 Hz |
| SLL Loop Bandwidth | 0.5 Hz |
| Modulation type | BOC(1, 1) |

From the results shown in Figure 11, it is observed that the developed semi-analytic technique is able to effectively capture the behavior of the system. For high CN values, a good agreement between theoretical and simulation results is found. However, for CN lower than 22 dB-Hz theoretical and simulation results start diverging. This is more clear in parts a) and c) of the figures. For such low CN values, the loop is no longer working in the linear region of the discriminator input/output function. Thus, the theoretical model is unable to capture the behavior of the loop that is losing lock. The semi-analytic technique implemented in the SATLSim MT toolbox is able to effectively describe the non-linear behavior of the loop requiring only limited computation resources.

## CONCLUSIONS

In this chapter, the development of fast semi-analytic techniques using MT has been analyzed. In the semi-analytic framework, the knowledge of the system under analysis is exploited to reduce the computational load and complexity that full Monte Carlo simulations would require. In this way, the strengths of both analytical and Monte Carlo methods are effectively combined.

Two examples of semi-analytic techniques have been thoroughly analyzed and used to illustrate the two main configurations developed within the semi-analytic framework. The first example illustrates the sequential configuration where simulations and the analytical model are used sequentially. This type of configuration provides increased precision with respect to full Monte Carlo simulations, particularly when the quantities to be estimated assume small values. In addition to this, a significant reduction in terms of computational complexity is achieved. In the example considered, full Monte Carlo simulations require the implementation of a full transmission/reception chain including the interaction between two different systems, DVB-T and GNSS. This requirement led to a significant computational and development complexity. The considered semi-analytic approach is an effective solution for alleviating those requirements.

The second example considered the closed-loop approach and specific focus was devoted to the SATLSim MT toolbox. This toolbox has been developed for the analysis of digital tracking loops and fully exploits the flexibility of the MT programming language. The code has been organized in functions that can be easily replaced by different MT modules. In this way, different loop components such as discriminators, loop filters and NCO models can be integrated in the SATLSim toolbox. The efficiency of semi-analytic techniques and the reduced development time enabled by the MT language are an effective tool for the analysis of complex communications systems

## REFERENCES

1. W. H. Tranter, K. S. Shanmugan, T. S. Rappaport, K. L. Kosbar, Principles of Communication Systems Simulation with Wireless ApplicationsPrentice Hall, January 2004

2.   M. C. Jeruchim, P. Balaban, K. S. Shanmugan, Simulation of communication systemsKluwer Academic/Plenum Publishers, new york edition, 2000

3.   F. M. Gardner, J. D. Baker, Techniques. Simulation, of. Models, Signals. Communication, John. Processes, JohnWiley & Sons, 2003

4.   M. Jeruchim, Techniques for estimating the bit error rate in the simulation of digital communication systemsIEEE Journal on Selected Areas in Communications21153170January 1984

5.   M. Pent, L. Lo, G. Presti, D'Aria, G. De Luca, B. E. R. Semianalytic, by evaluation, for simulation, nonlinear. noisy, channels. I. E. E. bandpass, IEEE Journal on Selected Areas in Communications613441January 1988

6.   M.T. Core, R. Campbell, P. Quan, and J. Wada. Semianalytic BER for PSK. IEEE Transactions on Wireless Communications, 8(4):1644-1648, April 2009.

7.   A. R. Golshan, Post-correlator modeling for fast simulation and joint performance analysis of GNSS code and carrier tracking loops. In Proc. of the ION/NTM (National Technical Meeting), 312318Monterey, CA, January 2006

8.   J. S. Silva, P. F. Silva, A. Fernandez, J. Diez, J. F. M. Lorga, correlator. Factored, A. model, for. Solution, flexible. fast, G. N. S. S. realistic, simulations. receiver, Proc. In, I. O. N. G. N. S. S. of, 267. pages, 268, Worth. T. X. Forth, September 2007

9.   D. Borio, P. B. Anantharamu, G. Lachapelle, simulations. Semi-analytic, extension. An, unambiguous. B. O. C. to, tracking, In Proc. of the ION/ITM (International Technical Meeting), 10231036San Diego, CA, January 2010

10.  Daniele Borio, Pratibha Anantharamu, and Gérard Lachapelle. SATLSim: a semi-analytic framework for fast GNSS tracking loop simulations.GPS Solutions152011427431

11.  J. M. Geist, Computer generation of correlated gaussian random variablesProceedings of the IEEE675862863may 1979

12.  A. R. Golshan, Loss of lock analysis of a firstorder digital code tracking loop and comparison of results to analog loop theory for BOC and NRZ signals. In Proc. of the ION/NTM (National Technical Meeting), 299305San Diego, CA, January 2005

13.  J. W. Betz, Effect of narrowband interference on GPS code tracking accuracy. In Proc. Of ION/NTM, 1627Anaheim, CA, January 2000

14.  D. Borio, G. N. S. S. acquisition, the. in, of. presence, wave. continuous,

I. E. E. interference, IEEE Transactions on Aerospace and Electronic Systems4614760January 2010

15. T. Van Slyke, W. Kuhn, B. Natarajan, Measuring interference from a UWB transmitter in the GPS l1 band. In Proc. of the IEEE Radio and Wireless Symposium, 887890Orlando, FL, March 2008

16. A. Simsky, T. De Wilde, D. Mertens, E. Koitsaly, J. , M. Sleewaegen, First field experience with L5 signals: DME interference reality check. In Proc. of ION/GNSS, 2937Savannah, GA, September 2009

17. D. Borio, S. Savasta, L. Lo, Presti, On the DVB-t coexistence with galileo and GPS system. In Proc. of the 3rd ESA Workshop on Satellite Navigation User Equipment Technologies (NAVITEC), Noordwijk, The Netherlands, December 2006

18. M. Wildemeersch, A. Rabbachin, E. Cano, J. Fortuny, Interference assessment of DVB-t within the GPS l1 and galileo e1 band. In Proc. of the 5th ESA European Workshop on GNSS Signals and Signal Processing (NAVITEC), 18Noordwijk, The Netherlands, December 2010

19. M. Steven, Kay, Fundamentals of Statistical Signal Processing, 2 Detection Theory, volume 2. Prentice Hall, 1rt edition, February 1998

20. ETSI.Digital video broadcasting (DVB); framing structure, channel coding and modulation for digital terrestrial television, 2006EN 300 744.

21. E. D. Kaplan, C. Hegarty, Understanding. G. P. S. editors, Principles, Artech. Applications, 2nd. House, edition, November 2005

22. H. L. Van Trees, Estimation. Detection, Theory. Modulation-Part,. , 1st. Wiley-Interscience, edition, September 2001

23. Borio. A. Daniele, theory statistical, G. N. S. S. for, acquisition. signal, Phd thesis, Politecnico di Torino, April 2008

24. W. John, Betz, Effect of partial-band interference on receiver estimation of C/N0. In Proc. of the ION/NTM, 817828Long Beach, CA, January 2001

25. J. Fortuny-Guasch, M. Wildemeersch, D. Borio, of. D. V. B. Assessment-T, on. G. N. S. S. impact, acquisition, performance. tracking, In Proc. of the ION/GNSS, 347356San Diego, CA, January 2011

26. P. Anantharamu, D. Borio, G. Lachapelle, Sub-carrier shaping for BOC modulated GNSS signalsEURASIP Journal on Advances in Signal Processing2011133 EOF

27. M. S. Hodgart, P. D. Blunt, A. dual, receiver. estimate, binary. of, carrier. . B. O. C. offset, signals. modulated, navigation. global, systems. satellite, Electronics Letters, 4316877878August 2007

28. S. A. Stephens, J. B. Thomas, Controlled-root formulation for digital phase-locked loopsIEEE Transactions on Aerospace and Electronic Systems3117895january 1995

29. A. J. V. Dierendonck, P. Fenton, T. Ford, Theory and performance of narrow correlator spacing in a GPS receiverNAVIGATION: the Journal of The Institut of Navigation, 393265283Fall 1992

# Chapter 4

# KINEMATICAL ANALYSIS AND SIMULATION OF HIGH-SPEED PLATE CARRYING MANIPULATOR BASED ON MATLAB

Ke Wang, Jiping Zhou

School of Mechanical Engineering, University of Yangzhou, Yangzhou, China

## ABSTRACT

In order to construct the more effective kinematics method for industry, by taking a high-speed plate handing robot as an example, the structure and parameters of the robot linkages are analyzed, and the standard Denavit-Hartenberg method is applied to establish the coordinates and the kinematic equation of the linkages. Depending on the graphics and matrix calculation ability of Matlab especially including the Robotics Toolbox, the handling robot has been modeled and its kinematics, inverse kinematics and the trajectory planning have been simulated. Therefore, the correctness of kinematic equation has been verified, meanwhile, the functions of displacement, velocity, acceleration and trajectory of all the joints are also obtained. In a further step, this has verified the validity of all the structure parameters and provided a reliable basis for the theoretical research on the design, dynamics analysis and trajectory planning of the manipulator control system.

## INTRODUCTION

As a typical representative and a main technical means of information technology and advanced manufacturing technology, the complete set of automatic stamping processing line has become a high technology to which the developed countries have paid much attention. And its development level has become one of the most important standards to measure a nation's technical development. It has been extensively applied in the industry of mechaniccal manufacturing, nuclear, aerospace, energy and transportation, petroleum chemistry, building, electronic and etc. [1]. But, one key aspect of the automatic stamping processing line is to develop a mechanical manipulator with the characteristic of the high-speed and dynamic transmission.

At present, the sheet metal forming production line is operated at the high productive rate of 15 - 25 SPM. Removing the time spent by the press stamping operations, there is only about 2 s left to the robot manipulator. In such a short period of time, the manipulator has to operate at a high speed of 200 - 250 m/min, so that the operations such as loading, moving and unloading can be fulfilled. And this has put forward higher requirements to the mechanical structure, material friction characteristics, and structural dynamic characteristics of the robot manipulator.

Taking a high-speed plate carrying manipulator as the research object, this paper firstly analyzes the structure and connecting rod parameters; then adopts the standard D-H method [2] to establish the kinematics equation; and then discusses the positive and inverse kinematics algorithm, and the trajectory planning problems; finally in the environment of Matlab, the kinematics model is built to take kinematics simulation by using of Robotics Toolbox [3]. In simulation process, we can not only directly observe the robot motion, but also get the required data in the graphic form. Therefore, the virtual performace of the product can be tested in the conceptual design stage, so as to improve the design performance, reduce design cost and decrease product development time.

# THE STRUCTURE DESIGN AND LINK PARAMETERS OF HIGH-SPEED PLATE CARRYING MANIPULATOR

## The Design Requirements of High-Speed Plate Carrying Manipulator

### The Workplace of High-Speed Plate Carrying Manipulator

The traditional stamping processing method relies on a stand-alone manual which is inefficient, inaccurate and insecurity. So it has already become increasingly unsuited to the requirements of modern mass production, especially when the requirements of the sheet metal processing annual output exceeds thousands tons, this contradiction is more prominent. The current stamping equipment is toward the trend of automation, sets and online development. With the rapid development of China's market economy, especially in the coastal areas where the shortage of skilled workers is severe, and labor costs rises sharply, many machinery manufacturers have the urgent demand for manufacturing automation, and require machine and equipment manufacturing industry to provide users with a complete set of on-line technical services to improve production efficiency and reduce labor costs. Therefore, the development of the technology of sheet metal production line is even more important. One of the core technologies of sheet metal stamping equipment line is the design of high-speed plate carrying manipulator. As shown in **Figure 1**, it is high-speed plate carrying manipulator in the application of stamping processing complete sets of equipment on-line system.

## The Operation Process of High-Speed Plate Carrying Manipulator

According to the composition of the stamping process sets of on-line system and the role of the high-speed plate carrying manipulator, the working cycle of a manipulator contains nine action process, as shown in Figure **2**:

1) From the point of origin, the left electromagnetic valve is energized after pressing the start button, then the manipulator moves to the left. It won't stop until it encounters the left limit switch.

2) Simultaneously the manipulator begins to drop after the decreased electromagnetic valve opens, then, it won't stop until it encounters the lower limit switch.

3) At the same time the clamp electromagnetic valve is energized, then the manipulator is clamped.

4) After the rise electromagnetic valve is energized, the manipulator begins to rise, then it won't stop until it encounters the rise limit switch.

5) Simultaneously the manipulator moves to the right after the right electromagnetic valve is energized, then, it won't stop until it encounters the right limit switch.

6) Simultaneously the manipulator begins to drop after the decreased electromagnetic valve is energized, then, it won't stop until it encounters the lower limit switch.

7) At the same time the clamp electromagnetic valve is opened, then, the manipulator is opened.

8) After the rise electromagnetic valve is energized, the manipulator begins to rise, then, it won't stop until it encounters the rise limit switch.

Double sets unstacker unit Feeding manipulator Press 1 Transfer manipulator 1 Press 2 Transfer manipulator 2 Press 3 Transfer manipulator 3

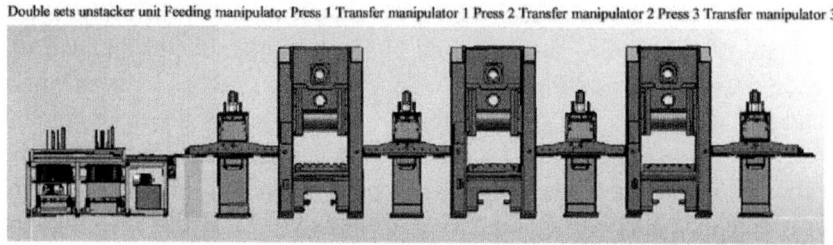

**Figure 1**: The composition of the stamping processing sets of on-line system.

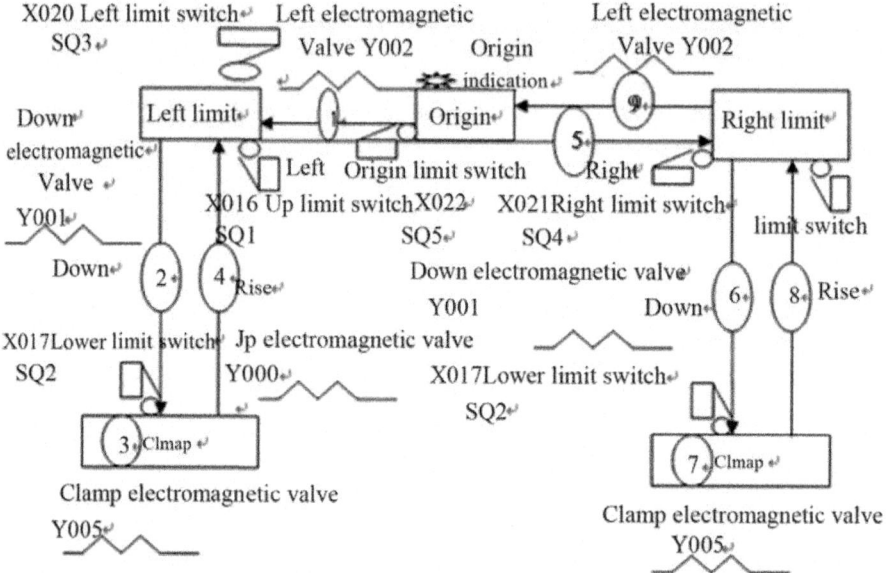

**Figure 2**: The operation process of the manipulator.

9)   Simultaneously the manipulator moves to the origin after the left electromagnetic valve is energized, then, it won't stop until it encounters the left limit switch.

So far, the manipulator after 9 step action completes a cycle of its movement, and then, it continues the cycle work.

Based on the above analysis, we have developed a three-dimensional model of the manipulator as shown in Figure 3, which can be used to analyze the dynamic characteristics of the manipulator.

## The Design Parameters of High-Speed Plate Carrying Manipulator

### *D-H Transformation*

In order to describe the translational and rotational relationships between adjacent rods, Denavit and Hartenberg (1955) have proposed a matrix method to establish the possessed coordinate system for each rod in the linkage chain. This method is to establish a homogeneous transformation matrix for each of the join bar coordinate, which represents the relationship of the previous coordinate system of the bar, and the principle [4,5] is detailed as follows:

OXYZ: A fixed reference coordinate system of the fixed coordinates, is called the coordinate system.

$O_iX_iY_iZ_i$: Fixed connected with the member bar of number I of the robot, the origin of coordinate is at the center point of the joint of the I + 1th.

Identify and establish each coordinate system by following three rules.

1):    $Z_{i-1}$ axis along the motion shaft of the join of the i th.

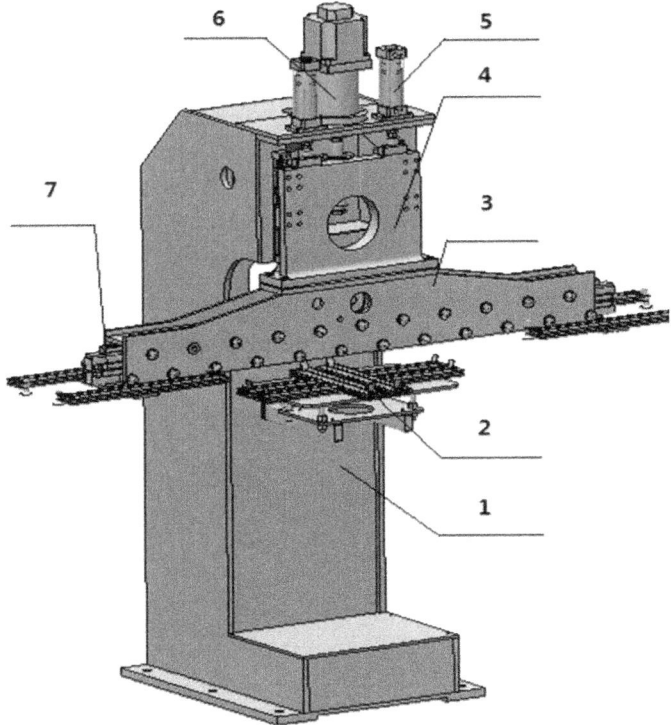

**Figure 3**: The assembly model of manipulator, 1-body, 2- feeder, 3-beam, 4-slipway, 5-balance cylinder, 6-servo motor, 7-X axis transmission system.

2):    $X_i$ axis vertical the axis of $Z_i$ and $Z_{i-1}$ and point to the direction of away from the $Z_{i-1}$ axis.

3):    $Y_i$ axis according to the requirements of the right hand coordinate system to establish.

Meanwhile the notation of D-H of the rigid bar depends on the four parameters of the connecting rod.

The angle of two connecting rod $\theta_i$ : $X_{i-1} \to X_i$ around the corner of $Z_{i-1}$.

The distance of two connecting rod $d_i$ : $X_{i-1} \to X_i$ along the distance of $Z_{i-1}$.

The length of the connecting rod $a_i$: $Z_{i-1} \to Z_i$ along the distance of $X_{i-1}$.

The torsion angle of the connecting rod: $a_i$ : $Z_{i-1} \to Z_i$ around the corner of $X_i$.

For rotational joints, $\theta_i$ is the joint variables, the rest are joint parameters (remain unchanged); for mobile joints, $d_i$ is the joint variables, the rest are joint parameters.

## The Design Coordinate of Manipulator

High-speed plate carrying manipulator is mainly composed of vertical pillars (body), horizontal arm (beam), sliding table (Y axis transmission system) and base. Horizontal arm mounted on the machine body level can move around and can move up and down along the vertical pillars. By reference to the high-speed plate carrying manipulator in Figure 3, we establish the coordinate system of manipulator in Figure 4 based on the above analysis.

According to the D-H parameters method, four parameters are defined for each link: the connecting rod angle $\theta_i$, the distance of two connecting rod $d_i$, the length of the connecting rod $a_i$, the connecting rod torsion Angle $a_i$, the D-H parameters of manipulator as is shown in **Table 1**.

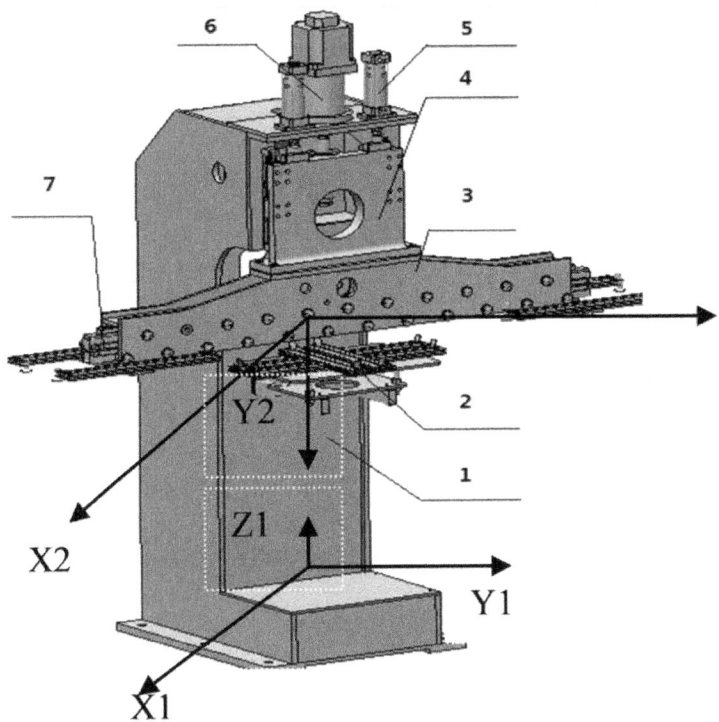

**Figure 4**: The D-H coordinate of high-speed plate carrying manipulator.

**Table 1**: The D-H parameters and joint variables of manipulator

| Joint | $\theta_i$ | $d_i$ | $a_i$ | $d_i$ |
|-------|-----|-------|-------|-------|
| 1 | 0 | $d_1$ | 0 | −90 |
| 2 | 0 | $d_2$ | 0 | 0 |

# The Kinematics Simulation Algorithm of Manipulator

## The Kinematics of Manipulator

The so-called kinematics problem [4,5] is given to the manipulator of each joint variable, and then obtains the position and posture of the end of the actuator, and its essence is to establish the kinematics equations. For the solution of kinematics equation, this paper uses homogeneous transformation matrix $^{i-1}_i A$ to describe the coordinate system of i relative to the position and pose of the coordinate system of the i − 1, this is the general formula for transformation matrix[14].

$$
{}^{i-1}A_i = \begin{bmatrix} \cos\theta_i & -\sin\theta_i\cos\alpha_i & \sin\theta_i\sin\alpha_i & a_i\cos\theta_i \\ \sin\theta_i & \cos\theta_i\cos\alpha_i & -\cos\theta_i\sin\alpha_i & a_i\sin\theta_i \\ 0 & \sin\alpha_i & \cos\alpha_i & d_i \\ 0 & 0 & 0 & 1 \end{bmatrix}
$$

(1)

Now, putting the D-H parameters and joint variables of manipulator substitution of the **Table 1** in (1), we get the homogeneous transformation matrix ${}^0T_i$ of the i coordinate system relative to the position and posture of the base coordinate system, expressed as:

$$
{}^0T_i = {}^0A_1\,{}^1A_2 \cdots {}^{i-1}A_i
$$

(2)

Especially, when $i = 2$, $T = {}^0T_2$ can be obtained, it determines the position and posture of the end of the manipulator relative to the base coordinate system, the matrix of T can be expressed as:

$$
T = {}^0T_2 = {}^0A_1{}^1A_2 = \begin{bmatrix} 1 & 0 & 0 & 0 \\ 0 & 0 & 1 & d_2 \\ 0 & -1 & 0 & d_1 \\ 0 & 0 & 0 & 1 \end{bmatrix}
$$

(3)

Note:

$$
{}^0A_1 = \begin{bmatrix} 1 & 0 & 0 & 0 \\ 0 & 0 & 1 & 0 \\ 0 & -1 & 0 & d_1 \\ 0 & 0 & 0 & 1 \end{bmatrix}, \quad {}^1A_2 = \begin{bmatrix} 1 & 0 & 0 & 0 \\ 0 & 1 & 0 & 0 \\ 0 & 0 & 1 & d_2 \\ 0 & 0 & 0 & 1 \end{bmatrix}
$$

## The Inverse Kinematics Issue of Manipulator

The inverse kinematics issue of manipulator is defined as follows: with the known the position and posture of the end of the actuator, we need to solve the variables of each joint of the manipulator, here in our case the variables are $d_1$ and $d_2$.

The target point of the movement of the manipulator center is ${}^0P = [0 \ Y_0 \ Z_0 \ 1]^T$. By using the transitional joint, the manipulator's center point can coincide with the target point, and the target point can be expressed as: ${}^2P = [0 \ -d_1 \ d_2 \ 1]^T$, then we can establish the following equation:

$$
{}^0P = {}^0T_2\,{}^2P
$$

(4)

Combining (3) and (4) and solving the kinematics equation, we can get the following equation.

$$d_2 = Y_0/2 \qquad (5)$$

$$d_1 = Z_0/2 \qquad (6)$$

## The Kinematics Simulation of Manipulator

### *Use Link and Robot Functions to Establish the Manipulator's Object*

1)    Before the manipulators simulation, firstly input the manipulator's parameters by calling the function Link:

$$L = \left( [\text{alpha}, A, \text{theta}, D, \text{sigma}], \text{standard} \right)$$

(Note: alpha, $A$, theta, $D$ represent the variables of $\alpha_i$, $\alpha_i$, $\theta_1$ and $d_1$ respectively; "sigma" represents the joint types: 0 for the rotational joint, and 1 for the transitional joint; "standard" is for the standard D-H parameters. The function robot is used to create a manipulator object by using the Link function in the format of Robot (Link...). The commands for creating the high-speed plate carrying manipulators is expressed as follows:

$$L_1 = \left( [-pi/2 \ 0 \ 0 \ 0.08 \ 1], \text{'standard'} \right);$$
$$L_2 = \left( [0 \ 0 \ 0 \ 0.9 \ 1], \text{'standard'} \right);$$
$$r = \text{robot} \left( \{L_1 \ L_2\}, \text{'Manipulator'} \right);$$
$$\text{plot}(r)$$
$$\text{drivebot}(r)$$

2)    We can immediately see the three-dimensional view of the manipulator which can be used in the form manually for driving through the slider in the chart to drive the movement of the manipulator, which is just like the actual control of the manipulator [3]. As shown in **Figure 5**, it has brought great convenience for manipulator's teaching and training.

## The Test of Kinematic Model

In the Figure 5, by moving each slide in the bar to move the manipulator, the first joint movement can change the height of the manipulator in the vertical direction. The second joint can change the length of the manipulator in horizontal direction, So that we can obtain different position and posture. By adjusting the slides and the joint variables, we can get the approximate model corresponding to the actual structure in Figure 3, as shown in Figure 6.

In order to verify the correctness of the kinematics Equation (3), the geometric parameters and the joint variables of each rod of the manipulator are put into the kinematic equations to solve the setting position and posture of the coordinate system of the end link rod in relation to the base coordinate system. And then the corresponding coordinate values are input to the manipulator's trajectory planning, and the actual end position and posture information are compared with the results from solving the equation. Two groups of joint variable values are randomly chosen:

$$q_1 = \begin{bmatrix} 1.0053 & 0.50265 \end{bmatrix}$$
$$q_2 = \begin{bmatrix} -0.50265 & -0.50265 \end{bmatrix}$$

The error is very small through the comparative analysis of the actual value and set point, which explains that the kinematics equation and the model are reliable and consistent.

## The Simulation of Trajectory Planning

According to the requirements of the task, trajectory planning will designate in advance the robot's operating procedure and action process. The simulation method could describe in details the movement process of the industrial robot [6-10]. Planning can be made in the space of both joint and operation. We would introduce two main motions [4, 5]: 1) PTP, Point-To-Point motion; 2) CP, Continuous-Path. With regard to continuous-path, not only the initial and final points of the manipulator need to be set, but also some other points between the two points (called path points) which must move along some specific path (path constraint) need to be indicated.

The design of the high-speed plate carrying manipulator adopts a point-to-point trajectory planning. Set A as the starting point, moving to B where a task is finished, and then set B as a starting point, moving to C where a certain task is completed, and then set C as a starting point, moving to point C to complete the preset tasks, and continue to move, and so on. Here, the move from point A to point B and from point B to point C does not exist any point between them, and there is not any requirement for the known path of the movement. Therefore, we can take this above planning as PTP planning.

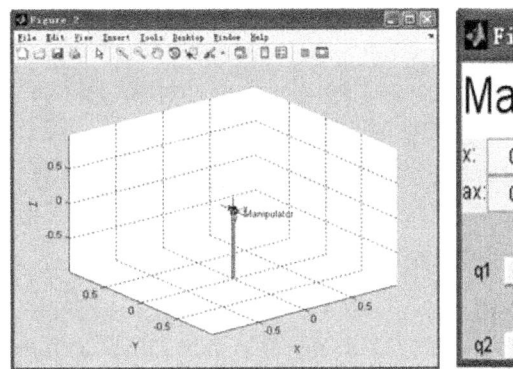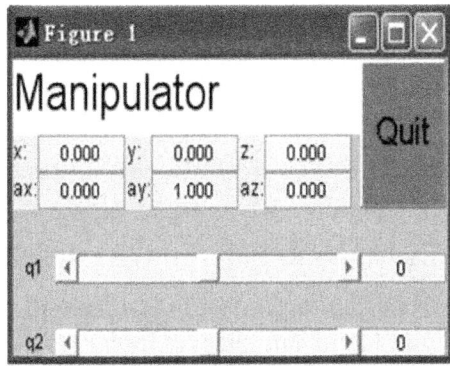

**Figure 5**: The three-dimensional diagram and slide block control chart of the high-speed plate carrying manipulator.

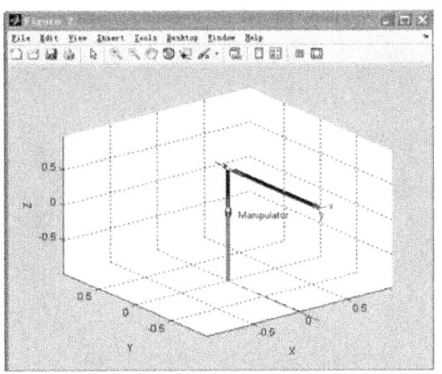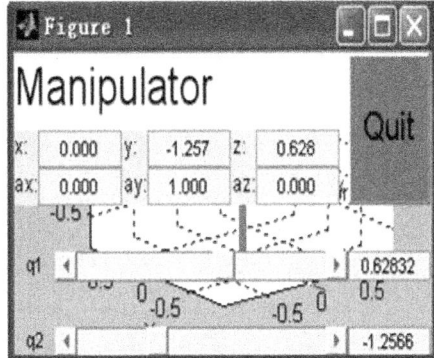

**Figure 6**: A state of high-speed plate carrying manipulator's three-dimensional diagram and slide block control chart.

The starting point is set to be $q_0 = [0 \quad 0]$, ending point $q_1 = [0.62832 \quad -1.2566]$, and between the two points the initial and final speeds are both zero, with the time of movement time: $t = 2$ s, the relative program as follows:

$q_0 = [0 \quad 0]; q_1 = [0.62832 \quad -1.2566]$

$t = [0: \quad 0.1: \quad 2]$

$q = \text{jtraj}(q_0, \quad q_1, \quad t)$

$\text{plot}(r, \quad q)$

By conducting the above program, we can observe the whole process of the high-speed plate carrying manipulator moving from Figures 5 and 6. We can also draw, by way of the function of $[q, \quad qd, \quad qdd] = \text{jtraj}(q_0, \quad q_1, \quad t), \text{plot}(t, \quad q(:, \quad i))$

each joint's displacement curve, as shown in **Figure** 7(a) and (b). (Note: q represents the displacement, i represents the joint Numbers). Also we can draw speed curve as shown in Figures 7(c)-(d), acceleration curve as shown in Figures 7(e)-(f), through the corresponding functions:

$\text{plot}(t, \ qd(:, \ i)), \text{plot}(t, \ qdd(:, \ i))$

## The Analysis Based on the Results of Simulation

From Figures 7(a) and (b), it can be seen that the displacement of the mobile joint 1 gradually changes from zero to 0.62832 m, movable joint 2 displacement from zero change to 1.2566 m; from Figures 7(c) and (d) we

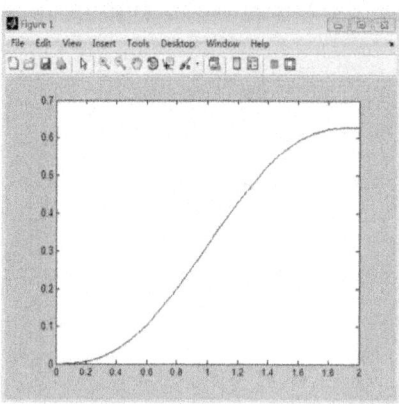

(a) The displacement curve of joint 1

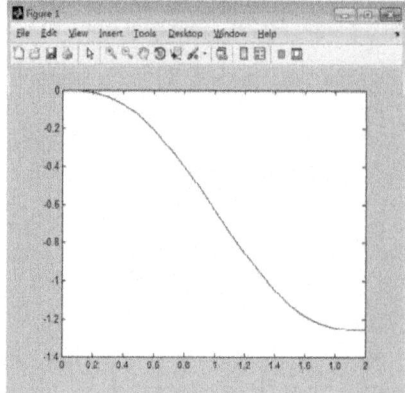

(b) The displacement curve of joint 2

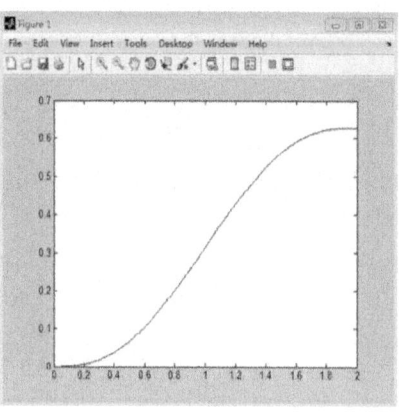

(c)The velocity curve of joint 1

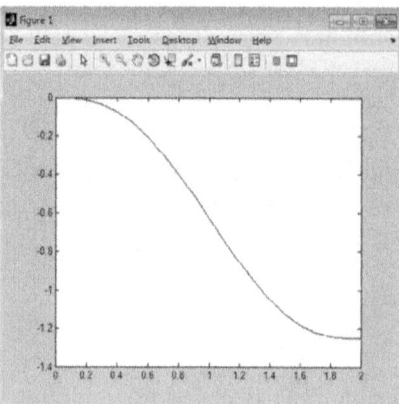

(d) The velocity curve of joint 2

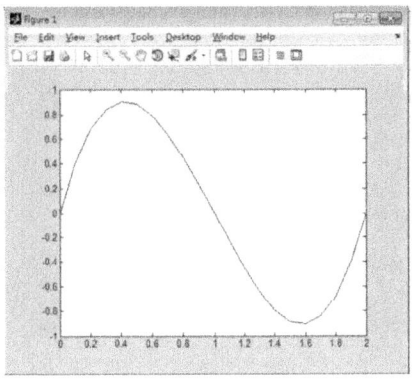 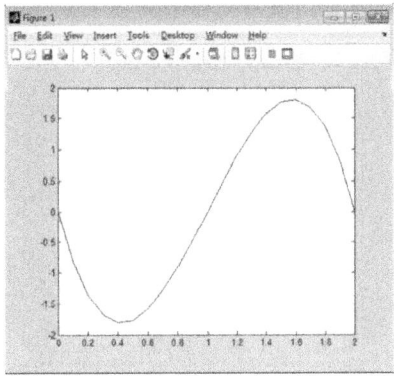

(e) The acceleration curve of joint 1          (f) The acceleration curve of joint 2

**Figure 7**: Displacement, velocity and acceleration curves of high-speed plate carrying manipulator.

can obtain that the initial and final speeds of the joints 1, 2 are zero, the maximum velocity appearing at the intermediate time t = 1 s; from Figures 7(e)-(f), it can be seen, the initial and final acceleration speeds are zeros, two maximum values in motion, with one positive and the other negative. The manipulator's displacement curve is smooth, and the curve of velocity and acceleration is continuous. Therefore, it is concluded that: in the process, the manipulator runs smoothly, and the whole body vibrates in a normal scope, which thus explains the validity of the design.

## CONCLUSIONS

The paper carries out a kinematical analysis of a highspeed plate carrying manipulator which is designed by using the module function of Robotics Toolbox in the Matlab. Based on this, we have achieved the following conclusions in three aspects: 1) the structure and parameters of the plate carrying manipulator have been designed; 2) the kinematics model has been established with the standard D-H method, and its positive and inverse kinematics have been analyzed; 3) by using Robotics Toolbox in Matlab, the kinematics model has been set up in order to make the study of the motions of manipulator easier and more apparent, and to conduct a kinematics simulation for the manipulator for obtaining the displacement, velocity and acceleration curve of each manipulator joint. Thus, this proves the validity of the design of manipulator structure parameters, as well as provides the reliable basis for the theoretical analysis in terms of the design, dynamics analysis, and trajectory planning of control system of the manipulator.

# REFERENCES

1.  Q. Wang, X. J. Hu and L. Z. Li, "The Kinematic Analysis and Simulation of the Humanoid Welding Manipulator Based on the MATLAB," Materials Science and Engineering College of Hefei University of Technology, 2011.

2.  J. J. Craig, "Introduction to Robotics," Mechanical Industry Press, Beijing, 2006.

3.  P. I. Corke, "A Robotics Toolbox for MATLAB," IEEE Robotics and Automation Magazine, Vol. 3, No. 1, 1996, pp. 24-32. doi:10.1109/100.486658

4.  X. S. Jiang, "Introduction to Robotics," Liaoning Science and Technology Press, Shenyang, 1994.

5.  Z. X. Cai, "The Robotics," Tsinghua University Press, Beijing, 2000.

6.  J. X. Luo and G. Q. Hu, "The Kinematic Simulation Research of Robot Based on the MATLAB," Journal of Xiamen University, JCR Science Edition, Vol. 44, No. 5, 2005, pp. 640-644.

7.  J. Han and L. Hao, "Trajectory Planning and Simulation of Robot in Joint Coordinate System," Journal of Nanjing University of Science and Technology, Vol. 24, No. 6, 2000, pp. 540-543.

8.  Z. X. Wang and W. X. Fan, "The Kinematics Analysis and Simulation of Industrial Robot Based on the Matlab," Mechanical and Electrical Engineering, Vol. 29, No. 1, 2012, pp. 33-37.

9.  The MathWorks Inc., "Matlab the Language of Technical Computing Version 6," The MathWorks Inc., 2002.

10. P. I. Corke, "Robotics Toolbox for Matlab (Release 7.1) [EB/OL]," 2002. http://www.cat.csiro.au/cmst/staff/pic/robot

# Chapter 5

# DYNAMIC ANALYSIS OF ELASTICALLY SUPPORTED CRACKED BEAM SUBJECTED TO A CONCENTRATED MOVING LOAD

Hasan Ozturk[1] , Zeki Kiral[1] , Binnur Goren Kiral[1]

[1]Department of Mechanical Engineering, Dokuz Eylul University 35397, Buca, Izmir, Turkey, hasan.

## ABSTRACT

This study deals with the dynamic behavior of a cracked beam subjected to a concentrated force traveling at a constant velocity. Dynamic analyses for a hinged-hinged cracked beam resting on elastic supports under the action of a moving load are carried out by the finite element method. For the beam having rectangular cross-section, element formulation for crack element is developed by using the principles of fracture mechanics. In the numerical analysis, Newmark integration method is employed in order to calculate the dynamic response of the beam. The effects of crack depth, crack location, elastic support and load velocity on the dynamic displacements calculated for different locations on the beam are investigated. The results related to the dynamic response of the beam are presented in 3D graphs.

## INTRODUCTION

Recently, the dynamic analysis of the engineering structures subjected to moving loads has gained great importance. The engineering structures with moving loads often come out in buildings, bridges, railways and cranes. Moreover, the cracks can be seen in engineering structures due to reasons like erosion, corrosion, fatigue or accidents. The presence of a crack can not only cause a local variation in the stiffness, but many also affect the dynamic behavior of the entire structure to a considerable extent. In this context, dynamic behavior of an engineering structure subjected to moving loads is affected with the presence of a crack.

Many investigations on the dynamic behaviour of the different isotropic structures subjected to moving loads have been carried out. Olsson (1991) has presented the fundamentals of the moving load problem for an isotropic beam. Rao (2000) has studied the dynamic response of multi-span Euler-Bernoulli beam to moving loads. Wang (1997) has investigated the forced vibration of multi-span Timoshenko beams. The effects of span number, rotary inertia, and shear deformation on the maximum moment, maximum deflection, and critical load velocity were investigated in this study. The moving force identification for a Timoshenko beam model has been given and the results, which obtained for an Euler-Bernoulli beam model, have been compared by Law and Zhu (2000). Esmailzadeh and Ghorashi (1997) have analyzed the effects of shear deformation, rotary inertia and the length of load distribution on the vibration of the Timoshenko beam subjected to a traveling mass. Hino et al. (1985) has investigated the vibration of a beam subjected to a moving load by using the Galerkin finite element formulation.

The deterministic and random vibration analyses of a nonlinear beam resting on an elastic foundation subjected to a moving load have been performed by Chang and Liu (1996). Thambiratnam and Zhuge (1996) have developed a simple procedure based on the finite element method for treating the dynamic analysis of beams on an elastic foundation subjected to moving point load, where the foundation was modelled by springs of variable stiffness. A method for the dynamic analysis of elastic beams subjected to dynamic loads induced by the arbitrary movement of a spring-mass-damper system has been presented by Lin and Trethewey (1990). In this study, the governing equations have solved with a Runge-Kutta integration scheme to obtain the dynamic response for both the supporting beam and the moving system. Kidarsa et al. (2008) have proposed a new numerical integration scheme in force-based elements, which enables to computing the internal force history at any location along a structural member under the effect of moving load. Gören (Kıral and Kıral 2009, 2008) and Kıral (2009)have studied the dynamic response of a symmetric laminated composite beam subjected to a concentrated moving load with constant velocity considering the effect of different boundary conditions, foundation stiffness and damping.

As seen from the aforementioned references, all studies on the moving load problem are related to the intact beams. Because of the practical importance of the subject, effects of cracks on the dynamic behaviors of beams have been the subject of many investigations. Unfortunately, literature research reveals that much less study investigating the dynamic behavior of cracked beam subjected to a moving load and elastic foundation is present in the published literature. A surface crack on a beam section has introduced a local flexibility to the structural member in the study which has been carried out by Gounaris and Dimarogonas (1988). In this study, a finite element model for a cracked prismatic beam has been developed. Ostachowicz and Krawczuk (1990) have studied the forced vibrations of the beam and the effects of the crack locations and sizes on the vibrational behavior of the structure were discussed. Karaagac, Ozturk and Sabuncu (2009) have investigated the effects of crack ratios and positions on the fundamental frequencies and buckling loads of slender cantilever Euler beams with a single-edge crack both experimentally and numerically using the finite element method, based on energy approach. Zheng and Kessissoglou (2004) have studied the natural frequencies and mode shapes of a cracked beam using the finite element method. In this study, an "overall additional flexibility matrix", instead of the "local additional flexibility matrix", has been added to the flexibility matrix of the corresponding intact beam element to obtain the total flexibility matrix, and therefore the stiffness matrix. Yokoyama and Chen (1998) have investigated the vibration characteristics of a uniform Bernoulli-Euler beam with a single edge crack using a modified line-spring model.Law and Zhu (2006) have studied the dynamic behavior of a beam with a breathing crack subject to moving loads. For dynamic analysis, the equations of motion in matrix form for a cracked beam subjected to a moving load with constant velocity have been formulated using Hamilton's principle and the assumed mode method by Lee and Ng (1994). In this study, the beam has modeled as two separate beams divided by the crack. Mahmoud and Abou Zaid (2002) have developed an iterative modal analysis approach to determine the effect of transverse cracks on the dynamic behavior of simply supported undamped Bernoulli-Euler beams subject to a moving mass. Lin and Chang (2006) have developed an analytical method to present the dynamic response of a cracked cantilever beam subject to a concentrated moving load. The cracked beam system has modeled as a two-span beam and each span of the continuous beam is assumed to obey Euler-Bernoulli beam theory. Yang et al. (2008) have investigated the free and forced vibration of slender functionally graded material (FGM) beams with open edge cracks under a combined action of an axial compression and a concentrated transverse moving load. Kargarnovin et al. (2012) have presented dynamic response of a delaminated composite beam under the action of moving oscillatory mass with the Poisson's effect, shear deformation and rotary inertia. Esen (2015)has presented a combined plate element for the analysis of transverse and longitudinal vibrations of a thin plate which carries a load moving along an arbitrary trajectory with variable velocity.

In this study, the dynamic behavior of a cracked beam subjected to a concentrated force traveling at a constant velocity is investigated by using the finite element method. The end conditions of the beam are assumed to be hinged-hinged. The effect of the elastic support is included in the study by using the linear springs distributed to the nodes of the beam. For the rectangular cross-section beam, a crack element is modeled by using the principles of fracture mechanics. The Newmark integration method, which is frequently used in the structural dynamic analyses, is employed in order to calculate the dynamic response of the beam. The effects of crack depth, crack location, elastic support and load velocity on the dynamic displacements of the beam are investigated. The results are presented in 3D graphs.

## CALCULATION OF THE LOCAL FLEXIBILITY

A crack on a beam introduces considerable local flexibility due to the strain energy concentration in the vicinity of the crack tip under load. The idea of an equivalent spring i.e. a local compliance is used to quantify, in a macroscopic way, the relation between the applied load and the strain concentration around the tip of the crack (Karaagac et al., 2009). A beam element of rectangular cross-section has an edge crack with a tip line parallel to the $z$-axis, i.e. with a uniform depth. A generalized loading is indicated by three general forces: an axial force $P_1$, shear force $P_2$ and bending moment $P_3$ as seen in Figure 1 (a).

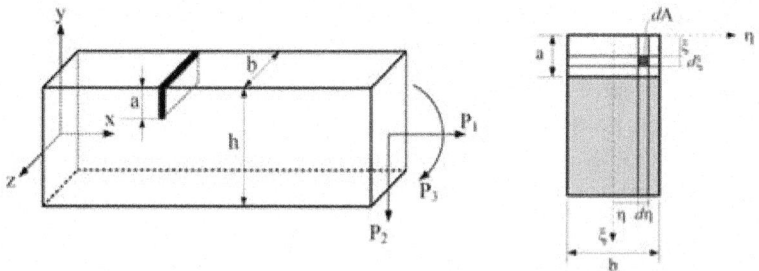

**Figure 1:** Schematic view of a cracked beam under generalized loading conditions.

In this work, the cross section of the beam is assumed to be rectangular. The additional strain energy due to the existence of a crack can be expressed as (Karaagac et al., 2009; Zheng and Kessissoglou, 2004):

$$\Pi_c = \int_{A_c} G \, dA$$

$$(1)$$

Where $G$ is the strain energy release rate function and $A_c$ is the effective cracked area. The strain energy release rate function $G$ can be expressed as (Zheng and Kessissoglou, 2004)

$$G = \frac{1}{E'}\left[\left(K_{I1} + K_{I2} + K_{I3}\right)^2 + K_{II1}^2\right]$$

$$(2)$$

In this expression, $E' = E$ for plane stress problem and $E' = E/(1-²)$ for plane strain problem (Karaagac et al., 2009; Zheng and Kessissoglou, 2004), where $E$ denotes the modulus of elasticity and denotes the poissons's ratio. $K_{In}$ and $K_{IIn}$ (n=1,2,3) are the stress intensity factors of the two modes of fracture (opening and sliding types) corresponding to generalized loading $P_n$, respectively. The stress intensity factor $K_{I2}$ is ignored due to the Euler beam theory and $K_{II}$ is given as (Ibrahim et al., 2013; Shen and Pierre, 1994).

$$K_{I1} = \frac{P_1}{bh}\sqrt{\pi\xi}F1\left(\frac{\xi}{h}\right), \quad K_{I3} = \frac{6P_3}{bh^2}\sqrt{\pi\xi}F2\left(\frac{\xi}{h}\right), \quad K_{II2} = \frac{P_2}{bh}\sqrt{\pi\xi}F3\left(\frac{\xi}{h}\right)$$

(3)

where

$$F1(s) = \sqrt{\frac{\tan\left(\frac{\pi s}{2}\right)}{\left(\frac{\pi s}{2}\right)}} \frac{0.752 + 2.02s + 0.37\left[1-\sin\left(\frac{\pi s}{2}\right)\right]^3}{\cos\left(\frac{\pi s}{2}\right)}$$

(4)

$$F2(s) = \sqrt{\frac{\tan\left(\frac{\pi s}{2}\right)}{\left(\frac{\pi s}{2}\right)}} \frac{0.923 + 0.199\left[1-\sin\left(\frac{\pi s}{2}\right)\right]^4}{\cos\left(\frac{\pi s}{2}\right)}$$

(5)

$$F3(s) = \frac{1.122 - 0.561s + 0.085s^2 + 0.18s^3}{\sqrt{1-s}}$$

(6)

$Fn(s=(/h)$ represents the correction function which takes into account the finite dimensions of the beam and takes particular forms for different geometry and loading modes. It is worth noting that $a$ is the final crack depth while $($ is the crack depth during the process of penetration from zero to the final depth.

The elements of the overall additional flexibility matrix $c_{ij}$ can be expressed as (Karaagac et al., 2009; Zheng and Kessissoglou, 2004).

$$c_{ij} = \frac{\partial \delta_i}{\partial P_i} = \frac{\partial^2 \Pi_C}{\partial P_i \partial P_j} \quad (i, j = 1,2,3...)$$

(7)

Substituting Eq. ($^3$- $^6$) in Eqn 7 yields the general equation for the local compliances as follows (considering that all $K$'s are independent of $($; $($ see Figure 1 (b) ):

$$c_{ij} = \frac{b}{E'}\frac{\partial^2}{\partial P_i \partial P_j} \int_0^a \left\{ \left[\frac{P_1}{bh}\sqrt{\pi\xi}F1\left(\frac{\xi}{h}\right) + \frac{6P_3}{bh^2}\sqrt{\pi\xi}F2\left(\frac{\xi}{h}\right)\right]^2 + \frac{P_2^2}{b^2h^2}\pi\xi F3^2\left(\frac{\xi}{h}\right) \right\}d\xi$$

$$(i, j = 1,2,3...)$$

(8)

where $c_{ij}$ is the local flexibility matrix:

$$c_{ij} = \begin{vmatrix} c_{11} & c_{12} & c_{13} \\ c_{21} & c_{22} & c_{23} \\ c_{31} & c_{32} & c_{33} \end{vmatrix}$$

(9)

# CRACKED BEAM MODEL AND ENERGY EQUATIONS

In this study, a finite element model is constituted to represent a cracked beam element of length d and the crack is located at a distance $d_1$ from the left end of the element as shown in Figure 2. The element is then considered to be split into two segments by the crack. The left and right segments are represented by non-cracked sub elements. The crack represents net ligament effect created by loadings. This effect can be related to the deformation of the net ligament through the compliance expressions ($c_{ij}$) by replacing the net ligament with a fictitious spring connecting both faces of the crack (Yokoyama and Chen, 1998; Ibrahim et al., 2013).

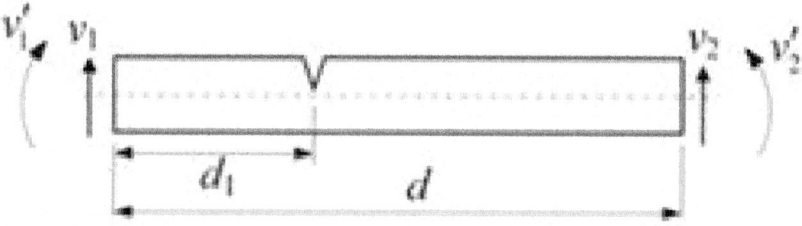

**Figure 2:** Crack locations in crack element.

The spring effects are introduced to the system by using the local flexibility matrix given by Eq. (9). The cracked element has two nodes with two degrees of freedom in each node. They are denoted as lateral bending displacements $(v_1, v_2)$ and slopes $(v'_1, v'_2)$,

For $0 \leq x \leq d_1$

$$v_1(x) = a_1 + a_2 x + a_3 x^2 + a_4 x^3$$

$$v'_1 = \frac{dv_1}{dx}$$

(10A)

For $d_1 \leq x \leq d$

$$v_2(x) = a_5 + a_6 x + a_7 x^2 + a_8 x^3$$

$$v'_2 = \frac{dv_2}{dx}$$

(10B)

The coefficients *a* of the polynomials can be expressed uniquely in terms of the boundary conditions and the local flexibility concept at the crack location. Eventually, the following expressions are obtained for a cracked element:

For lateral bending,

$$v_1(0) = q_1, \quad v_1'(0) = q_2$$
$$v_2(d) = q_3, \quad v_2'(d) = q_4$$

$$(11)$$

At the crack location $d_1$, the flexibility concept requires:

For lateral bending:

Continuity of the vertical displacement,

$$v_1(d_1) = v_2(d_1)$$

$$(12A)$$

Discontinuity of the cross-sectional rotation (slope),

$$v_2'(d_1) = v_1'(d_1) + c_{33} M_1(d_1)$$

$$(12B)$$

Where $M_1(d_1) = EIv_1''|_{x=d_1}$ and $I$ denotes the area moment of inertia of the beam cross section.

Continuity of bending moment,

$$M_1(d_1) = M_2(d_1)$$

$$(12C)$$

Continuity of shear force,

$$S_1(d_1) = S_2(d_1)$$

$$(12D)$$

By considering Eq. (10) which describes the displacement for the left and right parts of the element and rearranging Eqs. (11) and (12), the nodal displacement can be expressed in matrix forms as

$$\{q_v\} = [D_v]\{a\}$$

$$(13)$$

where

$$\{q_v\} = \begin{bmatrix} q_1 & q_2 & 0 & 0 & 0 & 0 & q_3 & q_4 \end{bmatrix}^T$$

$$(13A)$$

$$\{a\} = \begin{bmatrix} a_1 & a_2 & a_3 & a_4 & a_5 & a_6 & a_7 & a_8 \end{bmatrix}^T$$

$$(13B)$$

The generalized displacement vector according to local reference coordinates can be expressed as:

$$q = \begin{bmatrix} u_1 & v_1 & v_1' & u_2 & v_2 & v_2' \end{bmatrix}$$

(14)

Energy equations should be expressed separately from the crack element and intact elements on the left side of the crack element.

The elastic potential energy U:

For intact elements on the left side of the cracked element,

$$U_L = \frac{1}{2}\left| \int_0^d EI(v_1'')^2 dx \right|$$

(15)

For the cracked element,

$$U_C = \frac{1}{2}\left( \int_0^{d_1} EI(v_1'')^2 dx \right) + \frac{1}{2}\left( \int_{d_1}^d EI(v_2'')^2 dx \right)$$

(15B)

For intact element on the right side of the cracked element,

$$U_R = \frac{1}{2}\left( \int_0^d EI(v_2'')^2 dx \right)$$

(15C)

Similarly, the kinetic energy $T$ of an element of length $d$ in an Euler beam is given as follows for the intact elements on the left side of the cracked element,

$$T_L = \frac{1}{2}\left[ \int_0^d \rho A(\dot{v}_1^2) dx \right]$$

(16A)

For the cracked element,

$$T_C = \frac{1}{2}\left[ \int_0^{d_1} \rho A(\dot{v}_1^2) dx \right] + \frac{1}{2}\left[ \int_{d_1}^d \rho A(\dot{v}_2^2) dx \right]$$

(16B)

For intact element on the right side of the cracked element,

$$T_R = \frac{1}{2}\left[ \int_0^d \rho A(\dot{v}_2^2) dx \right]$$

(17C)

## PROCEDURE FOR DYNAMIC ANALYSIS

For a hinged-hinged cracked beam with elastic supports under the action of a moving load shown in Figure 3, the overall mass and stiffness matrices are obtained by assembling the element matrices. The number of elements used in the finite element vibration analysis is 30.

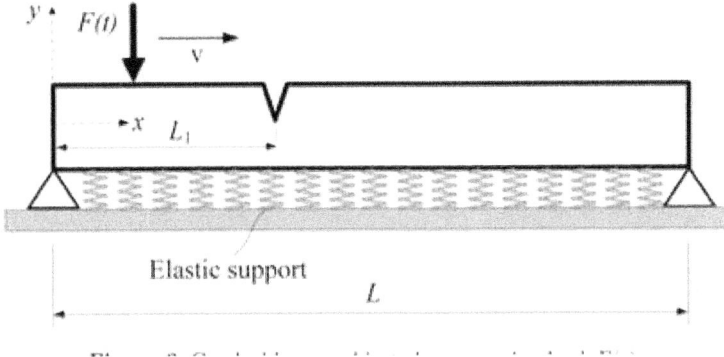

**Figure 3:** Cracked beam subjected to a moving load, F (t).

The dynamic response of the cracked beam is calculated by using the procedure described in the Newmark integration method which is widely used in structural dynamics. By using the overall mass $[M]$, damping $[C]$ and stiffness $[K]$ matrices for the beam, the governing equations of the system are written in the matrix form as

$$[M]\{\ddot{q}\}_t + [C]\{\dot{q}\}_t + [K]\{q\}_t = \{F\}_t \qquad (17)$$

Where $\{\ddot{q}\}_t$, $\{\dot{q}\}_t$ and $\{q\}_t$ are the nodal acceleration, velocity and displacement vectors at time $t$, respectively. $\{F\}_t$ is the global external excitation vector subjected to the beam at time $t$. The nodal external excitation vector at time t is constructed using the linear variation assumption of the force when it travels between two adjacent nodes as described byEq. (18).

$$\{f_t\}_i = \begin{cases} F_0 - t_v \dfrac{F_0}{t_{node}} & \text{force is moving away from the node } i \\[2ex] t_v \dfrac{F_0}{t_{node}} & \text{force is approaching to the node } i \end{cases} \qquad (18)$$

where $F_0$ denotes the magnitude of the moving load, $t_{node}$ denotes the time interval to travel between two nodes and $t_v$ is the local time, which has the value between 0 and $t_{node}$. If the force is on the $i^{th}$ node, $t_v$ equals to zero, and if the force is on one of the adjacent nodes, $t_v$ equals to $t_{node}$. In the Newmark method, the integration constants are defined as follows (Clough and Penzien, 1993),

$$a_0 = \frac{1}{\beta \Delta t^2}, \quad a_1 = \frac{\gamma}{\beta \Delta t}, \quad a_2 = \frac{1}{\beta \Delta t}, \quad a_3 = \frac{1}{2\beta} - 1,$$

$$a_4 = \frac{\gamma}{\beta} - 1, \quad a_5 = \left(\frac{\Delta t}{2}\right)\left(\frac{\gamma}{\beta} - 2\right), \quad a_6 = \Delta t(1 - \gamma), \quad a_7 = \gamma \Delta t \qquad (19)$$

where ($t$ is the time increment used in the numeric analysis and taken as ($t = T_{10}/20$ and $T_{10}$ is the 10th natural period of the beam. The integration parameters (are selected as 1/4 and 1/2, respectively, in order to obtain a stable solution. The effective stiffness matrix is calculated to perform the dynamic analyses as

$$\left[K_{ef}\right] = \left[K\right] + a_0 \left[M\right] + a_1 \left[C\right]$$

(20)

The effective load vector $F_{ef}$ is calculated for each time step as

$$\{F_{ef}\} = \{F\}_{t+\Delta t} + [M]\left(a_0\{q\}_t + a_2\{\dot{q}\}_t + a_3\{\ddot{q}\}_t\right) + [C]\left(a_1\{q\}_t + a_4\{\dot{q}\}_t + a_5\{\ddot{q}\}_t\right)$$

(21)

The nodal displacement, acceleration and velocity responses at time t+ (t can be obtained by using the following equations

$$\{q\}_{t+\Delta t} = \left[K_{ef}\right]^{-1}\{F_{ef}\}$$

(22)

$$\{\ddot{q}\}_{t+\Delta t} = a_0\left(\{q\}_{t+\Delta t} - \{q\}_t\right) - a_2\{\dot{q}\}_t - a_3\{\ddot{q}\}_t$$

(23)

$$\{\dot{q}\}_{t+\Delta t} = \{\dot{q}\}_t + a_6\{\ddot{q}\}_t + a_7\{\ddot{q}\}_{t+\Delta t}$$

(24)

In this study, the effect of damping on the dynamic response of the beam is not taken into consideration and the damping matrix $[C]$ is taken as zero. Moreover, the elastic support is modeled using the linear translation springs located on each node as seen in Figure 3. The linear springs have the effect on the vertical translational degree of freedom and the stiffness of each spring is added to the diagonal terms of the global elastic stiffness matrix. The elastic foundation parameter is used as the ratio between the equivalent spring coefficient of the elastic support and the hinged-hinged beam.

$$Elastic\ foundation\ parameter = \frac{\left(k_{eq}\right)_{\text{support}}}{48EI\ /\ L^3}$$

(25)

## NUMERIC ANALYSIS AND RESULTS

In this study, a computer code developed using the MATLAB 6.5 (Dabney and Harman, 1999) is used to calculate the dynamic response of a hinged-hinged cracked beam with elastic supports under the action of a concentrated moving load as shown in Figure 3. The effects of crack depth, crack location, elastic foundation parameter and load velocity on the dynamic response of the beam are presented in 3D graphs. The dimensions and material properties of the beam are given in Table 1.

**Table 1:** Properties of the beam.

| Properties | | | Quantity |
|---|---|---|---|
| Modulus of elasticity,E | | | 69 GPa |
| Mass density, $\rho$ | | | 2700 kg/m3 |
| Cross-section | h | 5 mm | |
| | b | 20 mm | |
| Beam length, L | | | 500 mm |

Figures from [4] to [6] show the effect of crack ratio ($a/h$) and crack location ($L_1/L$) on the first three natural frequencies of the cracked beam with different elastic foundation parameters. As can be seen from these figures, as the crack ratio (or depth) increases, the variations of the natural frequencies become significant. When the crack ratio is between 0 and 0.3, there is no considerable effect of the crack depth and location on the first natural frequency. But, the effects of these parameters towards the higher natural frequencies become evident even in smaller crack ratios. Moreover, the crack does not affect the natural frequencies when it is located at the particular points (nodes in mode shapes) of the beam length, since stresses in these points are so small. The maximum reduction occurs in the first, second and third natural frequency values, respectively for unsupported case. These reductions are 8.85%, 8.23% and 7.75%.However, for supported case, the maximum reduction occurs in the second, third and first natural frequency values, respectively namely as 7.72%, 7.65% and 4.28%.

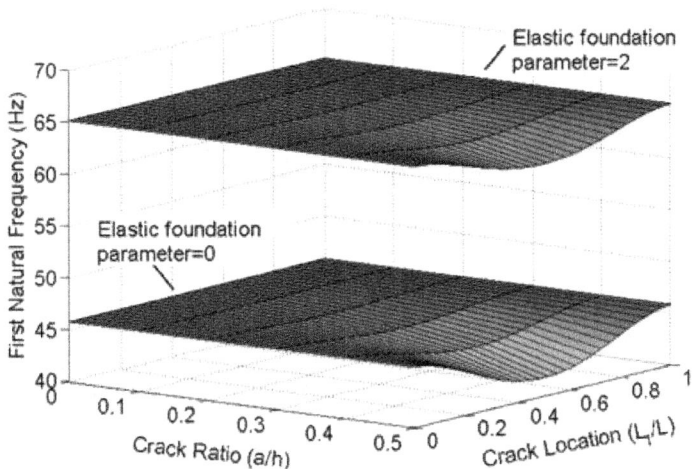

**Figure 4:** The effect of crack ratio and location on the first natural frequency of cracked beam with elastic support.

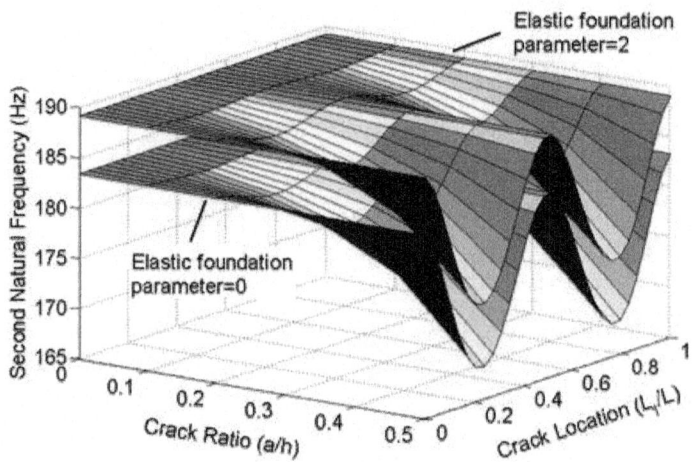

**Figure 5:** The effect of crack ratio and location on the second natural frequency of cracked beam with elastic support.

The maximum decreases in the natural frequencies occur when the crack location at the mid-point of the beam ($L_1/L=0.5$) for the first natural frequency, at the $L_1/L=0.24$ and $L_1/L=0.75$ for the second natural frequency and at the $L_1/L=0.138$, $L_1/L=0.5$ and $L_1/L=0.86$ for the third natural frequency. It is observed that the crack location is more effective at the higher natural frequencies while the crack ratio increases. As known fact is that when an elastic support or spring is added to the hinged-hinged beam, the natural frequencies increase. As seen in Figures 4,[5] and [6], while the first natural frequency increases significantly with increasing the elastic foundation parameter, this increment is greatly reduced towards the higher natural frequencies. The effects of the crack ratio and location on the natural frequencies are similar to unsupported case.

In this study, the dynamic response of the cracked beam is presented with respect to the nondimensional load velocity which is defined as:

$$\alpha = \frac{T_1}{\tau} \qquad (26)$$

where $T_1$ is the first natural period of the beam, and is the total time in which the moving load completes its movement on the beam. The moving load velocity increases as the nondimensional load velocity increase. In the numerical analyses, the cracked beam is discretized with 30 finite elements resulting in 31 nodes. The concentrated load is moved from 1st node to 31nd node with a constant velocity.

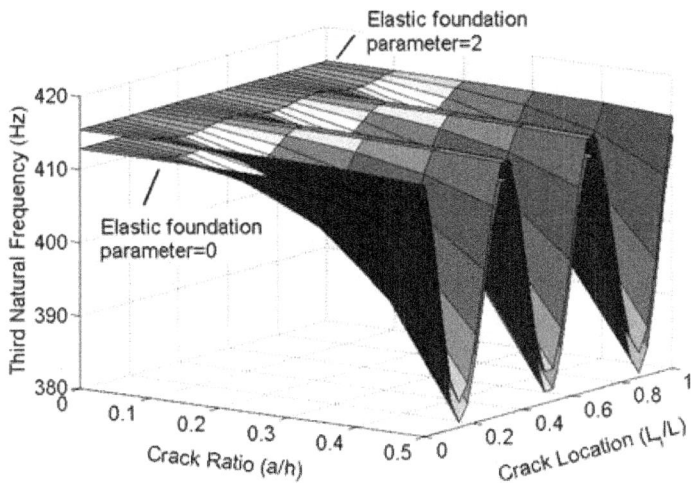

**Figure 6:** The effect of crack ratio and location on the third natural frequency of cracked beam with elastic support.

Figures 7-[11] present the effects of the elastic foundation parameter, crack ratio and crack location on the maximum displacement values calculated at the nodes of the cracked beam for different nondimensional load velocities. The dynamic displacement has the maximum value around the mid-point of the beam without being dependent on the nondimensional load velocity, crack ratio and crack location. When the nondimensional load velocity is 2, the maximum displacement is observed at the locations $0.5L$ and $0.66L$ for supported case.

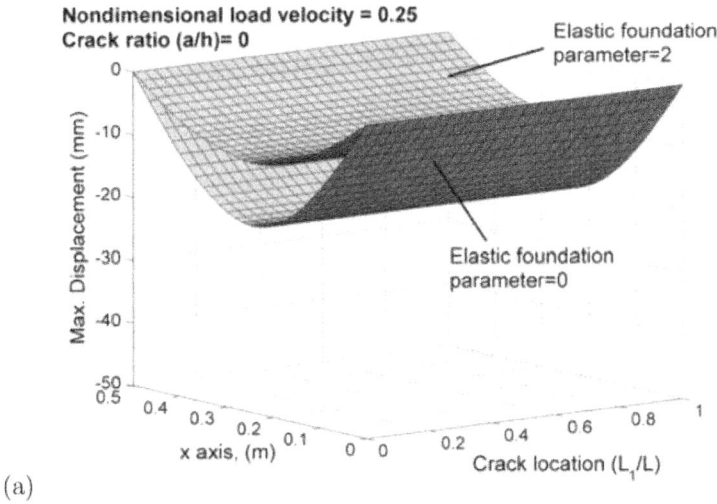

(a)

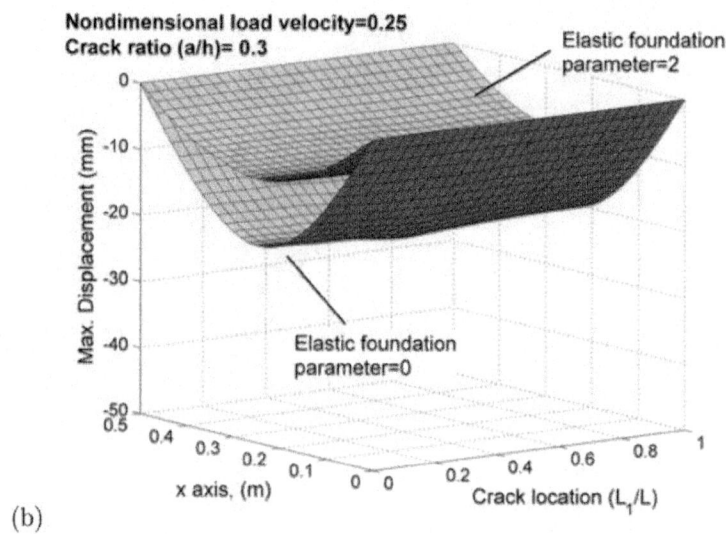

(b)

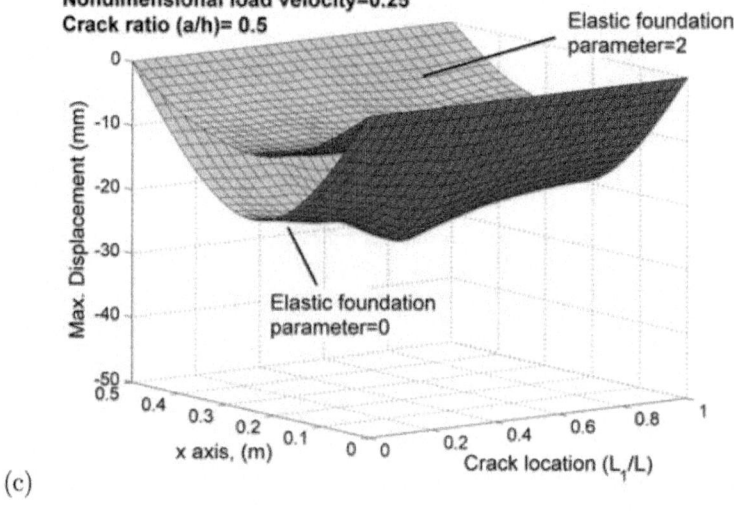

(c)

**Figure 7:** The effects of elastic foundation parameter, crack ratio and crack location on the maximum displacement at the nodes of the cracked beam for nondimensional load velocity $\alpha=0.25$.

The displacement decreases as expected when the elastic foundation parameter increases. As the crack ratio increases, considerable increment in the dynamic displacements appear as shown in figures, but this increment is more noticeable when the crack ratio is larger than the value of 0.3 and the crack location is between$L_1/L$=0.2 and$L_1/L$=0.8 . Similar to the natural frequency analysis, there is no strong influence of the crack location on the dynamic displacements until a certain crack depth. When the crack ratio is 0.5, the pattern of the maximum displacement values of the beam changes significantly with respect to the crack location as seen in figures from[7(c)] to [11(c)].

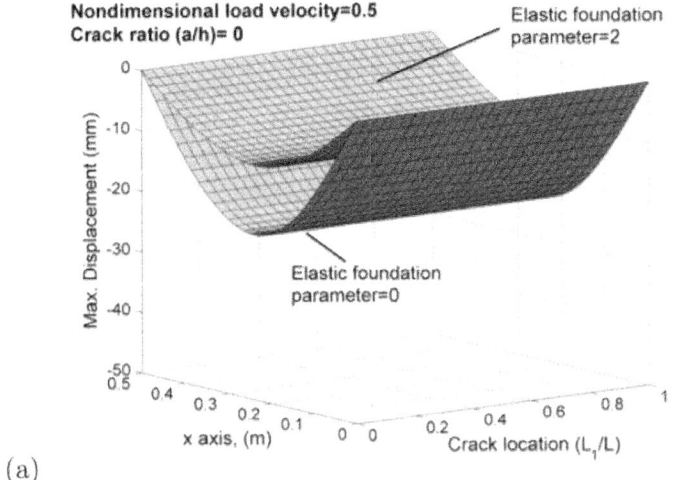

(a)

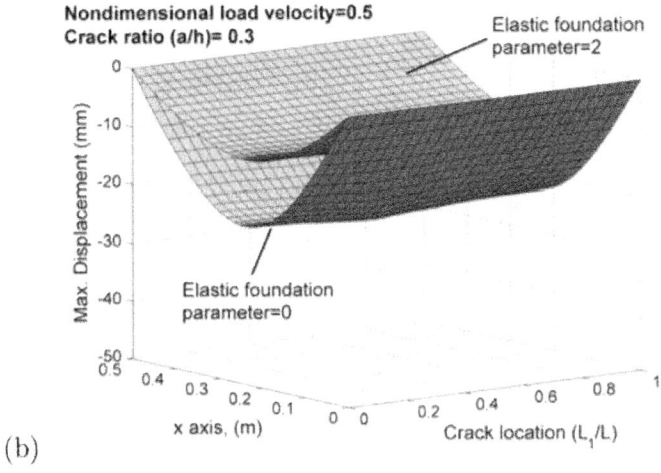

(b)

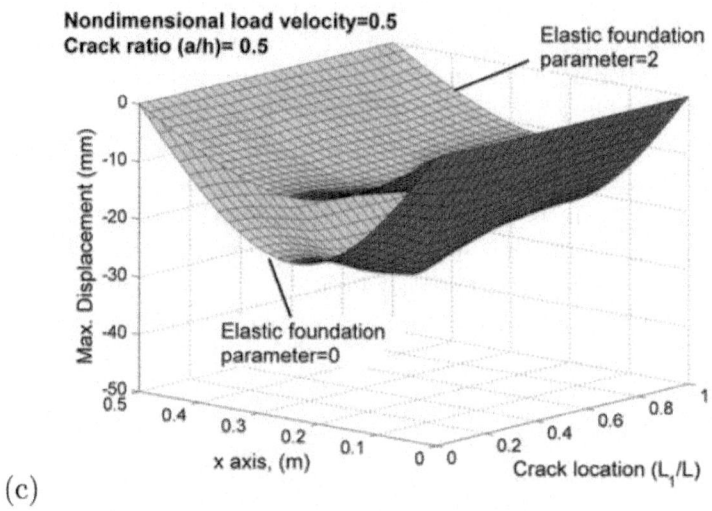

(c)

**Figure 8:** The effects of elastic foundation parameter, crack ratio and crack location on the maximum displacement at the nodes of the cracked beam for nondimensional load velocity $\alpha=0.5$.

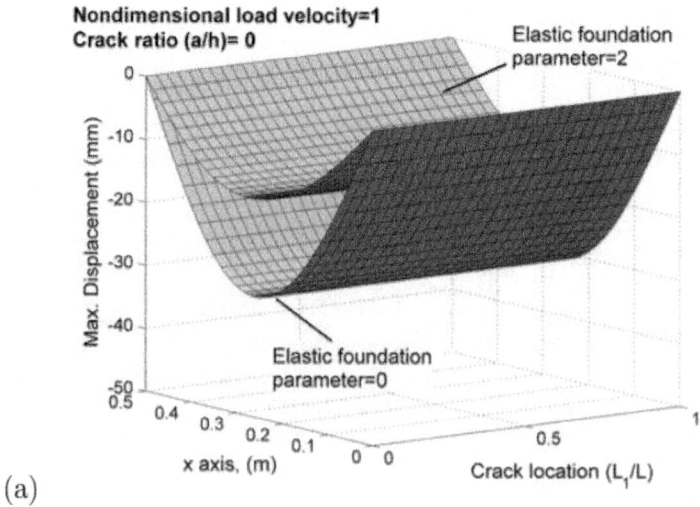

(a)

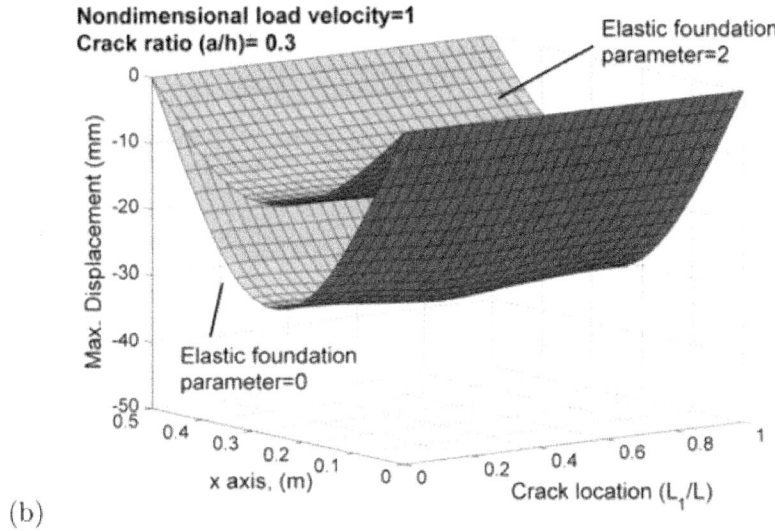

(b)

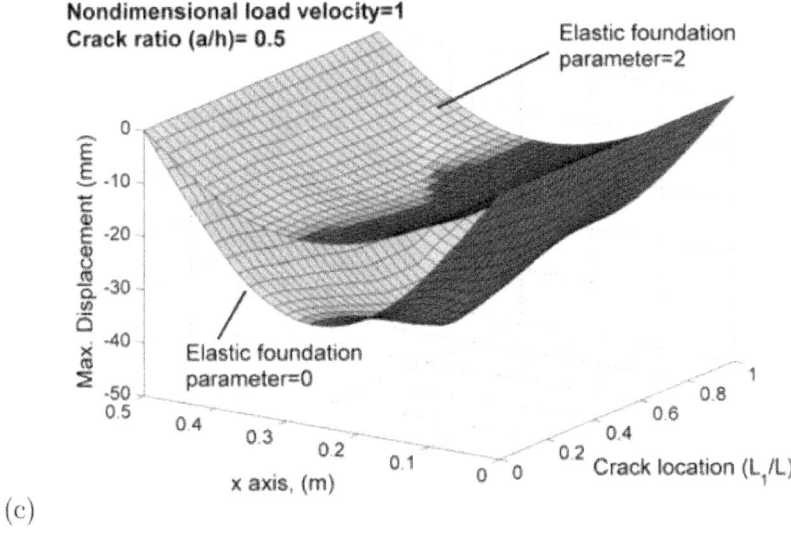

(c)

**Figure 9:** The effects of elastic foundation parameter, crack ratio and crack location on the maximum displacement at the nodes of the cracked beam for nondimensional load velocity $\alpha=1$.

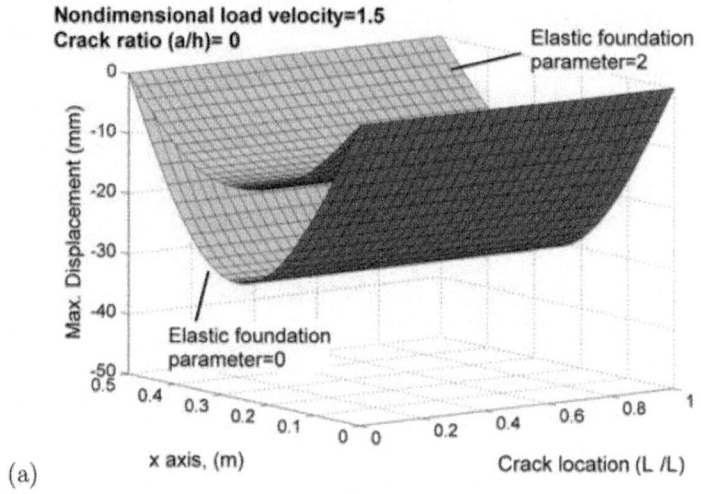

(a)

(b)

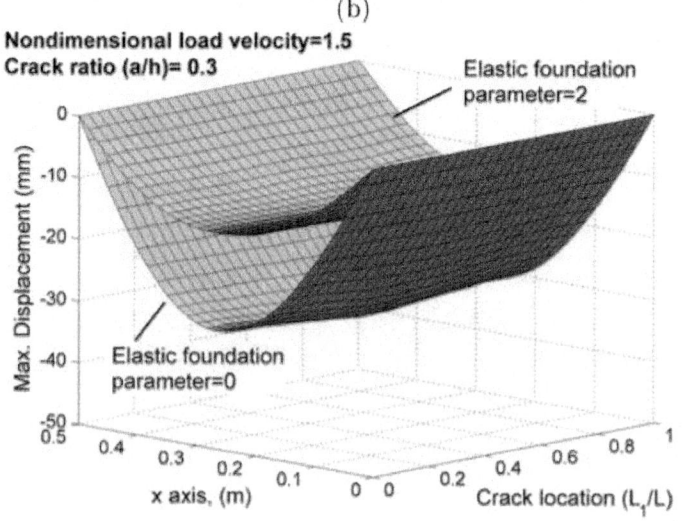

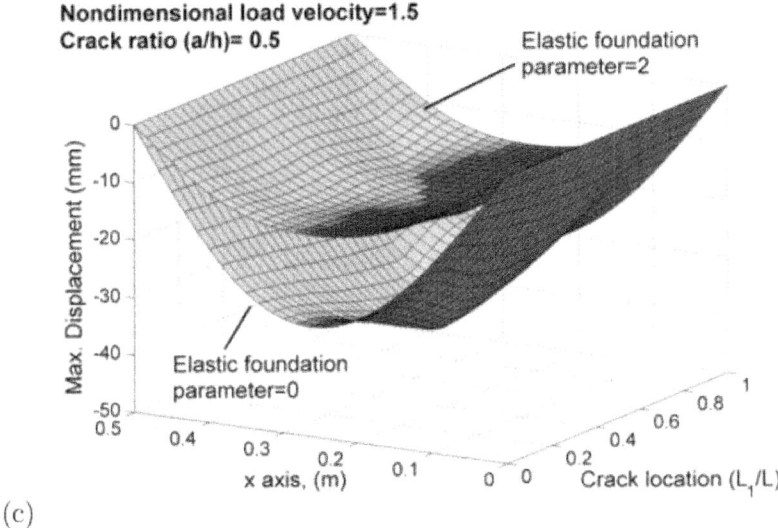

(c)

**Figure 10:** The effects of elastic foundation parameter, crack ratio and crack location on the maximum displacement at the nodes of the cracked beam for nondimensional load velocity $\alpha=1.5$.

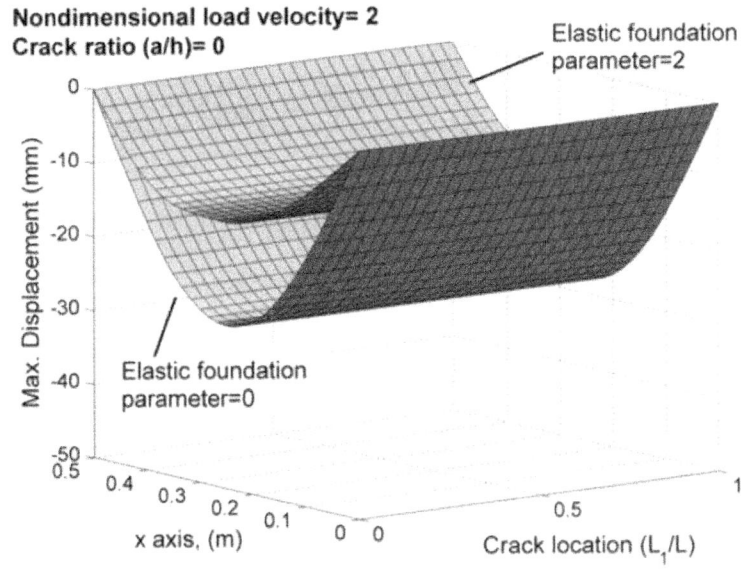

(a)

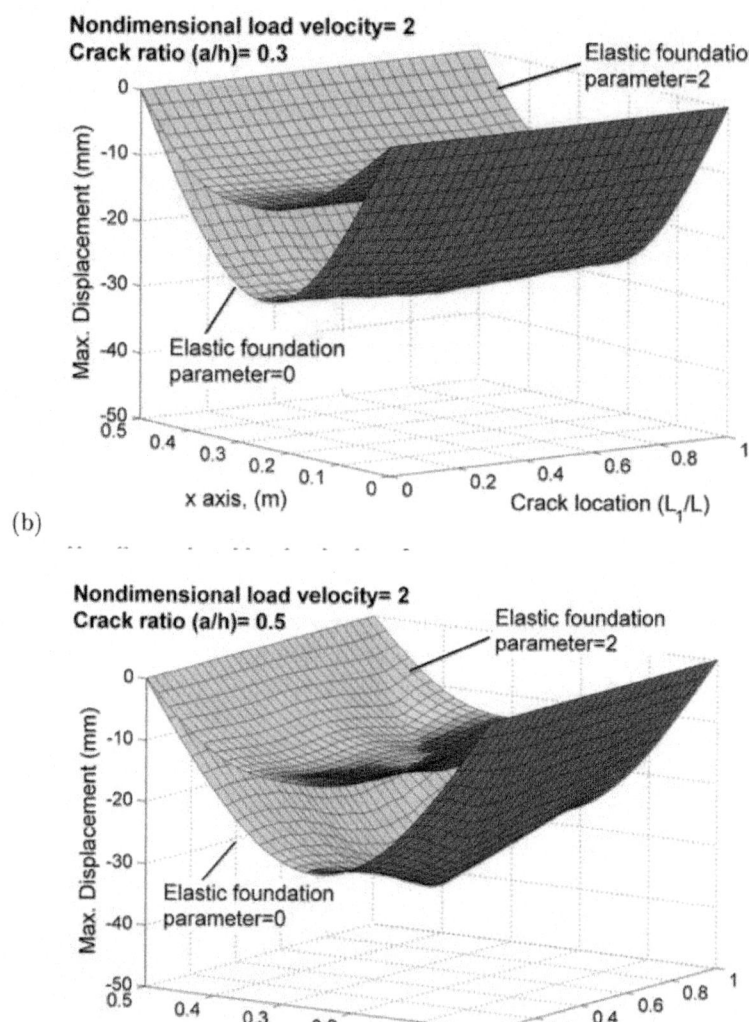

(b)

(c)

**Figure 11:** The effects of elastic foundation parameter, crack ratio and crack location on the maximum displacement at the nodes of the cracked beam for nondimensional load velocity α=2.

As can be seen from Figures 7-[11], it is observed that the dynamic displacement of the beam increases with nondimensional load velocity value approaching to 1. However, it decreases after nondimensional load velocity value of 1.5. This case can be distinctly observed when the crack ratio and elastic foundation parameter are equal to 0. Besides, Kiral (2009) have found that the critical nondimensional load velocity, at which the maximum mid-point displacements occur, is 1.2 for a hinged-hinged beam. As seen from Figures 7-[11], the maximum displacements along the beam occur symmetrically with respect to the mid-point of the beam until a crack ratio value of 0.3. After this value, this symmetric variation does not appear. As the elastic foundation parameter increases, the pattern of the maximum displacement values shows a difference according to the case without an elastic foundation due to the change in the structural stiffness. This situation can be seen more clearly in Figures 10 (c) and 11 (c) in which the nondimensional load velocity is 1.5 and 2.

The time-varying maximum displacements of the cracked beam calculated for the beam mid-point are given in 3D graphics in Figures from [12] to [16]. Figure 12 shows the dynamic displacements of the beam obtained at =0.25 for different sizes and locations of the crack. The time axis is normalized by dividing the elapsed time by the total load travel time for each case. It is seen from the Figures 12a, [12b] and [12c] that the dynamic displacements increase apparently as the crack depth increases. Similar to the maximum displacement of the beam analysis, the displacement changes with respect to the crack location $(L_1/L)$. These situations are shown inFigures 14 to [16]. At this low load velocity $(\alpha=0.25)$, the maximum mid-point displacements occur when the load is on the beam mid-point. In addition, Figure 12c shows that reduction in the mid-point displacements for the case that the crack is near to the mid-point of the beam, is bigger than those obtained when the crack is near to the end of the beam. The camel›s hump-like displacement pattern shown in Figure 12c is the clear reflection of this result. Figure 12 also shows that the elastic support reduces the dynamic displacements as expected. Figure 13 shows the mid-point dynamic displacements of the cracked beam for the nondimensional load velocity $\alpha=0.5$. For this load velocity, dynamic displacements increase and the time at which the maximum displacements are recorded shifts left. Effect of the crack location on the dynamic displacements is more apparent as seen in Figure 13c. But, the elastic support significantly suppresses the effect of the crack location on the dynamic displacements.

Figure 14 represents the mid-point dynamic displacements of the cracked beam for the nondimensional load velocity$\alpha=1$. For this velocity, dynamic displacements get larger and the maximum displacements are recorded when the load past the mid-span of the beam.Figure 15 shows that, for the nondimensional load velocity $\alpha=1.5$, the maximum displacements are still large and they are observed when the moving load is very close to the right end of the beam.Figure 16 presents the dynamic displacements for the nondimensional velocity $\alpha=2$. At this load velocity, the maximum dynamic displacements are smaller than those obtained for$\alpha=1.5$ and they occur when the moving load leaves the beam. The effects of the elastic support, crack size and crack location are still apparent.

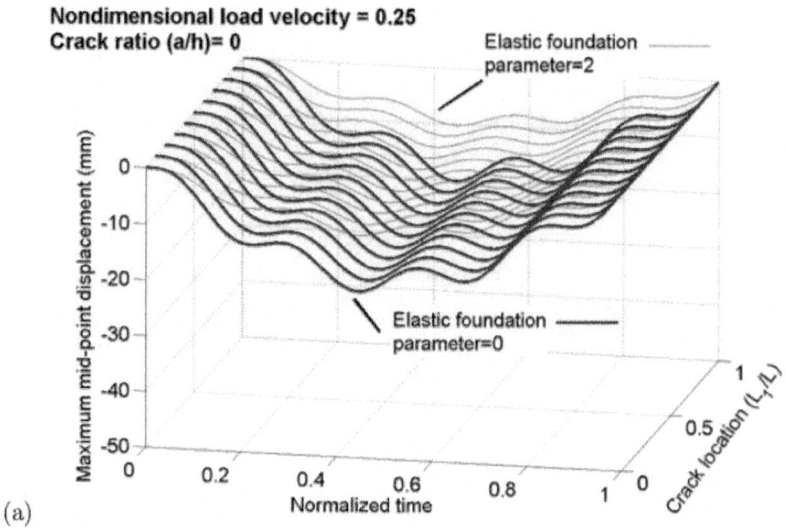

(a)

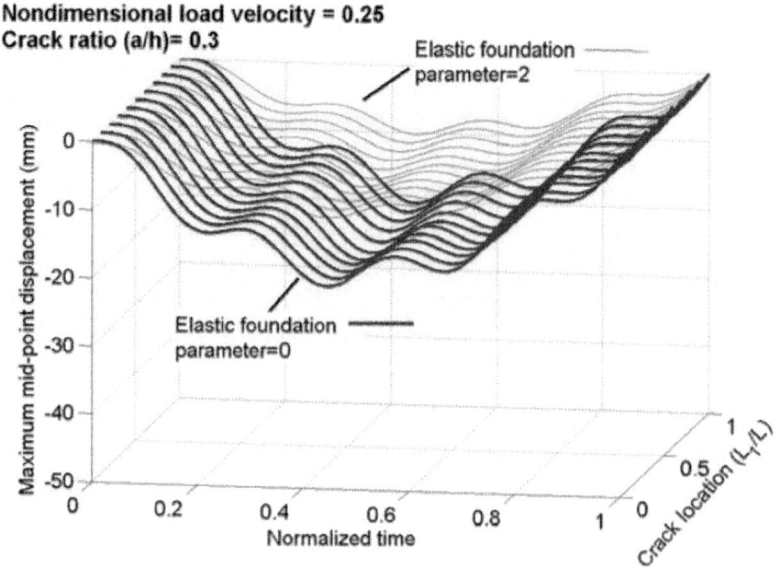

(b)

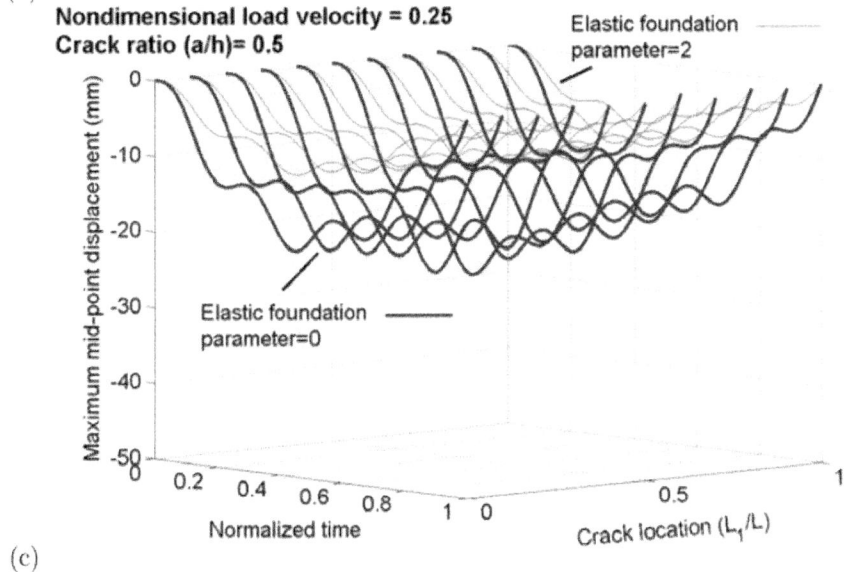

(c)

**Figure 12:** The mid-point dynamic displacements of the hinged-hinged cracked beam for nondimensional load velocity $\alpha$=0.25.

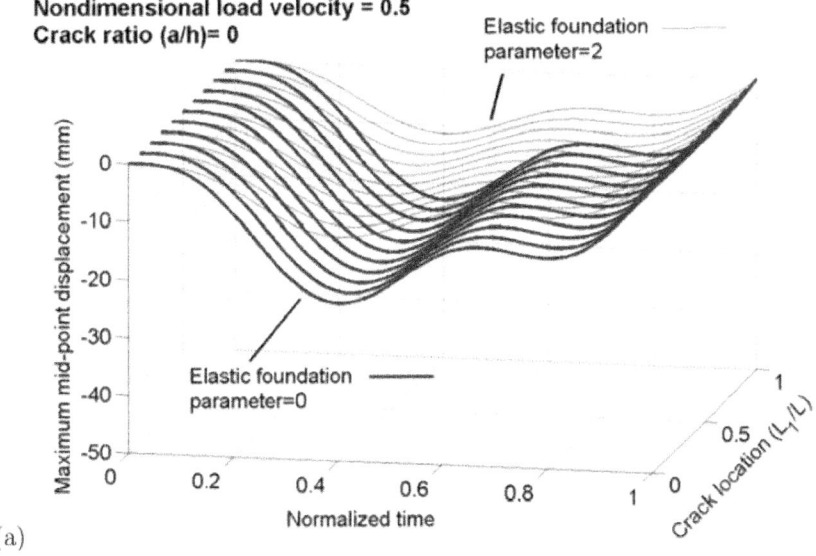

(a)

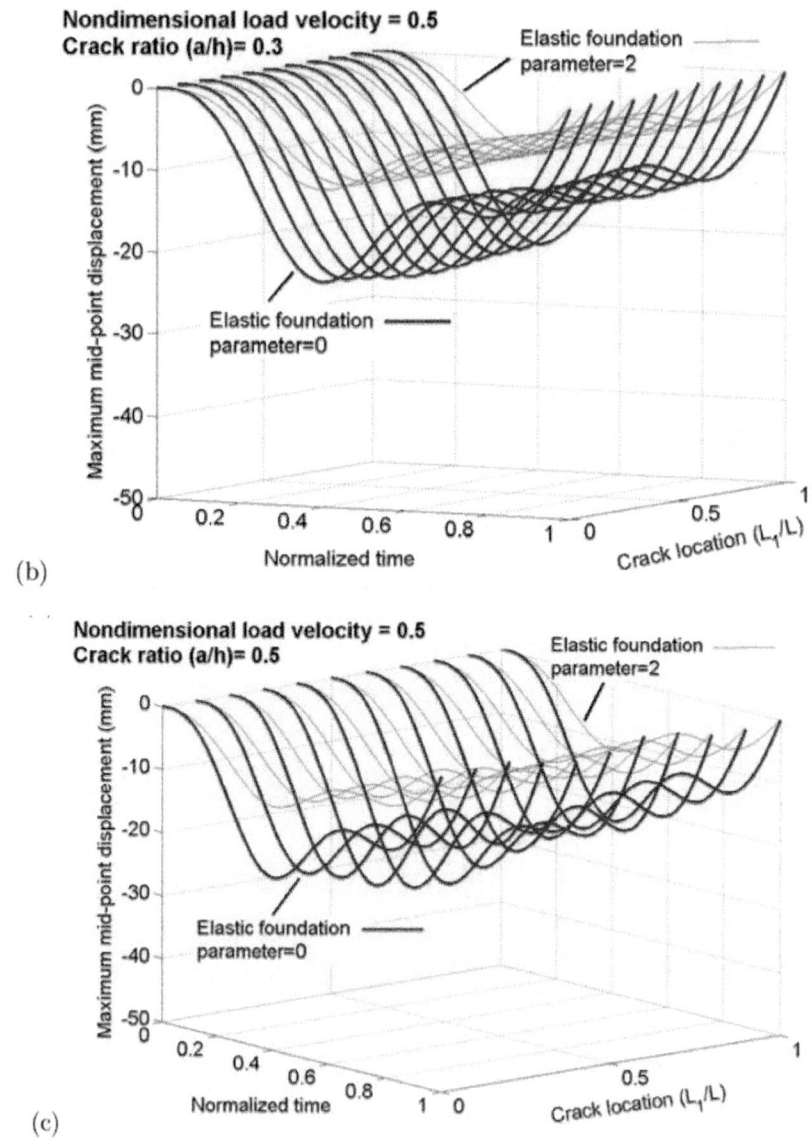

**Figure 13:** The mid-point dynamic displacements of the hinged-hinged cracked beam for nondimensional load velocity $\alpha$=0.5.

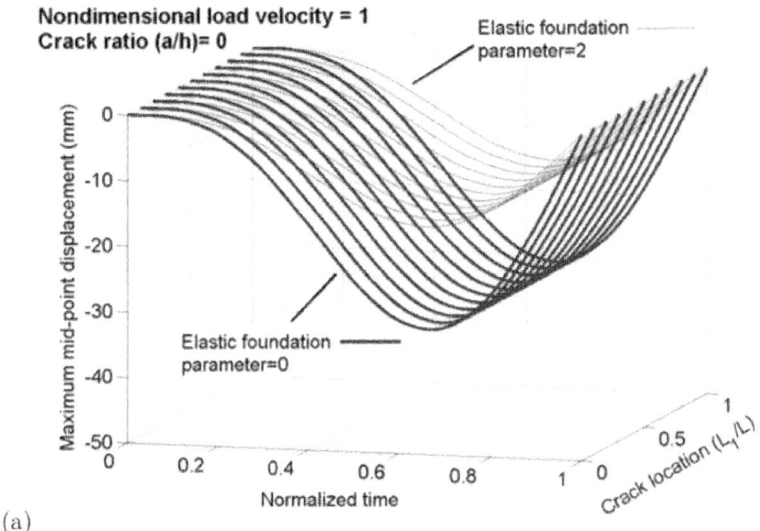

(a)

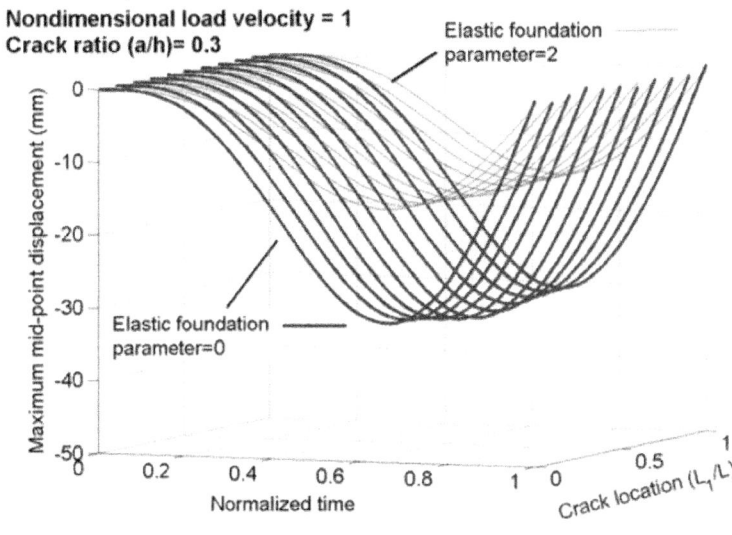

(b)

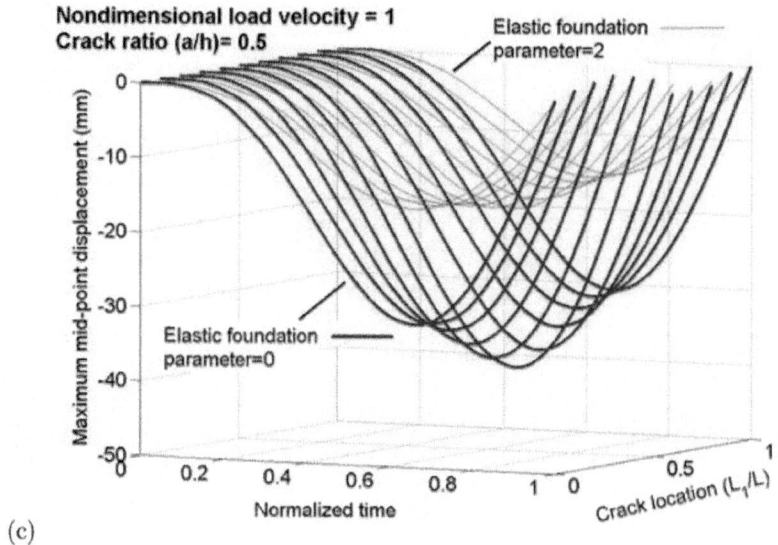

(c)

**Figure 14:** The mid-point dynamic displacements of the hinged-hinged cracked beam for nondimensional load velocity $\alpha=1$.

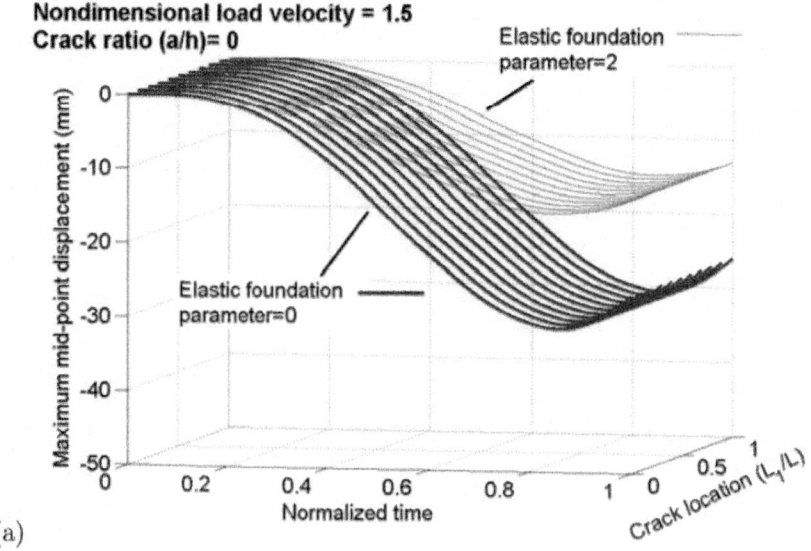

(a)

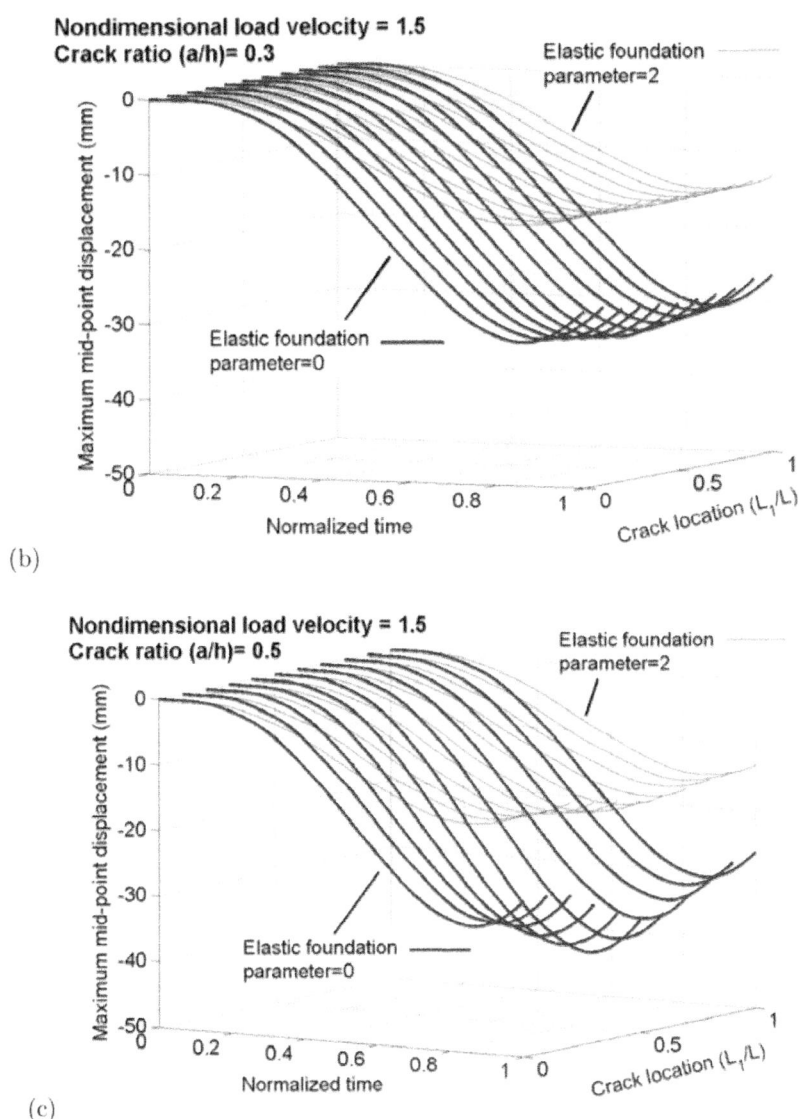

(b)

(c)

**Figure 15:** The mid-point dynamic displacements of the hinged-hinged cracked beam for nondimensional load velocity $\alpha=1.5$.

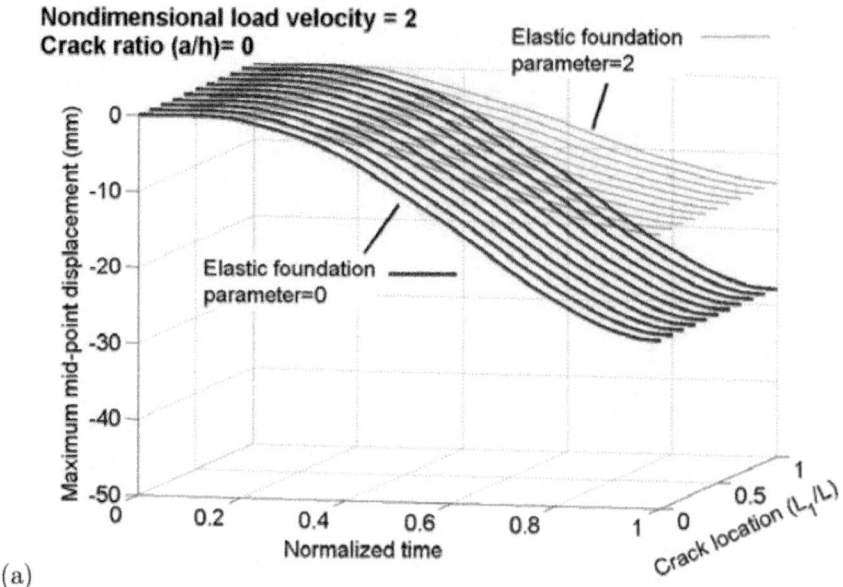

(a)

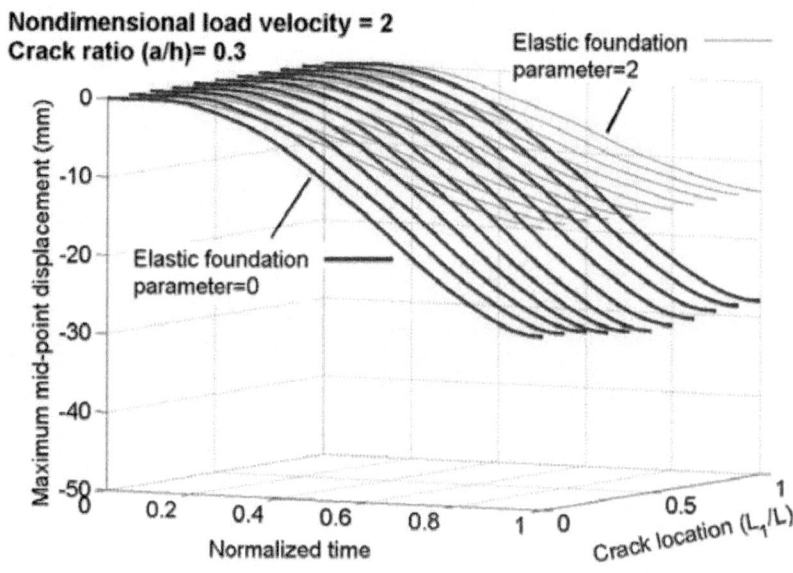

(b)

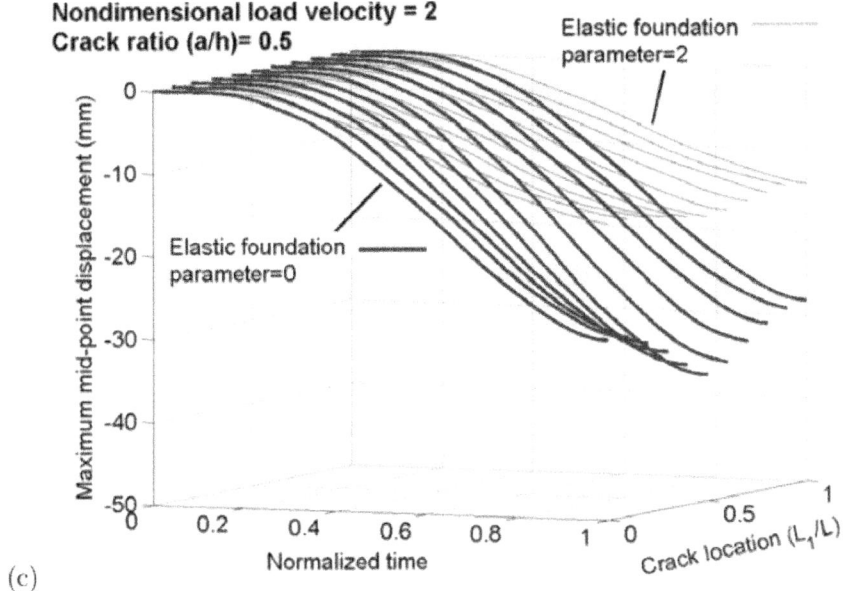

(c)

**Figure 16:** The mid-point dynamic displacements of the hinged-hinged cracked beam for nondimensional load velocity $\alpha$=2.

## CONCLUSIONS

In this study, dynamic displacements of a cracked beam subjected to a concentrated moving load with constant velocity are investigated by using the finite element method. The crack element and its local flexibility are obtained by using the fundamental knowledge of fracture mechanics. The depth and location of the crack and the moving load velocity are used as the parameters of the numerical analysis. Effect of the elastic support, which is modelled by linear springs, on the dynamic response of the cracked beam is also investigated. Based on the numerical results obtained in this study, the following conclusions are drawn,

- The natural frequencies of the cracked beam get smaller as the crack depth increases. Moreover, the crack does not affect the natural frequencies when it is located at the particular points (nodes in mode shapes) of the beam length.

- Reduction in the natural frequencies induced by the increasing crack size is more apparent at the lower natural modes without the elastic supports.

- Effect of the elastic support on the natural frequencies of the cracked beam reduces at the higher vibration modes.

- The crack, which is located near to the middle of the beam, has more influence on the increase in the dynamic displacement amplitudes.

- The amplitude of the dynamic displacements increases as the crack ratio increases for all load velocities. The elastic support helps to reduce the dynamic displacements of the beam.

- When the crack ratio is 0.5, the pattern of the maximum displacement values of the beam changes significantly with respect to the crack location.

- The maximum displacements along the beam occur symmetrically with respect to the mid-point of the beam until a crack ratio value of 0.3. After this value, this symmetric variation does not appear. As the elastic foundation parameter increases, the pattern of the maximum displacement values shows a difference according to the case without an elastic foundation.

- For with or without the elastic support case, the displacement pattern loses its symmetry as the load velocity increases.

- Velocity of the moving load has considerable influence on both the mid-point dynamic displacements and their occurrence time.

## REFERENCES

1.  Chang, T.P., Liu, Y.N. (1996). Dynamic finite element analysis of a nonlinear beam subjected to a moving load. International Journal of Solids and Structures 33(12): 1673-1688.

2.  Clough, R.W., Penzien, J. (1993). Dynamics of Structures. New York: McGraw-Hill.

3.  Dabney, J., Harman, T.L. (1999). Advanced Engineering Mathematics with MATLAB. USA: Brooks/Cole Publishing Company.

4.  Esen, I. (2015). A new FEM procedure for transverse and longitudinal vibration analysis of thin rectangular plates subjected to a variable velocity moving load along an arbitrary trajectory. Latin American Journal of Solids and Structures 12:808-830.

5.  Esmailzadeh, E., Ghorashi, M. (1997). Vibration analysis of a Timoshenko beam subjected to a traveling mass. Journal of Sound and Vibration 199: 615-328.

6.  Goren Kıral, B., Kıral, Z. (2009). Effect of elastic foundation on the dynamic response of laminated composite beams to moving loads.

Journal of Reinforced Plastics and Composites 28(8): 913-935.

7.  Gounaris, G., Dimarogonas, A. (1988). A finite element of a cracked prismatic beam for structural analysis. Computers & Structures 28(3): 309-313.

8.  Hino, J., Yoshimura, T., Ananthanarayana, N. (1985). Vibration analysis of non-linear beams subjected to a moving load using the finite element method. Journal of Sound and Vibration 100(4): 477-491.

9.  Ibrahim, A.M., Ozturk, H., Sabuncu, M. (2013). Vibration analysis of cracked frame structures. Structural Engineering and Mechanics 45(1): 33-52.

10. Karaagac, C., Ozturk, H., Sabuncu, M. (2009). Free vibration and lateral buckling of a cantilever slender beam with an edge crack: Experimental and numerical studies. Journal of Sound and Vibration 326(1-2): 235-250.

11. Kargarnovin, M.H., Ahmadian, M.T., Talookolaei, R.A.J. (2012). Dynamics of a delaminated timoshenko beam subjected to a moving oscillatory mass. Mechanics Based Design of Structures and Machines 40: 218-240.

12. Kidarsa, A., Scott, M.H., Higgins, C.C. (2008). Analysis of moving loads using force-based finite elements. Finite Elements in Analysis and Design 44: 214-224.

13. Kıral, Z. (2009). Damped response of symmetric laminated composite beams to moving load with different boundary conditions. Journal of Reinforced Plastics and Composites 28(20): 2511-2526.

14. Kıral, Z., Gören Kıral, B. (2008). Dynamic analysis of a symmetric laminated composite beam subjected to a moving load with constant velocity. Journal of Reinforced Plastics and Composites 27(1): 19-32.

15. Law, S.S., Zhu, X.Q. (2000). Study on different beam models in moving force identification. Journal of Sound and Vibration 234(4): 661-679.

16. Law, S.S., Zhu, X.Q. (2006). Vibration of a beam with a breathing crack subject to moving mass, In: Liu, G.R., Tan, V.B.C., Han, X. (eds) Computational Methods. Netherlands: Springer, 1963-1968.

17. Lee, H.P., Ng, T.Y. (1994). Dynamic response of a to a moving load cracked beam subject. Acta Mechanica 106: 221-230.

18. Lin, H.P., Chang, S.C. (2006). Forced responses of cracked cantilever beams subjected to a concentrated moving load. International Journal of Mechanical Sciences 48: 1456-1463.

19. Lin, Y.H., Trethewey, M.W. (1990). Finite element analysis of elastic beams subjected to moving dynamic loads. Journal of Sound and Vibration 136(2): 323-342.

20. Mahmoud, M.A., Abou Zaid, M.A. (2002). Dynamic response of a beam with a crack subject to a moving mass. Journal of Sound and Vibration 256(4): 591-603.

21. Olsson, M. (1991). On the fundamental moving load problem. Journal of Sound and Vibration 145(2): 299-307.

22. Ostachowicz, W.M., Krawczuk, M. (1990). Vibration analysis of a cracked beam. Computers & Structures 36(2): 245-250.

23. Rao, G.W. (2000). Linear dynamics of an elastic beam under moving loads. Journal of Vibration and Acoustics 122(3): 281-289.

24. Shen, M.H.H., Pierre, C. (1994). Free vibrations of beams with a single-edge crack. Journal of Sound and Vibration 170(2): 237-259.

25. Thambiratnam, D., Zhuge, Y. (1996). Dynamic analysis of beams on an elastic foundation subjected to moving loads. Journal of Sound and Vibration 198(2): 149-169.

26. Wang, R.T. (1997). Vibration of multi-span Timoshenko beams to a moving force. Journal of Sound and Vibration 207(5): 731-742.

27. Yang, J., Chen, Y., Xiang, Y., Jia, X.L. (2008). Free and forced vibration of cracked inhomogeneous beams under an axial force and a moving load. Journal of Sound and Vibration 312: 166-181.

28. Yokoyama, T., Chen, M.C. (1998). Vibration analysis of edge-cracked beams using a line-spring model. Engineering Fracture Mechanics 59(3): 403-409.

29. Zheng, D.Y., Kessissoglou, N.J. (2004). Free vibration analysis of a cracked beam by finite element method. Journal of Sound and Vibration 273: 457-475.

# Chapter 6

# THE DIFFERENTIAL QUADRATURE SOLUTION OF REACTION-DIFFUSION EQUATION USING EXPLICIT AND IMPLICIT NUMERICAL SCHEMES

Mohamed Salah, R. M. Amer, M. S. Matbuly

Department of Engineering Mathematics and Physics, Faculty of Engineering, Zagazig University, Zagazig, Egypt

## ABSTRACT

In this paper, two different numerical schemes, namely the Runge-Kutta fourth order method and the implicit Euler method with perturbation method of the second degree, are applied to solve the nonlinear thermal wave in one and two dimensions using the differential quadrature method. The aim of this paper is to make comparison between previous numerical schemes and detect which is more efficient and more accurate by comparing the obtained results with the available analytical ones and computing the computational time.

## INTRODUCTION

Thermal wave is reaction-diffusion equation that plays an ever-increasing role in the study of material parameters. It has been employed in optical investigations of solids, liquids and gases with photo-acoustic and thermal lens spectroscopy. Thermal waves have also been used to analyze the thermal and thermodynamic properties of materials and image thermal and material features within a solid sample [1].

In the past several decades, there has been greeting activity in developing numerical and analytical methods for the thermal wave equation. Due to the nonlinearity and complexity of such problems, only limited cases can be analytically solved [2-5]. Yan applied the projective Riccati equation method to solve Schrodinger equation in nonlinear optical fibers [2]. Then Mei, Zhang and Jiang employed the same method to get the exact solutions for some reaction-diffusion problems [3]. Abdusalam applied a factorization technique to find exact traveling wave solutions [4]. Chowdhury and Hashim obtained analytical solution for Cauchy reaction-diffusion problems using homotopy perturbation method [5]. Literature on the numerical solution of reaction-diffusion equations is sparse, and singular perturbation method has been applied to solve reaction-diffusion equations by Puri et al. in [6]. David, Curtis and John introduced time integration methods to solve thermal wave propagation [7]. Marcus applied finite difference method to study the dynamics of predator-prey interactions [8]. As well as, Chen et al. employed the finite element method to solve adjective reaction-diffusion equations [9]. Then Christos et al. also applied the same method to solve the problem with boundary layers [10]. Meral and Sezgin used this method and finite difference method with a relaxation parameter to solve nonlinear reaction-diffusion equation in one and two dimensions [11]. Recently, differential quadrature method has been efficiently employed in a variety of engineering problems [12]. Wu and Liu had introduced the generalization of the differential quadrature method to solve linear and nonlinear differential equations [13]. Kajal applied differential quadrature and Runge-Kutta method to solve thermal wave, a blow-up and a Brusselator chemical dynamics system [14]. Kajal achieved high accuracy. But there are some difficulties in the previous method which are explicit schemes used to update the solution using very small step size due to the limitation of stability condition that leads to more computational cost and lower efficiency. Therefore, Meral applied differential quadrature method and implicit Euler method with Newton method to solve one dimensional density dependent nonlinear reaction-diffusion equation [15]. Meral obtained stable solutions, and larger time steps could be used.

In this research, the thermal wave propagation model is solved by using two numerical methods to make comparison between them. In the first method, we used the hybrid technique method of Runge-Kutta fourth order method (RK4) and differential quadrature method (DQM). In the second method, we used the combined algorithm of DQM, Perturbation method of second degree and implicit Euler method. Perturbation method is used to avoid the nonlinear term. The obtained results are compared with the previous analytical ones to complete the comparison between previous different numerical schemes.

## NUMERICAL PROCEDURE OF THERMAL WAVE

Propagation of thermal waves through a rectangular plate, is governed by [14]:

$$\frac{\partial U}{\partial t} = \alpha \frac{\partial^2 U}{\partial x^2} + \beta \frac{\partial^2 U}{\partial y^2} + \gamma (U_{max} - U)U^2, t > 0, (0,0) < (x,y) < (a,b)$$

(1)

where: U is a temperature$\alpha$ and $\beta$ are diffusion parameters in direction of x and y, respectivelyγ is reaction parametera and b are plate dimensions in direction of x and y, respectivelyU$_{max}$ is a maximum temperature of the system.

Along the external boundaries, the temperatures can be described as:

$$a_1U(o,y,t)+b_1\frac{\partial U}{\partial x}\Big|_{(0,y,t)}=f_1(y,t) \tag{2-a}$$

$$a_2U(a,y,t)+b_2\frac{\partial U}{\partial x}\Big|_{(a,y,t)}=f_2(y,t) \tag{2-b}$$

$$a_3U(x,0,t)+b_3\frac{\partial U}{\partial y}\Big|_{(x,0,t)}=f_3(x,t) \tag{2-c}$$

$$a_4U(x,b,t)+b_4\frac{\partial U}{\partial y}\Big|_{(x,b,t)}=f_4(x,t) \tag{2-d}$$

where $a_i, b_i$ and $f_i, (i=1,4)$, are known functions.

Then initial temperature may be described as:

$$U(x,y,0)=g(x,y) \tag{3}$$

where $g(x,y)$ is a known function.

## Numerical Procedure Using First Method (RK4)

The main strategy is to employ DQM to reduce the problem to a system of ordinary differential equations then to apply RK4 to solve the reduced system as follows:

1) Discretize the spatial domain using Chebyshev-Gauss-Lobatto grid points [12], such as:

$$x_i=\frac{a}{2}\left[1-\cos\frac{(i-1)\pi}{N-1}\right],i=1,2,\cdots,N \tag{4-a}$$

$$y_j=\frac{b}{2}\left[1-\cos\frac{(j-1)\pi}{M-1}\right],j=1,2,\cdots,M \tag{4-b}$$

2) Apply the method of differential quadrature in terms of nodal temperature, such that:

$$\frac{\partial U}{\partial x}\Big|_{(x_i,y_j,t)}=\sum_{k=1}^{N}A_{ik}^{x}U(x_k,y_j,t) \tag{5-a}$$

$$\frac{\partial U}{\partial y}\Big|_{(x_i,y_j,t)}=\sum_{l=1}^{M}A_{jl}^{y}U(x_i,y_l,t) \tag{5-b}$$

$$\frac{\partial^2 U}{\partial x^2}\Big|_{(x_i,y_j,t)}=\sum_{k=1}^{N}B_{ik}^{x}U(x_k,y_j,t) \tag{5-c}$$

$$\left.\frac{\partial^2 U}{\partial y^2}\right|_{(x_i,y_j,t)} = \sum_{l=1}^{M} B_{jl}^y U(x_i,y_l,t)$$

(5-d)

where $A_{ij}^p$ and $B_{ij}^p$, $(p=x,y)$ are the first and second order weighting coefficients with respect to p [12].

3)   On sustainable substitution from Equations (5) into (1), one can reduce the problem to the following system of ordinary differential equations as:

$$\left.\frac{dU}{dt}\right|_{(x_i,y_j,t)} = q_{ij}\left[U(x_1,y_1,t),U(x_2,y_1,t),\cdots,U(x_N,y_M,t)\right],$$

or simply

$$\left.\frac{dU}{dt}\right|_{(x_i,y_j,t)} = q_{ij}(U_{11},U_{21},\cdots,U_{NM},t), i=1,N \text{ and } j=1,M.$$

(6)

where

$$q_{ij} = \alpha\sum_{k=1}^{N} B_{ik}^x U(x_k,y_j,t) + \beta\sum_{l=1}^{M} B_{jl}^y U(x_i,y_l,t) + \frac{\gamma}{\delta^2}(U_0 - U_{ij})U_{ij}^2,$$

$$i=1,N \text{ and } j=1,M.$$

(7)

4)   Update the temperature using RK4 such that [16]

$$U(x_i,y_j,t_0+\Delta t) = U(x_i,y_j,t_0) + \frac{1}{6}\left[H_1^{ij} + 2H_2^{ij} + 2H_3^{ij} + H_4^{ij}\right],$$

(8)

where

$$H_1^{ij} = \Delta t q_{ij}(U_{11},U_{21},\cdots,U_{NM},t_0)$$

(9-a)

$$H_2^{ij} = \Delta t q_{ij}\left(U_{11}+\frac{H_1^{11}}{2},U_{21}+\frac{H_1^{21}}{2},\cdots,U_{NM}+\frac{H_1^{NM}}{2},t_0+\frac{\Delta t}{2}\right)$$

(9-b)

$$H_3^{ij} = \Delta t q_{ij}\left(U_{11}+\frac{H_2^{11}}{2},U_{21}+\frac{H_2^{21}}{2},\cdots,U_{NM}+\frac{H_2^{NM}}{2},t_0+\frac{\Delta t}{2}\right)$$

(9-c)

$$H_4^{ij} = \Delta t q_{ij}\left(U_{11}+H_3^{11},U_{21}+H_3^{21},\cdots,U_{NM}+H_3^{NM},t_0+\Delta t\right)$$

(9-d)

where $\Delta t = t_1-t_0 = t_2-t_1 = \cdots = t_p-t_{p-1}$

## Numerical Procedure Using Second Method (Implicit Euler)

The main strategy is to apply perturbation method of second order [17,18] then applying DQ discretization to reduce the problem to a system of ordinary differential equations then applying implicit Euler method to transform the previous system to a system of linear algebraic equations as follows:

1)    We can solve

$$\frac{\partial U}{\partial t} = \alpha \frac{\partial^2 U}{\partial x^2} + \beta \frac{\partial^2 U}{\partial y^2} + \varepsilon \gamma (U_{max} - U) U^2,$$

(10)

subjected to the prescribed to boundary and initial conditions in Equations (2) and (3), assuming

$$U = U_o + \varepsilon U_1 + \varepsilon^2 U_2 + \cdots \varepsilon^n U_n$$

(11)

where $U_0, U_1, U_2$ are unknowns functions and $\varepsilon$ is a perturbation parameter.

The following condition is tested to ensure the convergence condition [19] in previous series in Equation (11).

$$\left| \frac{U_{i+1}}{U_i} \right| < 1 \text{ where } i = 0, 1, \cdots, n-1$$

(12)

2)    On sustainable substitution from Equation (11) into (10), one can reduce the problem to the following equation.

$$\frac{\partial}{\partial t}(U_o + \varepsilon U_1 + \varepsilon^2 U_2) = \alpha \frac{\partial^2}{\partial x^2}(U_o + \varepsilon U_1 + \varepsilon^2 U_2) + \beta \frac{\partial^2}{\partial y^2}(U_o + \varepsilon U_1 + \varepsilon^2 U_2)$$
$$+ \varepsilon \gamma (U_{max} - (U_o + \varepsilon U_1 + \varepsilon^2 U_2))(U_o + \varepsilon U_1 + \varepsilon^2 U_2)^2,$$

(13)

3)    Applying zero order perturbation method such that,

$$\frac{\partial U_0}{\partial t} = \alpha \frac{\partial^2 U_0}{\partial x^2} + \beta \frac{\partial^2 U_0}{\partial y^2}$$

(14)

Subjected to boundary and initial conditions in Equations (2) and (3), where differential quadrature method and implicit method are used to reduce Equation (14) to a system of linear algebraic equations such that By substitution of Equation (5) into (14) result that,

$$\left. \frac{dU_o}{dt} \right|_{(x_i, y_j, t)} = \alpha \sum_{k=1}^{N} B_{ik}^x U_o (x_k, y_j, t_i) + \beta \sum_{k=1}^{M} B_{jk}^y U_o (x_i, y_k, t_i)$$

(15)

$$\left. \frac{dU_o}{dt} \right|_{(x_i, y_j, t)} = q_{ij}(U_{0_{11}}, U_{0_{21}}, \cdots, U_{0NM}, t), i = 1, N \text{ and } j = 1, M.$$

(16)

$$U_0(x_i, y_j, t_0 + \Delta t) = U_0(x_i, y_j, t_0) + \Delta t \times q_{ij}(U_{0_{11}}, U_{0_{21}}, \cdots, U_{0NM}, t_0 + \Delta t),$$

(17)

4)    First order perturbation method is applied such that,

$$\frac{\partial U_1}{\partial t} = \alpha \frac{\partial^2 U_1}{\partial x^2} + \beta \frac{\partial^2 U_1}{\partial y^2} + \gamma (U_{max} - U_o) U_o^2$$

(18)

Subjected to the same boundary and initial conditions in Equations (2) and (3), reduced to the following algebraic system in equations

$$\left. \frac{dU_1}{dt} \right|_{(x_i, y_j, t)} = \alpha \sum_{k=1}^{N} B_{ik}^x U_1(x_k, y_j, t_i) + \beta \sum_{k=1}^{M} B_{jk}^y U_1(x_i, y_k, t_i) + \gamma (U_{max} - U_{oi,j,l}) U_{oi,j,l}^2$$

(20)

$$U_1\left(x_i,y_j,t_0+\Delta t\right)=U_1\left(x_i,y_j,t_0\right)+\Delta t\times q_{ij}\left(U_{1_{11}},U_{1_{21}},\cdots,U_{1NM},t_0+\Delta t\right),\tag{21}$$

5)    Also second order perturbation method is applied such that,

$$\frac{\partial U_1}{\partial t}=\alpha\frac{\partial^2 U_1}{\partial x^2}+\beta\frac{\partial^2 U_1}{\partial y^2}+\gamma\left(2U_{max}U_0U_1-3U_0^2U_1\right)\tag{22}$$

Subjected to the same boundary and initial conditions in Equations (2) and (3), reduced to the following algebraic system

$$\left.\frac{dU_2}{dt}\right|_{\left(x_i,y_j,t\right)}=\alpha\sum_{k=1}^{N}B_{ik}^{x}U_1\left(x_k,y_j,t_i\right)+\beta\sum_{k=1}^{M}B_{jk}^{y}U_1\left(x_i,y_k,t_i\right)+\gamma\left(2U_{max}U_{0i,jj}U_{1i,jj}-3U_{0i,jj}^2U_{1i,jj}\right)\tag{23}$$

$$U_2\left(x_i,y_j,t_0+\Delta t\right)=U_2\left(x_i,y_j,t_0\right)+\Delta t\times q_{ij}\left(U_{2_{11}},U_{2_{21}},\cdots,U_{2NM},t_0+\Delta t\right),\tag{24}$$

Finally, the series solution can be written as

$$U_{numerical}=\lim_{\varepsilon\to 1}U=U_o+U_1+U_2\tag{25}$$

We carry on previous procedure until the specified time is reached.

# RESULTS AND DISCUSSIONS FOR ONE DIMENSION ANALYSIS

To ensure the accuracy of the proposed numerical techniques, the thermal wave propagating model is solved using presented methods and compared with the available analytical solution [14, 20].

Consider a one-dimensional problem of thermal wave propagation along x-direction as $\gamma=1,\beta=a_3=a_4=b_1=b_2=b_3=b_4=f_3=f_4=0\ \alpha=a_1=a_2=U_{max}=1,$.

While

$$f_1(t)=\frac{1}{1+\exp(-t/2)},f_2(t)=\frac{1}{1+\exp\left(\left(1-t/\sqrt{2}\right)/\sqrt{2}\right)},$$

$$g(x)=\frac{1}{1+\exp\left(x/\sqrt{2}\right)},0\le x\le 1.\tag{26}$$

The exact solution for such problem can be obtained as [20]:

$$U_{exact}(x,t)=\frac{1}{1+\exp\left(\left(x-t/\sqrt{2}\right)/\sqrt{2}\right)},t>0,0\le x\le 1\tag{27}$$

To validate the accuracy of numerical results, the following errors [16] are computed,

$$\text{Root mean square of errors}=R.M.S.\text{of errors}=\sqrt{\sum_{i=1}^{NM}\left(U_{numerical}-U_{exact}\right)^2\bigg/NM}\tag{28-a}$$

$$\text{Max. absolute error}=\max_{\substack{1\le i\le N\\1\le j\le M}}\left|U_{numerical}\left(x_i,y_j,t_k\right)-U_{exact}\left(x_i,y_j,t_k\right)\right|\tag{28-b}$$

## Numerical Results of First Method (RK4)

For the numerical computation, the time domain is limited to $0 \leq t \leq 20$ and $N = 7$. The efficiency of presented techniques is tested by CPU time required when the computation reaches to t = 20 s. Two time step sizes of 0.001 and 0.005 are used for layer marching in the time direction. The numerical results of these two cases are listed respectively in Tables 1 and 2. From these two tables, it can be seen that RK4 method can achieve high accuracy at very small step size at $\Delta t = 0.001$ with $R.M.S. \text{ of errors} < 4 \times 10^{-9}, \text{Max. absolute error} < 3 \times 10^{-9}$ and at $\Delta t = 0.005$ with $R.M.S. \text{ of errors} < 1.4 \times 10^{-8}, \text{Max. absolute error} < 1.0 \times 10^{-8}$. Moreover, as shown in Figure 1, as $\Delta t$ increased slightly to 0.00515 the stability condition will not achieved and the oscillation will occurs in the period $6 \leq t \leq 25$. On the other hand the efficiency is very small as the CPU time required to reach t = 20 s is much larger.

**Table 1**: R.M.S., absolute errors, time steps and CPU times given by 4-stage Runge-Kutta at $\Delta t = 0.001$

| t = 30.0 | | N = 5 | N = 8 | N = 11 | N = 15 |
|---|---|---|---|---|---|
| Max.absolute error by implicit Euler | | 0.0025 | 6.9707E−004 | 2.4820E−004 | 0.0045 |
| Root mean square of errors by implicit Euler | | 0.0015 | 4.0116E−004 | 1.3981E−004 | 0.0023 |
| CPU time | | 0.018 | 0.018 | 0.02 | 0.042 |

**Table 2**: R.M.S., absolute errors, time steps and CPU times given by 4-stage Runge-Kutta at $\Delta t = 0.005$.

| Time (s) | Time steps | R.M.S. error | Max. absolute error | CPU time (s) |
|---|---|---|---|---|
| 0.5 | 101 | 1.4898E−008 | 1.0284E−008 | 0.029 |
| 1.0 | 201 | 1.4445E−008 | 1.0047E−008 | 0.037 |
| 1.5 | 301 | 8.5098E−009 | 6.5970E−009 | 0.05 |
| 2.0 | 401 | 2.6387E−009 | 1.5218E−009 | 0.05 |
| 2.5 | 501 | 6.8062E−009 | 5.1014E−009 | 0.062 |
| 5.0 | 1001 | 5.0520E−010 | 3.4387E−010 | 0.1 |
| 10.0 | 2001 | 1.3932E−011 | 1.0817E−011 | 0.27 |
| 20.0 | 4001 | 9.7838E−013 | 6.7968E−013 | 1.13 |

## Numerical Results of Second Method (Implicit Euler)

In the obtained results the advantage of using an implicit scheme has been observed. Stability problems are not encountered due to the use of implicit time integration step and larger time increments can be used, e.g. for t = 30,

$\Delta t = 3.0$ can be taken. Table 3 shows the maximum absolute errors and root mean square of errors for a fixed time (t = 30.0) for various numbers of grid points. The accuracies by using N = 8, 11 are almost the same and there is a drop for N = 15. From the table, DQM is observed to give very good accuracy with a small number of grid points. For N = 15, the drop of accuracy is due to the ill-conditioned Vandermonde-system obtained after the DQM discretization, which is the known nature of DQM for large N [15]. Tables 4-6 give the comparison of the DQM solution with the exact solution in terms of maximum absolute error and root mean square of errors for small time levels and for the times tending to steady-state, respectively. The computations are carried out with N = 11 and it is seen to be enough to obtain the solution with five digits accuracy at steady-state. Moreover, Figure 2 shows the absolute error at different times and locations. Also convergence condition in Equation (12) is tested achieving higher accuracy at second order perturbation method as shown in Figure 3.

## RESULTS AND DISCUSSIONS FOR TWO DIMENSIONS ANALYSIS

Consider also a simple two dimensional problem with $\alpha = \beta = a_1 = a_2 = a_3 = a_4 = 1$, $\gamma = b_1 = b_2 = b_3 = b_4 = f_1 = f_2 = f_3 = f_4 = 0$ and

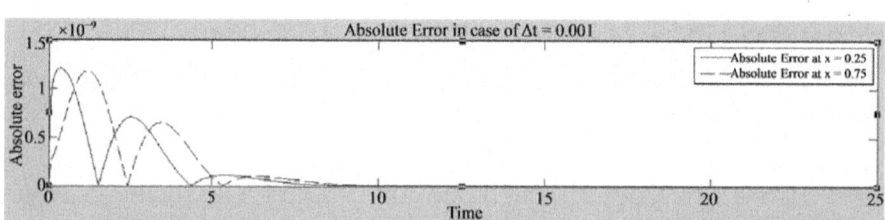

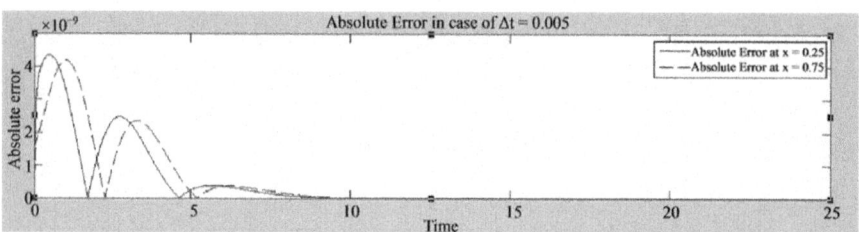

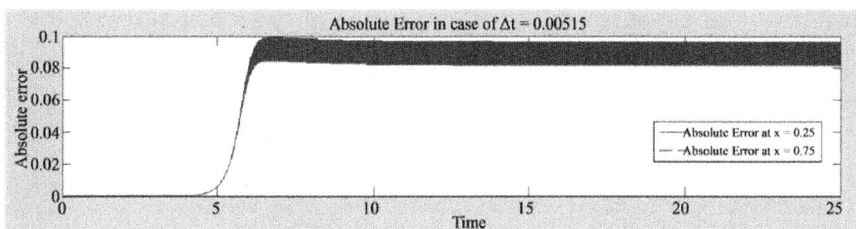

To show the effect of oscillation on figure we graph the absolute error in range $10 \leq t \leq 11$

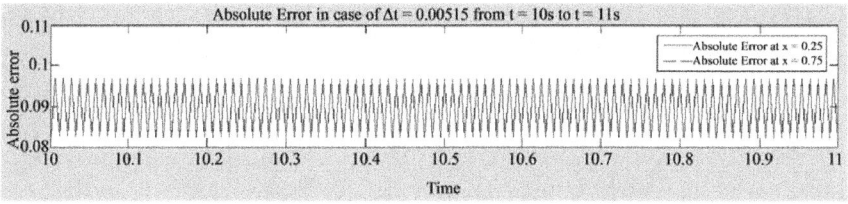

**Figure 1**: Absolute errors at different times and locations.

**Table 3**: R.M.S., absolute errors and CPU times given by implicit euler at $\Delta t = 3$.

| t = 30.0 | N = 5 | N = 8 | N = 11 | N = 15 |
|---|---|---|---|---|
| Max.absolute error by implicit Euler | 0.0025 | 6.9707E−004 | 2.4820E−004 | 0.0045 |
| Root mean square of errors by implicit Euler | 0.0015 | 4.0116E−004 | 1.3981E−004 | 0.0023 |
| CPU time | 0.018 | 0.018 | 0.02 | 0.042 |

**Table 4**: R.M.S., absolute errors and CPU times given by implicit euler at $\Delta t = t$.

| N = 11 | t = 0.01 | t = 0.1 | t = 0.5 | t = 1.0 |
|---|---|---|---|---|
| Max. absolute error by implicit Euler | 0.1320 | 0.0654 | 0.0129 | 1.0898E−004 |
| Root mean square of errors by implicit Euler | 0.0871 | 0.0391 | 0.0076 | 5.6590E−005 |
| CPU time | 0.019 | 0.015 | 0.015 | 0.011 |

**Table 5**: R.M.S., absolute errors and CPU times given implicit euler at $\Delta t = 2$.

| N = 11 | t = 6.0 | t = 12 | t = 22 | t = 34 |
|---|---|---|---|---|
| Max.absolute error by implicit Euler | 0.0220 | 0.0176 | 0.0012 | 0.0012 |
| Root mean square of errors by implicit Euler | 0.0134 | 0.0089 | 6.8117E−004 | 6.8188E−004 |
| CPU time | 0.021 | 0.017 | 0.024 | 0.038 |

**Table 6**: R.M.S., absolute errors and CPU times given by implicit euler at $\Delta t = 4$

| N = 11 | t = 16 | t = 36 | t = 60 | t = 84 |
|---|---|---|---|---|
| Max.absolute error by implicit Euler | 0.0093 | 2.7649E−004 | 7.0708E−005 | 7.0708E−005 |
| Root mean square of errors by implicit Euler | 0.0050 | 1.6835E−004 | 4.0801E−005 | 4.0801E−005 |
| CPU time | 0.024 | 0.024 | 0.029 | 0.029 |

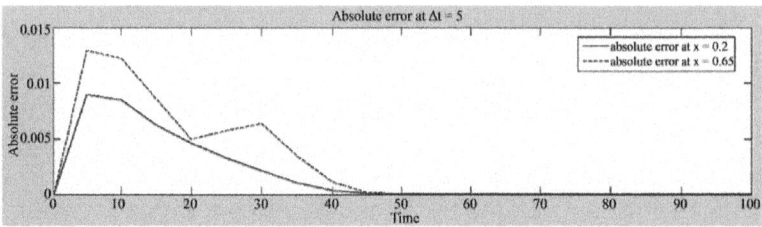

**Figure 2**: Absolute errors at different times and locations.

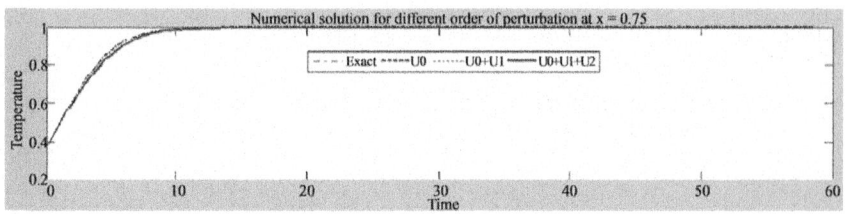

**Figure 3**: Satisfying convergence conditions.

$g(x,y) = \sin \pi x \sin \pi y, 0 \le x, y \le 1$

which can be solved exactly as [21]: $U(x,y,t) = e^{-2\pi^2 t} \sin \pi x \sin \pi y$. The design of the numerical scheme is extended to two dimensions.

Table 7 shows that for $\Delta t = 0.01, N = 5, M = 4$, the obtained results agree with the analytical ones [21] in both methods. Also Figure 4 shows that absolute error for the hybrid method absolute error <0.01, and for implicit Euler, the absolute error $0.6 \times 10^{-3}$.

# CONCLUSION

Throughout this study, thermal wave propagation model which is the type of reaction-diffusion equations is solved by using DQM for space discretization and two different time-integration schemes. Moreover, one can use a small number of discretization points, which lead to higher accuracy. Also for the nonlinear wave equation, the use of DQM with non-uniform grid discretization increases the accuracy and stability of solution. The resulting system of ordinary differential equations is solved by using two different time integration schemes in order to make comparison between two methods and detect which of them is better.

**Table 7:** Root mean square of errors for two dimensional thermal wave propagation at $\Delta t = 0.01$, N = 5, M = 4

| Time | 0.01 | 0.04 | 0.08 | 0.16 | 0.4 | 1.00 |
|---|---|---|---|---|---|---|
| Root mean square of errors by hybrid method | 0.0027 | 0.0060 | 0.0054 | 0.0021 | 3.9994E-005 | 5.1513E-010 |
| Root mean square of errors by implicit Euler | 0.0369 | 0.0143 | 0.0025 | 0.0013 | 7.4271E-005 | 2.9596E-009 |

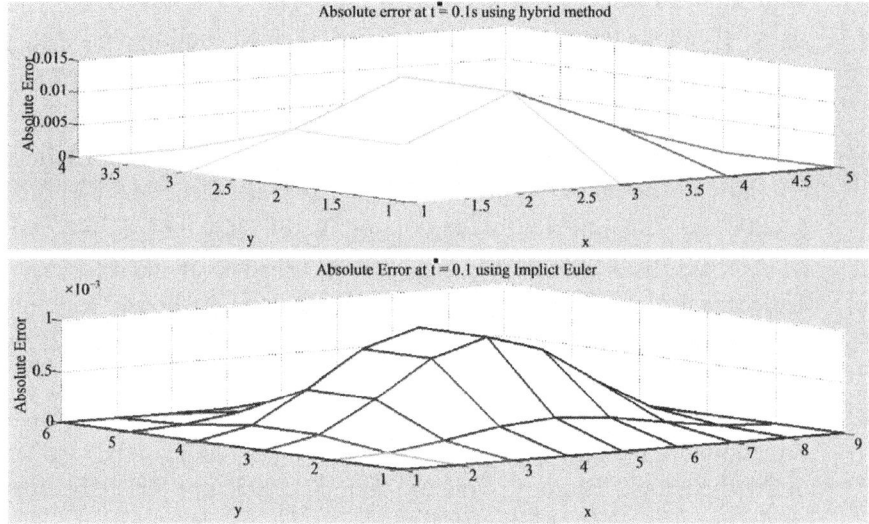

**Figure 4.** Absolute error at t = 0.1 s and different locations of x, y.

The numerical results obtained in this paper ensure that the problems have small desired time to reach it. Thus they have very small step size which is preferred and use RK4 to solve the system of ordinary differential equations in order to decrease the computational time. On the other hand, the problems which have high desired time to reach it, thus have large incremental time (stiff problems) which are preferred and use implicit Euler with perturbation method to solve the system of ordinary differential equations.

# REFERENCES

1.   J. Opsal, A. Rosencwaig and D. L. Willenborg, "Thermal-Wave Detection and Thin-Film Thickness Measurements with Laser Beam Deflection," Applied Optics, Vol. 22, No. 20, 1983, pp. 3169-3176. http://dx.doi.

org/10.1364/AO.22.003169

2.    Z. Y. Yan, "Generalized Method and Its Application in the Higher-Order Nonlinear Schrodinger Equation in Nonlinear Optical Fibres," Chaos Solitons and Fractals, Vol. 16, No. 5, 2003, pp. 759-766. http://dx.doi.org/10.1016/S0960-0779(02)00435-6

3.    J. Mei, H. Zhang and D. Jiang, "New Exact Solutions for a Reaction-Diffusion Equation and a Quasi-Camassa Holm Equation," Applied Mathematics E-Notes, Vol. 4, 2004, pp. 85-91.

4.    E. S. Fahmy and H. A. Abdusalam, "Exact Solutions for Some Reaction Diffusion Systems with Nonlinear Reaction Polynomial Terms," Applied Mathematical Sciences, Vol. 3, No. 11, 2009, pp. 533-540.

5.    M. S. H. Chowdhury and I. Hashim, "Analytical Solution for Cauchy Reaction-Diffusion Problems by Homotopy Perturbation Method," Sains Malaysiana, Vol. 39, No. 3, 2010, pp. 495-504.

6.    S. Puri and K. Wiese, "Perturbative Linearization of Reaction-Diffusion Equations" Journal of Physics A: Mathematical and General, Vol. 36, No. 8, 2003, pp. 2043-2054.http://dx.doi.org/10.1088/0305-4470/36/8/303

7.    L. D. Ropp, N. J. Shadid and C. C. Ober, "Studies of the Accuracy of Time Integration Methods for Reaction-Diffusion Equations," Journal of Computational Physics, Vol. 194, No. 2, 2004, pp. 544-574. http://dx.doi.org/10.1016/j.jcp.2003.08.033

8.    R. G. Marcus, "Finite-Difference Schemes for Reaction-Diffusion Equations Modeling Predator-Prey Interactions in MATLAB," Bulletin of Mathematical Biology, Vol. 69, No. 3, 2007, pp. 931-956. http://dx.doi.org/10.1007/s11538-006-9062-3

9.    B. Liu, M. B. Allen, H. Kojouharov, B. Chen, et al., "Finite-Element Solution of Reaction-Diffusion Equations with Advection" Computational Mechanics, 1996, pp. 3-12.

10.   X. Christos and L. Oberbroeckling, "On the Finite Element Approximation of Systems of Reaction-Diffusion Equations by p/hp Methods," Global Science, Vol. 28, No. 3, 2010, pp. 386-400.

11.   G. Meral and M. S. Tezer, "Solution of Nonlinear Reaction-Diffusion Equation by Using Dual Reciprocity Boundary Element Method with Finite Difference or Least Squares Method," Advances in Boundary Element Techniques, Vol. 3, 2008, pp. 317- 322.

12.   C. Shu, "Differential Quadrature and Its Application in Engineering," Springer Verlag, London, 2000. http://dx.doi.org/10.1007/978-1-4471-0407-0

13.  T. Y. Wu and G. R. Liu, "A Differential Quadrature as a Numerical Method to Solve Differential Equations," Computational Mechanics, Vol. 24, No. 3, 1999, pp. 197-205.http://dx.doi.org/10.1007/s004660050452

14.  V. Kajal, "Numerical Solutions of Some Reaction-Diffusion Equations by Differential Quadrature Method," International Journal of Applied Mathematics and Mechanics, Vol. 6, No. 14, 2010, pp. 68-80.

15.  G. Meral, "Solution of Density Dependent Nonlinear Reaction-Diffusion Equation Using Differential Quadrature Method," World Academy of Science, Engineering and Technology, Vol. 41, 2010, pp. 1178-1183.

16.  W. Y. Yang, W. Cao, T. Chung, J. Morris, et al., "Applied Numerical Methods Using Matlab," John Wiley & Sons, Hoboken, 2005. http://dx.doi.org/10.1002/0471705195

17.  J. S. Nadjafi and A. Ghorbani, "He's Homotopy Perturbation Method: An Effective Tool for Solving Nonlinear Integral and Integro-Differential Equations," Computers and Mathematics with Applications, Vol. 58, No. 11-12, 2009, pp. 2379-2390.http://dx.doi.org/10.1016/j.camwa.2009.03.032

18.  H. S. Prasad and Y. N. Reddy, "Numerical Solution of Singularly Perturbed Differential-Difference Equations with Small Shifts of Mixed Type by Differential Quadrature Method," American Journal of Computational and Applied Mathematics, Vol. 2, No. 1, 2012, pp. 46-52. http://dx.doi.org/10.5923/j.ajcam.20120201.09

19.  J. P. Hambleton and S. W. Sloan, "A Perturbation Method for Optimization of Rigid Block Mechanisms in the Kinematic Method of Limit Analysis," Computers and Geotechnics, Vol. 48, 2013, pp. 260-271. http://dx.doi.org/10.1016/j.compgeo.2012.07.012

20.  M. Bastani and D. K. Salkuyeh, "A Highly Accurate Method to Solve Fisher's Equation," Indian Academy of Sciences, Vol. 78, No. 3, 2012, pp. 335-346.

21.  E. Kreyszig, "Advanced Engineering Mathematics," John Wiley & Sons, Columbus, 2006

# Chapter 7

## FRACTIONAL HEAT CONDUCTION MODELS AND THERMAL DIFFUSIVITY DETERMINATION

Monika Žecová and Ján Terpák

Institute of Control and Informatization of Production Processes, Faculty of BERG, Technical University of Košice, 042 00 Košice, Slovakia

### ABSTRACT

The contribution deals with the fractional heat conduction models and their use for determining thermal diffusivity. A brief historical overview of the authors who have dealt with the heat conduction equation is described in the introduction of the paper. The one-dimensional heat conduction models with using integer- and fractional-order derivatives are listed. Analytical and numerical methods of solution of the heat conduction models with using integer- and fractional-order derivatives are described. Individual methods have been implemented in MATLAB and the examples of simulations are listed. The proposal and experimental verification of the methods for determining thermal diffusivity using half-order derivative of temperature by time are listed at the conclusion of the paper.

### INTRODUCTION

Unsteady heat conduction process, described by partial differential equation, was first formulated by Jean Baptiste Joseph Fourier (1768–1830). In 1807, he wrote an article "Partial differential equation for heat conduction in solids." The issue of heat conduction was addressed by other scientists as well, such as Adolf Fick (1829–1901) [1, 2], James Clerk Maxwell (1831–1879) [3–5], Albert Einstein (1879–1955) [6], Lorenzo Richards (1904–1993) [7], and Geoffrey Taylor (1886–1975) [8, 9].

The various analytical and numerical methods are used to solve the Fourier heat conduction equation (FHCE) [10, 11]. In the case of heat conduction in materials with nonstandard structure, such as polymers, granular and porous materials, and composite materials, a standard description is insufficient and required the creation of more adequate models with using derivatives of fractional-order [12–15]. The causes are mainly memory systems and ongoing processes [16–20], roughness, or porosity of the material [21–23] and also fractality and chaotic behavior of systems [24–28].

The more adequate models of processes subsequently require new methods to determine the parameters of these models. In the case of FHCE, the basic parameter of this equation is thermal diffusivity, which characterizes the dynamics of temperature changes in the substance. Measurement of thermal diffusivity can be realized by many ways. The latest methods for determining thermal diffusivity are mainly laser flash method[29, 30], Kennedy transient heat flow method [31–33], single rectangular pulse heating method [34], and thermal wave method [35, 36].

The issue of research and development methods and tools for processes modeling with using fractional-order derivatives is very actual, since it means a qualitatively new level of modeling. Important authors of the first articles were Fourier (1768–1830), Abel, Leibniz (1646–1716), Grünwald (1838–1920), and Letnikov (1837–1888). Mathematicians like Liouville (1809–1882) [37, 38] and Riemann (1826–1866) [39] made major contributions to the theory of fractional calculus. Nowadays the fractional calculus interests many scientists and engineers from different fields, such as mechanics, physics, chemistry, and control theory [40].

At the present time, there are a number of analytical [41–48] and numerical solutions of fractional heat conduction equation. In the case of numerical methods different methods are developed based on the random walk models [49–52], the finite difference method (FDM) [53–55], the finite element method [56–59], numerical quadrature [60–62], the method of Adomian decomposition [63, 64], Monte Carlo simulation [65,66], matrix approach [12, 13, 67], or the matrix transform method [68, 69]. The finite difference method is an extended method where an explicit [53, 70, 71], an implicit [54, 72–74], and a Crank-Nicolson scheme [55, 75] are used. For the Crank-Nicolson scheme, the literature describes the use of Grünwald-Letnikov definition only for a spatial derivative [73, 76–78].

## MODELS OF HEAT CONDUCTION PROCESSES

Heat conduction is a molecular transfer of thermal energy in solids, liquids, and gases due to the temperature difference. The process of heat conduction takes place between the particles of the substance to touch directly each other and has different temperature. Existing models of heat conduction processes are divided according to various criterions. We consider a division of models into two groups, namely models with using derivatives of integer- and fractional-order.

Models with using derivatives of integer order are the nonstationary and stationary models. Nonstationary models are described by Fourier heat conduction equation, where the temperature $T$ (K) is a function of spatial coordinate $x$ (m) and time $\tau$ (s). In the case of one-dimensional heat conduction, it has the following form [11, 79]:

$$\frac{\partial T(x,\tau)}{\partial \tau} = (\sqrt{a})^2 \frac{\partial^2 T(x,\tau)}{\partial x^2} \quad \text{for } 0 < x < L, \ \tau > 0, \tag{1}$$

$$T(0,\tau) = T_1, \quad T(L,\tau) = T_2 \quad \text{for } \tau > 0,$$

$$T(x,0) = f(x) \quad \text{for } 0 \le x \le L, \tag{2}$$

where $a = \lambda/(\rho c_p)$ is thermal diffusivity (m$^2$ ·s$^{-1}$), $\rho$ is density (kg·m$^{-3}$), $c_p$ is specific heat capacity (J·kg$^{-1}$·K$^{-1}$), and $\lambda$ is thermal conductivity (W·m$^{-1}$·K$^{-1}$).

Heat conduction model with using derivatives of fractional-order for various one-dimensional geometric cases was expressed by the following Oldham-Spanier equation [22]:

$$\frac{\partial T(x,\tau)}{\partial x} + \frac{1}{\sqrt{a}} \frac{\partial^{1/2}[T(x,\tau) - T_0]}{\partial \tau^{1/2}} + \frac{g[T(x,\tau) - T_0]}{x + R} = 0, \tag{3}$$

where $g$ is a geometric factor and $R$ is a radius of curvature.

In the case of one-dimensional heat conduction planar wall ($g = 0$),(3) will take the following form:

$$\frac{\partial^{1/2}[T(x,\tau) - T_0]}{\partial \tau^{1/2}} = -\sqrt{a} \frac{\partial T(x,\tau)}{\partial x}. \tag{4}$$

A more general formulation of the task for modeling not only one-dimensional heat conduction is based on the model in which, on the left-hand side of (1) instead of the first derivative with respect to time, the derivative of order $\alpha$ occurs; that is, we can find it in the form [42, 79]

$$\frac{\partial^\alpha u}{\partial \tau^\alpha} = (b)^2 \frac{\partial^2 u}{\partial x^2} \quad \text{for } 0 < x < L, \ \tau > 0, \tag{5}$$

$$u(0,\tau) = U_1, \quad u(L,\tau) = U_2 \quad \text{for } \tau > 0,$$

$$u(x,0) = f(x) \quad \text{for } 0 \le x \le L, \tag{6}$$

Where $b$ represents a constant coefficient with the unit m·s$^{-\alpha/2}$

Fractional-order models can also be described by the following equation, where $\alpha$ and $\beta$ are of arbitrary order [12, 13, 67]:

$$\frac{\partial^\alpha T(x,\tau)}{\partial \tau^\alpha} = (\sqrt{a})^2 \frac{\partial^\beta T(x,\tau)}{\partial |x|^\beta}. \tag{7}$$

# SOLUTIONS

One-dimensional heat conduction models using integer and fractional derivatives can be solved by analytical and numerical methods.

## Analytical Methods of Solution

Analytical methods can be used for solving problems in a bounded, semibounded, or unbounded interval.

Analytical solution of heat conduction model (1) for a bounded interval $\langle 0, L\rangle$ has the following shape [11, 42]:

$$T(x,\tau) = \sum_{n=1}^{\infty} \sin\left(\frac{n\pi x}{L}\right) \exp\left(-\left(\frac{n\pi \sqrt{a}}{L}\right)^2 \tau\right) c_k. \tag{8}$$

Analytical solution for a fractional diffusion-wave equation (5) has the form

$$T(x,\tau) = \sum_{n=1}^{\infty} \sin\left(\frac{n\pi x}{L}\right) E_\alpha\left(-\left(\frac{n\pi b}{L}\right)^2 \tau^\alpha\right) c_k. \tag{9}$$

For models (1) and (5),

$$c_k = \frac{2}{L} \int_0^L \left[f(\xi) - \frac{1}{L}(T_2 - T_1)\xi - T_1\right] \sin\frac{k\pi\xi}{L} d\xi. \tag{10}$$

We developed and derived the coefficient $ck$ for the form of the function $(\xi) = a_0 + a_1\xi + a_2\xi^2$, in order to implement simulations for different initial conditions (constant, straight line, and parabola). Coefficient $c_k$ has this final shape [79]:

$$c_k = \frac{2}{nL}\left[T_2 - a_0 - a_1 L + a_2\left(\frac{2}{n^2} - L^2\right)\right](-1)^k$$

$$- T_1 + a_0 - a_2\frac{2}{n^2}, \tag{11}$$

where $n = k\pi/L$.

## Numerical Methods of Solution

The best known numerical methods include finite element method, finite difference method, and boundary element methods.

Finite difference methods according to the type of differential expression can be divided into explicit, implicit, and Crank-Nicolson scheme.

Explicit Scheme. Explicit scheme for solving the heat conduction model defined by (1) in the case of homogeneous material has the form

$$T_{m,p} = MT_{m-1,p-1} + T_{m,p-1} - 2MT_{m,p-1} + MT_{m+1,p-1}, \qquad (12)$$

where module $M$ is determined by the relation

$$M = \left(\frac{\sqrt{a}}{\Delta x}\right)^2 \quad \Delta \tau \leq 0,5, \qquad (13)$$

and in the case of nonhomogeneous material, it has the following form:

$$T_{m,p} = M_{m-1}T_{m-1,p-1} + T_{m,p-1}$$
$$- (M_{m-1} + M_m) T_{m,p-1} + M_m T_{m+1,p-1}, \qquad (14)$$

where module $M_m$ is

$$M_m = \left(\frac{\sqrt{a_m}}{\Delta x}\right)^2 \quad \Delta \tau \leq 0,5. \qquad (15)$$

Implicit Scheme. In the case of the implicit scheme, the temperature at a given point is calculated for a homogeneous body according to the following formula:

$$-MT_{m-1,p} + (1 + 2M) T_{m,p} - MT_{m+1,p} = T_{m,p-1}, \qquad (16)$$

and for a nonhomogeneous body, it has the following formula:

$$- M_{m-1}T_{m-1,p} + (1 + M_{m-1} + M_m) T_{m,p}$$
$$- M_m T_{m+1,p} = T_{m,p-1}. \qquad (17)$$

Crank-Nicolson Scheme. For a homogeneous body, it has the form

$$T_{m,p} = \frac{M}{2} \left(T_{m-1,p} - 2T_{m,p} + T_{m+1,p}\right)$$
$$+ \frac{M}{2} \left(T_{m-1,p-1} - 2T_{m,p-1} + T_{m+1,p-1}\right) + T_{m,p-1} \qquad (18)$$

and for a nonhomogeneous body it has the form

$$T_{m,p} = \frac{1}{2} \left(M_{m-1}T_{m-1,p} - (M_{m-1} + M_m) T_{m,p}\right.$$
$$\left. +M_m T_{m+1,p}\right)$$
$$+ \frac{1}{2} \left(M_{m-1}T_{m-1,p-1} - (M_{m-1} + M_m) T_{m,p-1}\right.$$
$$\left. +M_m T_{m+1,p-1}\right)$$
$$+ T_{m,p-1}. \qquad (19)$$

Numerical Methods of Fractional-Order. For solving numerical methods of fractional-order, we use Grunwald-Letnikov ¨ definition with using the principle of "short memory" [16, 80]:

$$\frac{\partial^\alpha T(x,\tau)}{\partial \tau^\alpha} = \frac{\sum_{j=0}^{N_f} bc_j T(x, \tau - j\Delta\tau)}{\Delta\tau^\alpha},$$

(20)

where $L$ is the "length memory," $\tau$ is the time step, and the value of $N(f)$ will be determined by the following relation:

$$N(f) = \min\left\{\left[\frac{\tau}{\Delta\tau}\right], \left[\frac{L}{\Delta\tau}\right]\right\},$$

$$bc_0 = 1, \qquad bc_j = \left(1 - \frac{1+\alpha}{j}\right) \cdot bc_{j-1},$$

where $j \geq 1$.

(21)

Explicit Scheme. Explicit scheme for the heat conduction model using derivative of fractional-order (5) for a homogeneous material has the form

$$T_{m,p} = MT_{m-1,p-1} - \sum_{j=1}^{N_f} bc_j T_{m,p-j}$$

$$- 2MT_{m,p-1} + MT_{m+1,p-1}$$

(22)

and for a nonhomogeneous material it has the form

$$T_{m,p} = M_{m-1} T_{m-1,p-1} - \sum_{j=1}^{N_f} bc_j T_{m,p-j}$$

$$- (M_{m-1} + M_m) T_{m,p-1} + M_m T_{m+1,p-1}.$$

(23)

Implicit Scheme. Fractional shape for a homogeneous body is given by the following relation:

$$- MT_{m-1,p} + (1 + 2M) T_{m,p} - MT_{m+1,p}$$

$$= -\sum_{j=1}^{N_f} bc_j T_{m,p-j}$$

(24)

and for a nonhomogeneous body, it has the following relation:

$$- M_{m-1}T_{m-1,p} + (1 + M_{m-1} + M_m) T_{m,p}$$

$$- M_m T_{m+1,p} = -\sum_{j=1}^{N_f} bc_j T_{m,p-j}.$$

(25)

Crank-Nicolson Scheme. The fractional shape for a homogeneous body has the form and for a nonhomogeneous body it has the form

$$T_{m,p} = \frac{M}{2} \left( T_{m-1,p} - 2T_{m,p} + T_{m+1,p} \right)$$

$$+ \frac{M}{2} \left( T_{m-1,p-1} - 2T_{m,p-1} + T_{m+1,p-1} \right)$$

$$- \sum_{j=1}^{N_f} bc_j T_{m,p-j}$$

(26)

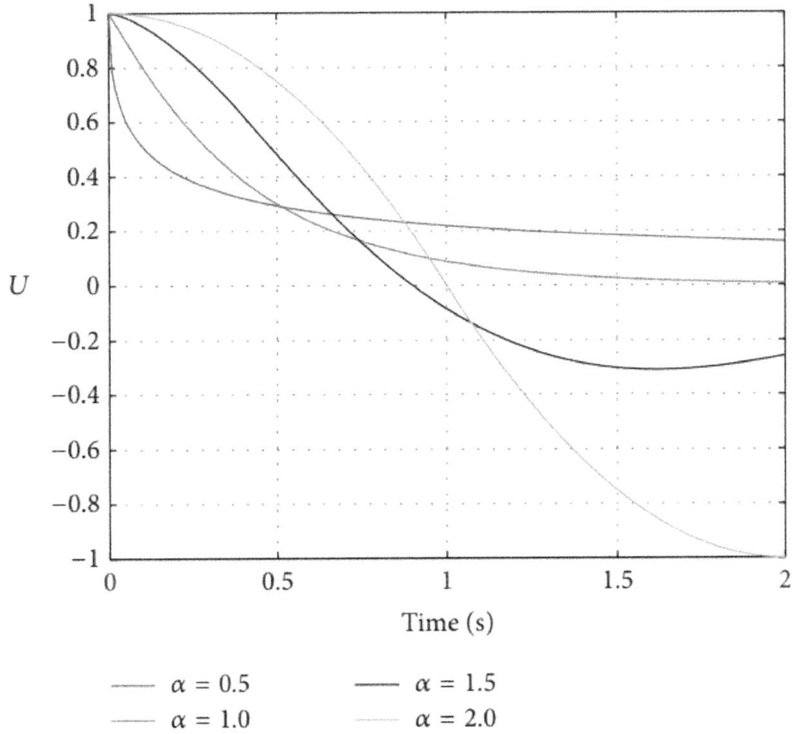

**Figure 1:** Comparison of analytical method for four different derivatives of temperature according to the time.

$$T_{m,p} = \frac{1}{2}\left(M_{m-1}T_{m-1,p} - (M_{m-1} + M_m)T_{m,p}\right.$$

$$\left. +M_m T_{m+1,p}\right)$$

$$+ \frac{1}{2}\left(M_{m-1}T_{m-1,p-1} - (M_{m-1} + M_m)T_{m,p-1}\right.$$

$$\left. +M_m T_{m+1,p-1}\right)$$

$$- \sum_{j=1}^{N_f} bc_j T_{m,p-j}.$$

(27)

## SIMULATIONS

Implementation of the one-dimensional heat conduction model was realized in the programming environment MATLAB. Two toolboxes for the one-dimensional heat conduction model with using integer- and fractional-order derivatives have been created. All implemented functions are published at Mathworks, Inc., MATLAB Central File Exchange as Heat Conduction Toolbox and Fractional Heat Conduction Toolbox [81, 82].

Simulations of heat conduction model for analytical solution have been implemented for four different derivatives temperatures according to time, namely, for the derivative order of 0.5, 1, 1.5, and 2 (Figure 1). The model input parameters were set as follows: initial temperature in the shape of parabolic function $f(x) = 2x - x^2$, boundary condition of the 1st kind for $U_1 = U_2 = 0$, total time simulation 2 s, time step 0.01 s, number of items' sum 100, distance 2 m, number of points 21, and coefficient for material properties 1 m·s$^{-\alpha/2}$.

Simulation with a heat conduction model for explicit, implicit, and Crank-Nicolson scheme was performed with the time step 0.01 s and order of the derivative of 1.5. Input parameters of the model were chosen as follows: initial value $U(x) = 0$, boundary condition of the 1st kind for $U_1 = U_2 = 1$, total time simulation 2 s, time step 0.01 s, number of items' sum 100, distance 2 m, number of points 21, and coefficient for material properties 1 m·s$^{-3/4}$

From the numerical methods, we have chosen Crank Nicolson scheme, in which we can see what effect a different order of the derivative has on the temperatures course (Figure 2).

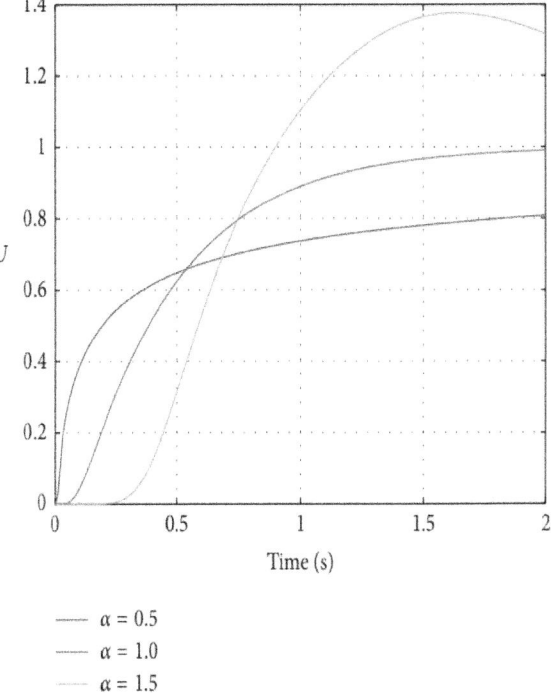

— $\alpha = 0.5$
— $\alpha = 1.0$
— $\alpha = 1.5$

**Figure 2:** Crank-Nicolson scheme for derivatives of 0.5, 1, and 1.5.

In Figure 3, we see the comparison of courses of individual numerical methods and analytical solution.

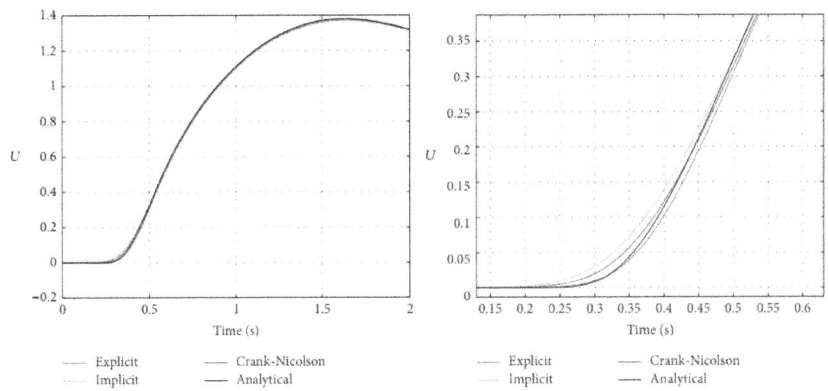

**Figure 3:** Comparison of analytical solution and numerical methods for the derivative of 1.5.

## PROPOSAL METHOD FOR THERMAL DIFFUSIVITY DETERMINATION

The method is based on the method of calculation of heat flows:

$$i_Q = \sqrt{c_p \rho \lambda \cdot_0} D_\tau^{1/2} g(\tau), \quad g(\tau) = T_w(\tau) - T_0. \tag{28}$$

Determination of the heat flow $i_Q$ is possible in two ways: namely,

(i)     from the gradient of the two measured temperatures $(T_1, T_2)$,

$$i_Q = -\lambda \frac{d}{dx} T_1(\tau), \tag{29}$$

(ii)     from the half-order derivative of one measured temperature $(T_1)$,

$$i_Q = \frac{\lambda}{\sqrt{a}} \frac{d^{1/2}}{d\tau^{1/2}} [T_1(\tau) - T_0]. \tag{30}$$

Share of half-order derivative and gradient of temperature is proportional to the square root of the thermal diffusivity:

$$\sqrt{a} = \frac{(d^{1/2}/d\tau^{1/2})[T_1(\tau) - T_0]}{-(d/dx) T_1(\tau)}. \tag{31}$$

$$\sqrt{a} = \frac{\Delta\tau^{-1/2} \sum_{j=0}^{N(f)} bc_j [T_{1,p-j} - T_0]}{\Delta x^{-1} [T_{2,p} - T_{1,p}]}. \tag{32}$$

For the numerical calculation of the first derivative of temperature according to the coordinate, respectively, temperature gradient (31) is sufficient to measure two temperatures (Figure 4).

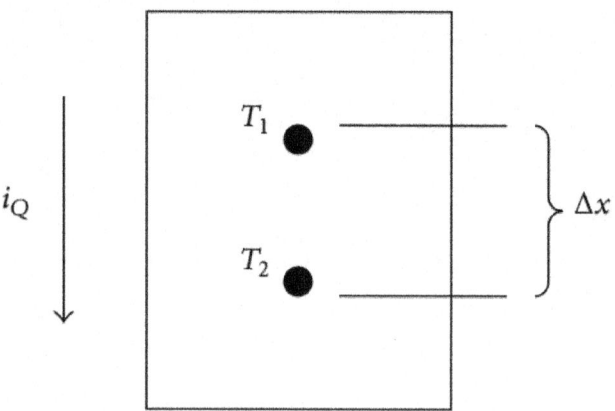

**Figure 4:** Measured temperatures.

The calculation of thermal diffusivity is based on the ratio half-order derivative of temperature according to the time to the temperature gradient (Figure 5) which is observed based on the values of two neighbouring temperatures in space obtained from simulations.

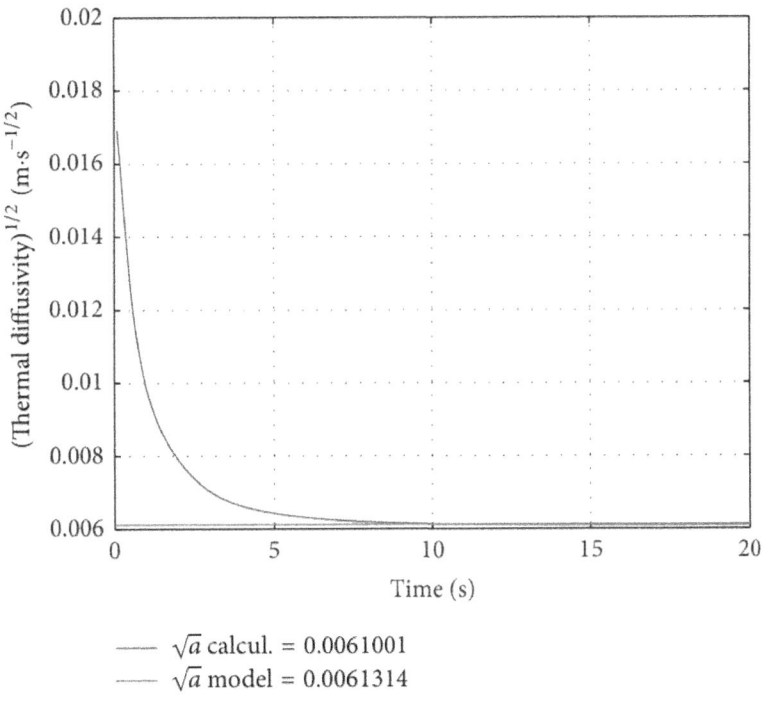

$\sqrt{a}$ calcul. = 0.0061001
$\sqrt{a}$ model = 0.0061314

**Figure 5:** Rate of half-order temperature derivative to the temperature gradient.

More previous values of temperature in time are used for the calculation of the half-order derivative, as in the case of the first derivative, which uses only one previous value [83].

The method was tested on the model using Crank-Nicolson scheme on a brass sample. The value of thermal diffusivity for a brass is $3.7594 \times 10^{-5}$ $m^2 \cdot s^{-1}$. The initial temperature of simulation was determined on 20∘ C, boundary condition of the 1st kind for 20∘ C and 100∘ C, with a time step of the simulation 0.01 s. Input parameters of the brass: density 8,400 $kg \cdot m{-}3$, specific heat capacity 380 $J \cdot kg^{-1} \cdot K^{-1}$, and thermal conductivity 120 $W \cdot m^{-1} \cdot K^{-1}$.

In Figure 6, we can see the effect of time step to calculate the square root of thermal diffusivity

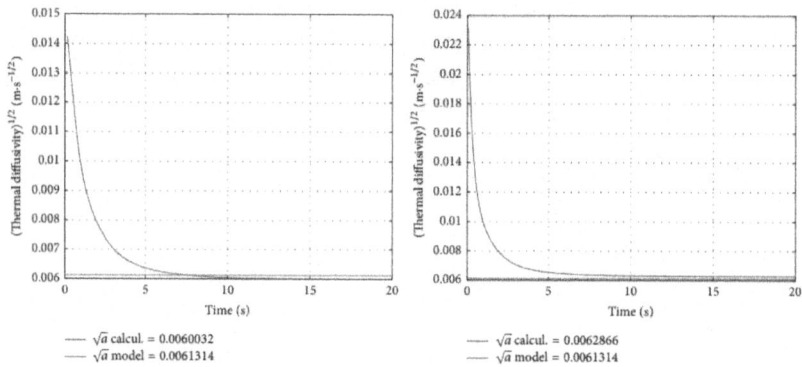

**Figure 6:** Crank-Nicolson scheme for a time step of 0.15 s and 0.05 s.

The calculation accuracy of determining the value of the square root of thermal diffusivity depends on the number of previous values of temperatures in time and also from the selected time step. Reducing the number of previous values of temperatures leads to higher inaccuracy of calculation.

## EXPERIMENTAL VERIFICATION

The method has been verified on the experimental measurements. Measurements were carried out on the devices HT10XC and HT11C. Module HT11C is a physical model of one-dimensional heat conduction [84]. It consists of a heating and cooling section between which is inserted the sample of material (Figure 7).

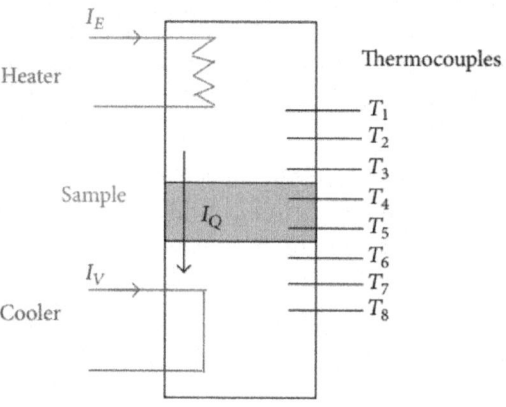

**Figure 7:** Scheme of HT11C.

Brass sample was used in the form of a cylinder with a diameter of 25 mm and a height of 30 mm. Contact areas of the sample were coated with a thin layer of thermal paste to minimize the transient thermal resistance. Module HT11C uses the thermocouples of type K in the temperature range from 0 to 133°C and the distance among them is 15 mm. The device HT10XC with HT11C module is connected via USB to a PC. The software that comes with the device allows setting conditions of the experiment and the measurement data saving to a file.

Experimental measurements which are referred to in this paper were carried out under the following conditions: namely, heater power 1.3 W, the water flow in the cooler 0.5 L/min, and the time step for recording of measured data 1.0 s. A unit jump in the heater power from 1.3 to 3.3 W was realized after stabilizing the temperatures. The transition from one steady state to another is shown in Figure 8.

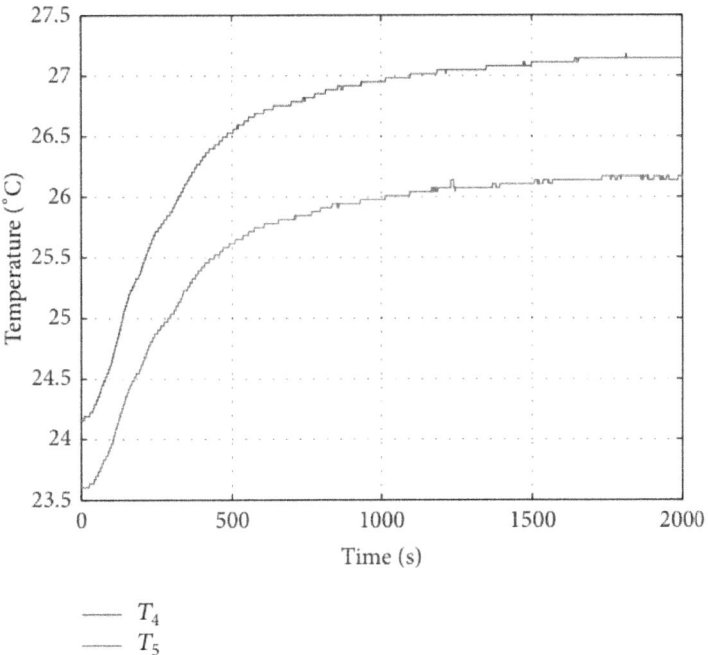

**Figure 8:** Experimental measurements of temperatures T4 and T5 for the brass

On Figure 9 is determined the square root of thermal diffusivity from the measured values of the device HT11C (Figure 8), that is, from the ratio of half-order derivative the temperature according to the time to the temperature gradient.

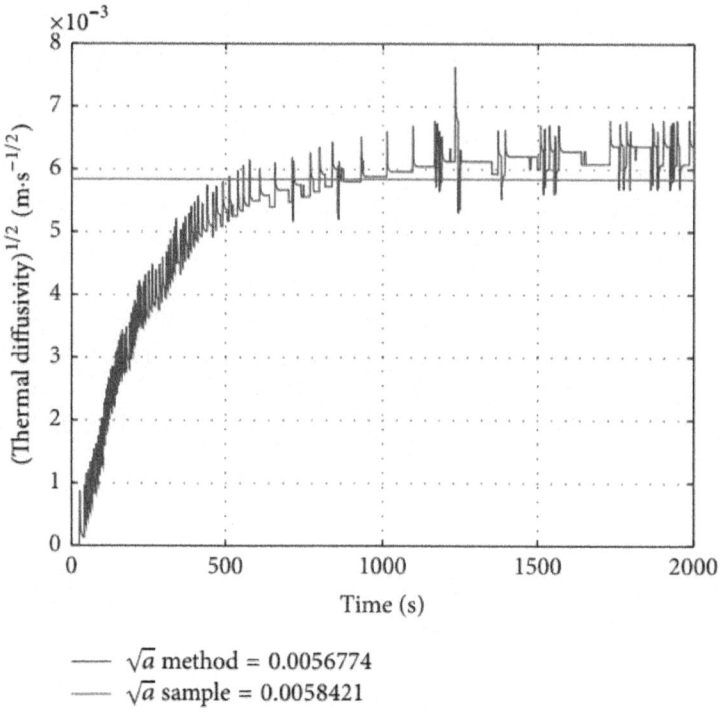

$\sqrt{a}$ method = 0.0056774
$\sqrt{a}$ sample = 0.0058421

**Figure 9:** Thermal diffusivity

The value of thermal diffusivity of the used brass sample for equipment HT11C was $3.2233 \times 10^{-5}$ m² ·s⁻¹ and it corresponds to the square root of thermal diffusivity 0.0056774 m·s⁻¹ᐟ². Brass sample was also measured on the device LFA [85] and the value of thermal diffusivity was $3.4130 \times 10^{-5}$ m² ·s⁻¹, which corresponds to the square root of thermal diffusivity 0.0058421 m·s⁻¹ᐟ². Calculated relative error between the measured values of the thermal diffusivity of the brass sample on HT11C and LFA is 5.5591% [79].

## CONCLUSION

Benefits of this work are mainly the developed analytical and numerical methods for solving one-dimensional heat conduction using integer and fractional derivatives, which are implemented in the form of libraries functions in MATLAB. Another benefit is the designed, implemented, and verified method of determining thermal diffusivity using the half-order derivative of temperature according to the time on the experimental equipment HT10XC with module HT11C.

# CONFLICT OF INTERESTS

The authors declare that there is no conflict of interests regarding the publication of this paper.

# ACKNOWLEDGMENTS

This work was supported by the Slovak Research and Development Agency under the Contract no. APVV-0482-11 and by Projects VEGA 1/0729/12, 1/0552/14, and 1/0529/15.

# REFERENCES

1.   A. Fick, "Ueber diffusion," Annalen der Physik, vol. 170, no. 1, pp. 59–86, 1855.

2.   A. Fick, "On liquid diffusion," Philosophical Magazine and Journal of Science, vol. 10, pp. 31–39, 1855.

3.   J. C. Maxwell, "Scientific worthies," Nature, vol. 8, no. 203, pp. 397–399, 1873.

4.   J. C. Maxwell, Treatise on Electricity and Magnetism, vol. 1, Oxford at the Clarendon Press, 2nd edition, 1881.

5.   J. C. Maxwell, An Elementary Treatise on Electricity, Clarendon Press, Oxford, UK, 2nd edition, 1888.

6.   A. Einstein, "Uber die yon der molekularkinetischen Fliissigkeiten snspendierten Teilchen," Annals of Physics, vol. 17, pp. 549–560, 1905, (English translation in Investigations on the Theory of Brownian Movement by Albert Einstein, edited with notes by R. Fiirth, Methuen, London, UK, pp. 1–35, 1926).

7.   L. A. Richards, "Capillary conduction of liquids through porous mediums," Journal of Applied Physics, vol. 1, no. 5, pp. 318–333, 1931.

8.   G. Taylor, "Dispersion of soluble matter in solvent flowing slowly through a tube," Proceedings of the Royal Society of London, Series A, vol. 219, pp. 186–203, 1953.

9.   T. N. Narasimhan, "Fourier's heat conduction equation: history, influence, and connections," Proceedings of the Indian Academy of Sciences, Earth and Planetary Sciences, vol. 108, no. 3, pp. 117–148, 1999.

10.  D. W. Hahn and M. N. Ozisik, Heat Conduction, John Wiley & Sons, 2012.

11.  P. V. O'Nell, Advanced Engineering Mathematics, Cengage Learning, Toronto, Canada, 2012.

12. I. Podlubny, A. Chechkin, T. Skovranek, Y. Chen, and B. M. Vinagre Jara, "Matrix approach to discrete fractional calculus II. Partial fractional differential equations," Journal of Computational Physics, vol. 228, no. 8, pp. 3137–3153, 2009.

13. I. Podlubny, T. Skovranek, B. M. Vinagre Jara, I. Petras, V. Verbitsky, and Y. Chen, "Matrix approach to discrete fractional calculus III: non-equidistant grids, variable step length and distributed orders,"Philosophical Transactions of the Royal Society A, vol. 371, no. 1990, pp. 1–15, 2013.

14. J. S. Leszczynski, An Introduction to Fractional Mechanics, Czestochowa University of Technology, 2011.

15. D. Sierociuk, A. Dzieliński, G. Sarwas, I. Petras, I. Podlubny, and T. Skovranek, "Modelling heat transfer in heterogeneous media using fractional calculus," Philosophical Transactions of the Royal Society A, vol. 371, no. 1990, Article ID 20120146, 2013.

16. I. Podlubny, Fractional Differential Equations, vol. 198 of Mathematics in Science and Engineering, Academic Press, San Diego, Calif, USA, 1999.

17. I. Podlubny, "Fractional-order models: a new stage in modelling and control," in Proceedings of the IFAC Conference: System Structure and Control, pp. 231–235, Nantes, France, July 1998.

18. S. Sakakibara, "Properties of vibration with fractional derivatives damping of order 1/2," JSME International Journal, Series C, vol. 40, no. 3, pp. 393–399, 1997.

19. P. J. Torvik and R. L. Bagley, "On the appearance of the fractional derivative in the behavior of real materials," Journal of Applied Mechanics, vol. 51, no. 2, pp. 294–298, 1984.

20. S. G. Samko, A. A. Kilbas, and O. I. Maričev, Fractional Integrals and Derivatives and Some of Their Applications, Nauka i technika, Minsk, Belarus, 1987.

21. K. B. Oldham, "Semiintegral electroanalysis: analog implementation," Analytical Chemistry, vol. 45, no. 1, pp. 39–47, 1973.

22. K. B. Oldham and J. Spanier, the Fractional Calculus: Theory and Applications of Differentiation and Integration to Arbitrary Order, Academic Press, New York, NY, USA, 1974.

23. K. B. Oldham and C. G. Zoski, "Analogue instrumentation for processing polarographic data," Journal of Electroanalytical Chemistry, vol. 157, no. 1, pp. 27–51, 1983.

24. B. Mandelbrot, "Some noises with I/f spectrum, a bridge between direct

current and white noise," IEEE Transactions on Information Theory, vol. 13, no. 2, pp. 289–298, 1967.

25.  A. Oustaloup, "From fractality to non-integer derivation through recursivity, a property common to these, two concepts: a fundamental idea for a new process control," in Proceedings of the 12th IMACS World Congress, pp. 203–208, Paris, France, 1988.

26.  A. Oustaloup, La Derivation non Entiere, Hermes, Paris, France, 1995, (French).

27.  L. O. Chua, L. Pivka, and C. W. Wu, "A universal circuit for studying chaotic phenomena," Philosophical Transactions of the Royal Society of London, Series A: Mathematical, Physical Sciences and Engineering, vol. 353, no. 1701, pp. 65–84, 1995.

28.  I. Petras, Fractional-Order Nonlinear Systems, Higher Education Press, Beijing, China, 2011.

29.  W. J. Parker, R. J. Jenkins, C. P. Butler, and G. L. Abbott, "Method of determining thermal diffusivity, heat capacity and thermal conductivity," Journal of Applied Physics, vol. 32, no. 9, p. 1961, 1961.

30.  A. M. Hofmeister, "Thermal diffusivity of garnets at high temperature," Physics and Chemistry of Minerals, vol. 33, no. 1, pp. 45–62, 2006.

31.  J. Madsen and J. Trefny, "Transient heat pulse technique for low thermal diffusivity solids," Journal of Physics E: Scientific Instruments, vol. 21, no. 4, pp. 363–366, 1988.

32.  C. S. McMenamin, D. F. Brewer, T. E. Hargreaves et al., "An improved analysis technique for heat pulse measurements," Physica B: Physics of Condensed Matter, vol. 194–196, no. 1, pp. 21–22, 1994.

33.  H. Bougrine, J. F. Geys, S. Dorbolo et al., "Simultaneous measurements of thermal diffusivity, thermal conductivity and thermopower with application to copper and ceramic superconductors," The European Physical Journal B, vol. 13, no. 3, pp. 437–443, 2000.

34.  K. Kobayasi, "Simultaneous measurement of thermal diffusivity and specific heat at high temperatures by a single rectangular pulse heating method," International Journal of Thermophysics, vol. 7, no. 1, pp. 181–195, 1986.

35.  B. Abeles, G. D. Cody, and D. S. Beers, "Apparatus for the measurement of the thermal diffusivity of solids at high temperatures," Journal of Applied Physics, vol. 31, no. 9, pp. 1585–1592, 1960.

36.  A. Kaźmierczak-Bałata, J. Bodzenta, and D. Trefon-Radziejewska, "Determination of thermal-diffusivity dependence on temperature of

transparent samples by thermal wave method," International Journal of Thermophysics, vol. 31, no. 1, pp. 180–186, 2010.

37. J. Liouville, "Memoire sur le calcul des differentielles a indices quelconques," Journal de l'École Polytechnique, vol. 13, p. 163, 1832.

38. J. Liouville, "Mémoire sur l'intégration de l'équation mx2+nx+pd2y/dx2+qx+rdy/dx + sy=0 à l'aide des différentielles à indices quelconques," Journal d l'Ecole Polytechnique, vol. 13, pp. 163–186, 1832.

39. B. Riemann, Versuch einer allgemeinen Auffassung der Integration und Differentiation: The Collected Works of Bernhard Riemann, edited by H. Weber, Dover, New York, NY, USA, 2nd edition, 1953.

40. S. Westerlund and L. Ekstam, "Capacitor theory," IEEE Transactions on Dielectrics and Electrical Insulation, vol. 1, no. 5, pp. 826–839, 1994.

41. R. Gorenflo, Y. Luchko, and F. Mainardi, "Wright functions as scale-invariant solutions of the diffusion-wave equation," Journal of Computational and Applied Mathematics, vol. 118, no. 1-2, pp. 175–191, 2000

42. O. P. Agrawal, "Solution for a fractional diffusion-wave equation defined in a bounded domain,"Nonlinear Dynamics, vol. 29, pp. 145–155, 2002.

43. F. Liu, V. V. Anh, I. Turner, and P. Zhuang, "Time fractional advection-dispersion equation," Journal of Applied Mathematics and Computing, vol. 13, no. 1-2, pp. 233–245, 2003.

44. F. Mainardi, Y. Luchko, and G. Pagnini, "The fundamental solution of the space-time fractional diffusion equation," Fractional Calculus & Applied Analysis, vol. 4, no. 2, pp. 153–192, 2001.

45. F. Huang and F. Liu, "The time fractional diffusion equation and the advection-dispersion equation," The ANZIAM Journal, vol. 46, no. 3, pp. 317–330, 2005.

46. W. Wyss, "The fractional diffusion equation," Journal of Mathematical Physics, vol. 27, no. 11, pp. 2782–2785, 1986.

47. W. R. Schneider and W. Wyss, "Fractional diffusion and wave equations," Journal of Mathematical Physics, vol. 30, no. 1, pp. 134–144, 1989.

48. J.-S. Duan, "Time- and space-fractional partial differential equations," Journal of Mathematical Physics, vol. 46, no. 1, Article ID 013504, 013504, 8 pages, 2005.

49. R. Gorenflo, F. Mainardi, D. Moretti, and P. Paradisi, "Time fractional diffusion: a discrete random walk approach," Nonlinear Dynamics, vol. 29, pp. 129–143, 2002.

50. R. Gorenflo and A. Vivoli, "Fully discrete random walks for space-time

fractional diffusion equations,"Signal Processing, vol. 83, no. 11, pp. 2411–2420, 2003.

51. F. Liu, S. Shen, V. Anh, and I. Turner, "Analysis of a discrete non-Markovian random walk approximation for the time fractional diffusion equation," The ANZIAM Journal, vol. 46, pp. 488–504, 2004/05.

52. Q. Liu, F. Liu, I. Turner, and V. Anh, "Approximation of the Lévy-Feller advection-dispersion process by random walk and finite difference method," Journal of Computational Physics, vol. 222, no. 1, pp. 57–70, 2007.

53. S. Shen and F. Liu, "Error analysis of an explicit finite difference approximation for the space fractional diffusion equation with insulated ends," The ANZIAM Journal, vol. 46, pp. 871–887, 2005.

54. P. Zhuang and F. Liu, "Implicit difference approximation for the time fractional diffusion equation,"Journal of Applied Mathematics & Computing, vol. 22, no. 3, pp. 87–99, 2006.

55. C. Tadjeran and M. M. Meerschaert, "A second-order accurate numerical method for the two-dimensional fractional diffusion equation," Journal of Computational Physics, vol. 220, no. 2, pp. 813–823, 2007.

56. J. P. Roop, "Computational aspects of FEM approximation of fractional advection dispersion equations on bounded domains in R2," Journal of Computational and Applied Mathematics, vol. 193, no. 1, pp. 243–268, 2006.

57. G. J. Fix and J. P. Roof, "Least squares finite-element solution of a fractional order two-point boundary value problem," Computers & Mathematics with Applications, vol. 48, no. 7-8, pp. 1017–1033, 2004.

58. V. J. Ervin and J. P. Roop, "Variational formulation for the stationary fractional advection dispersion equation," Numerical Methods for Partial Differential Equations, vol. 22, no. 3, pp. 558–576, 2006.

59. V. J. Ervin and J. P. Roop, "Variational solution of fractional advection dispersion equations on bounded domains in Rd," Numerical Methods for Partial Differential Equations, vol. 23, no. 2, pp. 256–281, 2007.

60. P. Kumar and O. P. Agrawal, "An approximate method for numerical solution of fractional differential equations," Signal Processing, vol. 86, no. 10, pp. 2602–2610, 2006.

61. O. P. Agrawal, "A numerical scheme for initial compliance and creep response of a system," Mechanics Research Communications, vol. 36, no. 4, pp. 444–451, 2009.

62. L. Yuan and O. P. Agrawal, "A numerical scheme for dynamic systems

containing fractional derivatives,"Journal of Vibration and Acoustics, vol. 124, no. 2, pp. 321–324, 2002.

63. S. Momani and Z. Odibat, "Numerical solutions of the space-time fractional advection-dispersion equation," Numerical Methods for Partial Differential Equations, vol. 24, no. 6, pp. 1416–1429, 2008.

64. H. Jafari and V. Daftardar-Gejji, "Solving linear and nonlinear fractional diffusion and wave equations by Adomian decomposition," Applied Mathematics and Computation, vol. 180, no. 2, pp. 488–497, 2006.

65. D. Fulger, E. Scalas, and G. Germano, "Monte Carlo simulation of uncoupled continuous-time random walks yielding a stochastic solution of the space-time fractional diffusion equation," Physical Review E—Statistical, Nonlinear, and Soft Matter Physics, vol. 77, no. 2, Article ID 021122, 2008.

66. M. M. Marseguerra and A. Zoia, "Monte Carlo evaluation of FADE approach to anomalous kinetics,"Mathematics and Computers in Simulation, vol. 77, no. 4, pp. 345–357, 2008.

67. I. Podlubny, "Matrix approach to discrete fractional calculus," Fractional Calculus & Applied Analysis, vol. 3, no. 4, pp. 359–386, 2000.

68. M. Ilić,, I. W. Turner, and D. P. Simpson, "A restarted Lanczos approximation to functions of a symmetric matrix," IMA Journal of Numerical Analysis, vol. 30, no. 4, pp. 1044–1061, 2010.

69. Q. Yang, I. Turner, F. Liu, and M. Ilić, "Novel numerical methods for solving the time-space fractional diffusion equation in two dimensions," SIAM Journal on Scientific Computing, vol. 33, no. 3, pp. 1159–1180, 2011.

70. S. B. Yuste and L. Acedo, "An explicit finite difference method and a new von Neumann-type stability analysis for fractional diffusion equations," SIAM Journal on Numerical Analysis, vol. 42, no. 5, pp. 1862–1874, 2005.

71. R. Lin, F. Liu, V. Anh, and I. Turner, "Stability and convergence of a new explicit finite-difference approximation for the variable-order nonlinear fractional diffusion equation," Applied Mathematics and Computation, vol. 212, no. 2, pp. 435–445, 2009.

72. F. Liu, P. Zhuang, V. Anh, and I. Turner, "A fractional-order im plicit difference approximation for the space-time fractional diffusion equation," ANZIAM Journal, vol. 47, pp. 48–68, 2006.

73. S. Chen, F. Liu, P. Zhuang, and V. Anh, "Finite difference approximations for the fractional Fokker-Planck equation," Applied Mathematical

Modelling, vol. 33, no. 1, pp. 256–273, 2009.

74.  T. A. Langlands and B. I. Henry, "The accuracy and stability of an implicit solution method for the fractional diffusion equation," Journal of Computational Physics, vol. 205, no. 2, pp. 719–736, 2005.

75.  C. Tadjeran, M. M. Meerschaert, and H.-P. Scheffler, "A second-order accurate numerical approximation for the fractional diffusion equation," Journal of Computational Physics, vol. 213, no. 1, pp. 205–213, 2006.

76.  E. Sousa, "Numerical approximations for fractional diffusion equations via splines," Computers and Mathematics with Applications, vol. 62, no. 3, pp. 938–944, 2011.

77.  C. Celik and M. Duman, "Crank-Nicolson difference scheme for the fractional diffusion equation with the Rieszfractional derivative," Journal of Computational Physics, vol. 231, no. 1, pp. 1743–1750, 2012.

78.  D. Wang, A. Xiao, and W. Yang, "Crank-Nicolson difference scheme for the coupled nonlinear Schrodinger equations with the Riesz space fractional derivative," Journal of Computational Physics, vol. 242, pp. 670–681, 2013.

79.  M. Zecova, Modelovanie procesov prenosu tepla vedenfm (Modeling of the heat conduction processes) [Ph.D. thesis], TUKE, Kosice, Slovakia, 2014, (Slovak).

80.  I. Petras, "Fractional derivatives, fractional integrals, and fractional differential equations in Matlab," inEngineering Education and Research Using MATLAB, pp. 239–265, 2011.

81.  M. Zecova and J. Terpak, Heat Conduction Toolbox—File Exchange—MATLAB Central, 2014,http://www.mathworks.com/matlabcentral/fileexchange/43146-heat-conduction-toolbox.

82.  M. Zecova and J. Terpak, Fractional Heat Conduction Toolbox, File exchange, MATLAB Central, 2014,http://www.mathworks.com/matlabcentral/fileexchange/45491-fractional-heat-conduction-toolbox.

83.  M. Zecova, J. Terpak, and L. Dorcak, "The modeling of heat conduction using integer-and fractional-order derivatives," in Proceedings of the 15th International Carpathian Control Conference (ICCC '14), pp. 710–715, Velke Karlovice, Czech Republic, May 2014.

84.  "HT10XC Computer Controlled Heat Transfer Teaching Equipment," 2014,http://discoverarmfield.com/en/products/view/ht10xc/computer-controlled-heat-transfer-teaching-equipment.

85.  "LFA 427-Laser Flash Apparatus (2000°C)-NETZSCH," 2013, http://www.netzsch.com/n87885.

# Chapter 8

# OPTIMIZATION OF BEARING LOCATIONS FOR MAXIMIZING FIRST MODE NATURAL FREQUENCY OF MOTORIZED SPINDLE-BEARING SYSTEMS USING A GENETIC ALGORITHM

Chi-Wei Lin

Department of Industrial Engineering and Systems Management, Feng Chia University, Taichung, Taiwan

## ABSTRACT

This paper has developed a genetic algorithm (GA) optimization approach to search for the optimal locations to install bearings on the motorized spindle shaft to maximize its first-mode natural frequency (FMNF). First, a finite element method (FEM) dynamic model of the spindle-bearing system is formulated, and by solving the eigenvalue problem derived from the equations of motion, the natural frequencies of the spindle system can be acquired. Next, the mathematical model is built, which includes the objective function to maximize FMNF and the constraints to limit the locations of the bearings with respect to the geometrical boundaries of the segments they located and the spacings between adjacent bearings. Then, the Sequential Decoding Process (SDP) GA is designed to accommodate the dependent characteristics of the constraints in the mathematical model. To verify the proposed SDP-GA optimization approach, a four-bearing installation optimazation problem of an illustrative spindle system is investigated. The results show that the SDP-GA provides well convergence for the optimization searching process. By applying design of experiments and analysis of variance, the optimal values of GA parameters are determined under a certain number restriction in executing the eigenvalue calculation subroutine. A linear regression equation is derived also to estimate necessary calculation efforts with respect to the specific quality of the optimization solution. From the results of this illustrative example, we can conclude that the proposed SDP-GA optimization approach is effective and efficient.

## INTRODUCTION

Spindle-bearing systems have been widely applied in mechanisms, such as machine tools, that need relative rotary motion to accomplish desired machining functions. Recently, to alleviate high-speed rotation, motorized spindles have been successfully developed and used in high-speed machine tools. This type of spindle is equipped with a built-in motor as an integrated part of the spindle shaft, eliminating the need for conventional power transmission devices. To achieve high cutting performance, the motorized spindle-bearing system must be designed deliberately since it is the main moving component in the machine tool.

The design process for a motorized spindle system usually begins with identification of the bearing types, configurations, preloading methods, and spindle motor systems that would be appropriate for the desired machining application [1] . Once the bearings and spindle system concept has been clarified, detailed product specifications are further used to define other design values of the spindle system. One example is the locations of bearings assembled on the spindle shaft. However, in the initial stage of the spindle-bearing system design, the engineers must ascertain the optimal values of design variables that maximize the first-mode natural frequency (FMNF) of the spindle-bearing system to avoid forced resonance from easily happening in machining operations [2] .

How the spindle shaft and bearing related design variables affect a system's static and dynamic characteristics has been intensively studied [3] -[7] . Recently, Lin and Tu [2] comprehensively explored the effects of design variables on the natural frequencies of a high-speed spindle system. The eight design variables they considered included material of spindle shaft, diameters and total length of the spindle shaft, bearing initial preload, spacings between the bearings of the front or rear bearing set, spacing between the middle line of the front and rear bearing sets, and spacing of the middle line of the front and rear bearing sets to the end of the cutting tool. Their results showed that the first two most important design factors among the eight were related to the locations of the bearings, which implied that the positions bearings installed on the spindle shaft must be determined carefully. There have been several papers dedicated to the optimization of locations or spans of bearings based on the stability of rotating machinery or machining from the practical viewpoints of machining experts and scholars [6] -[8] . However, there still lacks a research concentrated directly on the dynamics of spindle bearing systems.

To estimate natural frequencies of spindle systems analytically in the early detailed design stage, the finite element method (FEM) has been frequently adopted in modeling rotor dynamics, due to its flexibility in treating ever-complex spindle system designs, such as motorized spindles. Basically, the FEM model for the spindle systems of machine tools is similar to those developed in rotor dynamic literature [9] -[12] . However, the spindle shafts used in machine tools usually have smaller shaft diameters and bearings, and possess no disk-like components. As the FEM model is built, the system natural frequencies can then be solved as an eigenvalue problem of the dynamic model, which is equal to finding the roots of a high order polynomial equation [13]; therefore, the FMNF maximization problem generally appears nonlinear. Because derivatives of the objective function FMNF are difficult or even impossible to deduce, genetic algorithm (GA), an optimization approach without derivatives, is utilized to solve the optimization problem in this paper.

Genetic algorithm, first introduced by Holland in 1975 [14], is a modern metaheuristic method which can solve nonlinear optimization problems effectively. GA is a computational technique which simulates the evolution process based on the principles of natural genetics and natural selection [15]. In the academic field of engineering optimization, GA has been successfully applied in the researches such as identification of a spatial slider-crank mechanism [16] , optimal designs of pressure swing adsorption [17] , comparisons for damage detection on structures [18] , and rolling element bearing design [19] .

The main purpose of this paper is to demonstrate the application of GA on determining the optimal locations of bearings of motorized spindle shafts, which are the decisions expected to be made by design engineers when they start to plan assembling the bearings and spindle shaft after the spindle and bearing specifications have been specified during the concept design stage. In the following sections, the FEM model of a typical motorized spindle system is introduced first. Next, based on the FEM model, an optimization mathematical model constituted by an objective function and constraints represented with the design variables, i.e. the locations of bearings, is developed, and an appropriate GA-based optimal solutions searching approach is described. Then, to manifest the proposed GA, a spindle-bearing system design problem is illustrated. The mathematical model of the bearing location problem for the illustrative design is built first. To find the optimal solutions, a customized computer code is written in MATLAB by following the procedures of the proposed GA, where the GA parameters are determined by statistical analyses. The computer program is also used to investigate the influence of the parameters on the results. Finally, the primary conclusions and discussions of this paper are described.

## FEM MODEL OF SPINDLE-BEARING SYSTEMS

Without loss of generality, the proposed dynamic FEM model is developed based on the spindle-bearing system as shown in Figure 1, where two sets of angular-contact ball bearings (Set 1 and Set 2) are utilized. Figure 2 shows the essential mechanical models and the major independent design variables required to describe the dynamic properties of spindle-bearing systems [2], in which the integrated spindle-bearing system FEM model is a combination of the distributed spindle shaft FEM and the discrete bearing stiffness models. Those models can be represented as functions of spindle specifications, bearing specifications, initial preload, and bearing location. In this research, however, only the bearing location (highlighted in Figure 2) is retained as the design variables while the other factors are treated as design parameters.

To construct the spindle shaft FEM model, the spindle shaft is first discretized into a finite number of beam elements, and each node of the elements is assigned with four degrees of freedom, where, in sequence, the first two are assigned to the lateral directions and the remaining two are assigned to the angular directions. Associated with several specific shape functions, the kinetic and potential energy of each element can be obtained by integrating those of the cross-sections along the axis of the spindle shaft, and expressed as functions of the physical and geometrical properties of the element. Summing up the kinetic and potential energy of all elements, we can get the total energy of the spindle shaft, and by utilizing Lagrange's equation, the equations of motion (EOM) of the spindle shaft can be finally deduced. After combining the bearing stiffness with the EOM of the spindle shaft, the simple free vibration EOM for the motorized spindle-bearing system can be presented as [20]

$$M\ddot{q} + Kq = 0 \tag{1}$$

$$K = K_s + B \tag{2}$$

where $q$ in $R^{4(p+1)}$ is the global node displacement vector of the spindle shaft and $p$ is the element number; $M$ is the mass matrix of size $4(p+1) \times 4(p+1)$; $K$, comprising the spindle shaft and radial bearing stiffness matrices $K_s$ and $B$ as indicated in Equation (2), is the stiffness matrix of size $4(p+1) \times 4(p+1)$. The matrices $M$ and $K_s$ are all determined based on the dimensions and material of the spindle shaft, and the details of these matrices can be found in Appendix A, where the elements are modeled as Timoshenko beams. Note that no structural damping or axial forces are considered in the model.

The bearing stiffness matrix $B$ can be expressed by summing up the stiffness of all bearings as

$$B = \sum_{i=1}^{n_b} B_i \qquad (3)$$

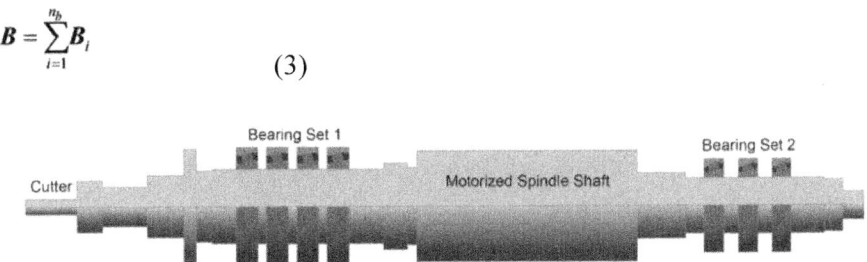

**Figure 1**: A general spindle-bearing system.

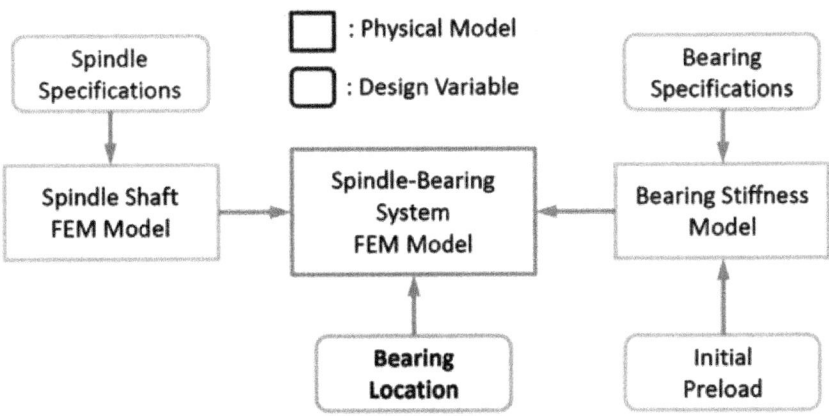

**Figure 2**: Mechanism of spindle-bearing system FEM model.

where $n_b$ is the total number of bearings in the system and $B_i$, with the same size as $K_s$, is the stiffness matrix contributed by bearing $i$. Usually the points of application of the bearings are arranged on the joint stations of spindle shaft FEM model [13]. If we neglect the axial stiffness contributed by the bearing as in the analysis of the spindle shaft, the entries of $B_i = [b_{lm}]$ can be written as

$$b_{lm} = \begin{cases} V & \text{if } l = m = 4j-3 \text{ or } 4j-2 \\ 0 & \text{otherwise.} \end{cases} \qquad (4)$$

where $l, m = 1, 2, \cdots, 4(p+1)$, $V$ is the radial stiffness of bearing $i$, and $j$ is the joint station which the bearing is acted on. The static radial stiffness $V$ of an angular-contact ball bearing can be represented by the bearing stiffness equation as a function of axial preload $(P_a)$, ball diameter $(d_b)$, number of balls $(n_a)$ and contact angle $(\theta)$ [21]:

$$V = c_a P_a^{1/3} n_a^{2/3} \sin^{2/3}\theta \cos\theta d_b^{1/3} \qquad (5)$$

Where $c_a$ is an empirical data decided by the experimental results?

In order to find the natural frequencies of the system, we relate the vibration problem to the generalized eigenvalue problem by substituting $q = he^{j\omega t}$ into Equation (1), which results in [13]

$$(M^{-1}K)h = \lambda h \qquad (6)$$

$$\lambda = \omega^2 \qquad (7)$$

where $\omega$ is the system natural frequency, $\lambda$ is the eigenvalue, and $h$ is the eigenvector.

To find a nonzero solution $h$ of Equation (6), we must have [22]

$$\det(M^{-1}K - \lambda I) = 0 \qquad (8)$$

which is the characteristic equation of $M^{-1}K$, and the roots of which are the eigenvalues of $M^{-1}K$. Expansion of Equation (8) by cofactors results in a $4(p+1)$th-degree polynomial equation in $\lambda$, i.e. the eigenvalues of $M^{-1}K$ are $\lambda = [\lambda_1, \lambda_2, \cdots, \lambda_{4(p+1)}]$, and therefore, the natural frequencies are $\omega = [\omega_1, \omega_2, \cdots, \omega_{4(p+1)}]$;

for each eigenvalue, there is a corresponding eigenvector.

## MATHEMATICAL MODEL OF THE BEARING LOCATION OPTIMIZATION

In this research, we define the locations of bearings as the design variables. Assume that there are $n_s$ sets of bearings to be installed on the spindle system, and for set $i$, $n_i$ identical bearings are used and, therefore, $n_b = \sum_{i=1}^{n_s} n_i$. Under this condition, the locations of bearings of set $i$ can be represented as

$$\tilde{x}_i = [\tilde{x}_{i,j}], \ j = 1, 2, \cdots, n_i \qquad (9)$$

where the real number $\tilde{x}_{i,j}$, named local design variable, indicates the coordinate of the middle line of bearing $j$ of the bearing set $i$. The lower the bearing sets numbered, the closer their locations are to the original point. Since the bearings are identical within a set, index $j$ can also be used to represent the sequence of the bearings such that the bearings with lower $j$'s are closer to the original point. Therefore, for bearing set $i$, $\tilde{x}_{i,1}$ and $\tilde{x}_{i,n_i}$ are the coordinates of the two end bearings respectively. The adjacent bearings of bearing $j$ are bearings $j-1$ and $j+1$ for $j = 2,3,\cdots,n_i-1$. The locations of all bearings $x$ can be formed by assembling all $\tilde{x}_i$ as

$$x = \left[ \tilde{x}_1 \ \tilde{x}_2 \ \cdots \ \tilde{x}_{n_s} \right] \qquad (10)$$

Besides the local variables, we define the global design variables as $x = \left[ x_1, x_2, \cdots, x_{n_b} \right]$, which are utilized in calculating natural frequencies after being converted into the corresponding numbers of joint stations in FEM, and for a local variable owning an index $(i,j)$, the index $k$ for its representing global variable can be evaluated by

$$k = \sum_{l=0}^{(i-1)} n_l + j \qquad (11)$$

with $n_0 = 0$.

Since the FMNF is the smallest among the natural frequencies of spindle-bearing systems, maximizing it draws the following objective function

$$\text{Max } \omega_1 \qquad (12)$$

where $\omega_1 = \min \omega$. As discussed in Section 2, $\omega$ is the square root of eigenvalue $\lambda$ and depends on $x$ since the matrix $K$ is a function of $x$.

There are two different kinds of location-related constraints for the design variables considered in this research. First, the locations of bearings are ordinarily constrained by, for example, the geometrical boundaries of the segments where they are located. This kind of constraint is set to avoid interference with other parts, and or by the requirements originated from the results of other analyses such as statics. Here we name those constraints as Type I constraints and represent them with

$$L_{i,j} \leq \tilde{x}_{i,j} \leq U_{i,j}, \quad i = 1,2,\cdots,n_s, \ j = 1,2,\cdots,n_i \qquad (13)$$

where $L_{i,j}$ and $U_{i,j}$ are the upper and lower limits of $\tilde{x}_{i,j}$ respectively.

The second type of constraints, Type II constraints, reflects the fact that the least amount of spacing between two bearings must be greater than a specified value such as the width of the bearing. Apparently, only the bearings within the same group would be involved in the same constraint equations of Type II constraints. If the smallest spacing of bearings $j$ and $j+1$ in the bearing set $i$ is recognized as $C_{i,j,j+1}$, the constraints can be written as

$$\tilde{x}_{i,j+1} - \tilde{x}_{i,j} \geq C_{i,j,j+1}, \quad i = 1,2,\cdots,n_s, \quad j = 1,2,\cdots,n_i - 1 \tag{14}$$

Ideally, the optimal solutions are obtained by identifying the ones possessing the maximum FMNF from the feasible solutions. However, since the objective function is complicated and nonlinear, it may be difficult or even impossible to derive an analytical or gradient-based method to solve the optimization problem. In this paper, a GA-based approach is constructed instead to search for the optimal solutions, which simultaneously satisfy all the constraints presented in Equation (13) and Equation (14).

## FORMULATION OF THE GENETIC ALGORITHM

As a randomized population-based search technique, GA utilizes genetic operators, i.e. crossover and mutation, inspired by natural evolution process to manipulate individuals in a population over generations to improve their fitness gradually based on the "survival of the fittest" strategies. The individuals in GA are likened to chromosomes which are most commonly represented as strings with an equal length of binary numbers formed by the coded design variables, and to each chromosome, there corresponds a value of the objective function, referred to as the fitness of the chromosome [23].

To encode the real variables $x$ as binary numbers for forming the chromosome, we adopt the simple binary representation scheme with $m$ bits [24]. The value of bit number $m$ is generally determined based on the required precision of the design variables. For each component $x_i$, by translating and scaling, we map its feasible region $[x_i^L, x_i^U]$ onto the interval ▮▮▮. The integers in the interval $[0, 2^m - 1]$ are then expressed as $m$-bit binary strings $y_i$, which defines the corresponding encoding variable of $x_i$ as

$$y_i = a_{m(i-1)+1} a_{m(i-1)+2} \cdots a_{mi} \tag{15}$$

where $a_{m(i-1)+j} \in \{0,1\}$, $j = 1,2,\cdots,m$, and by collecting all $y_i$'s, we can form $y = [y_1, y_2, \cdots, y_n]$, the design variables in binary format. As a result, the chromosome $S$ can be constituted by attaching the $m$-bit binary strings of all variables, end for end, in sequence as

$$S = y_1 y_2 \cdots y_{n_b} = \underbrace{a_1 a_2 \cdots a_m}_{y_1} \underbrace{a_{m+1} a_{m+2} \cdots a_{2m}}_{y_2} \cdots \underbrace{a_{m(n_b-1)+1} a_{m(n_b-1)+2} \cdots a_{n_c}}_{y_{n_b}}$$

(16)

where $n_c = n_b \times m$ is the length of chromosome $S$. $S$ can be converted back to $y$ with a similar reverse transformation.

When we perform the decoding tasks to convert the binary codes back to real numbers, the feasible regions of variables must be identified deliberately according to the constraints applied to them. For a simple problem with fixed upper and lower limits, we can conduct decoding on all variables simultaneously since they are independent to each other. However, for mutually dependent variables such as those considered in this paper, their values in new generations, yielded by genetic operations of GA, may violate the dependency constraints (such as Type II constraints) if we try to decode the variables parallel in time or fail to handle the dependencies of the variables properly. There are several methods to ensure variable dependent constraints satisfied during GA optimization. The most effective approach is to restrict the search to valid regions of the search space, i.e. the chromosome is decoded in such a way that invalid solutions are prevented. We can thereby avoid wasting effort by evaluating infeasible solutions [25].

In this research, to ascertain that the design variables satisfy all constraints after decoding from binary codes to real values, we propose the Sequential Decoding Process (SDP), in which the valid regions of variables are decided one by one, sequentially. Since the locations of bearings of two different sets are independent, we can simultaneously decode the variables within different bearing sets at the same time, and for each set of bearings, we decode the variables, from the first to the last, in sequence. The first step of SDP is converting the global binary variable $y$ to the local binary variable $\tilde{y}_{i,j}$ by using Equation (11). Then, for each bearing set $i$, $i = 1, 2, \cdots, n_s$, the binary $\tilde{y}_{i,j}$, with respect to global $y_k$, is decoded into real $\tilde{x}_{i,j}$ by using the standard binary decoding formula for $j$ from 1 to $n_i$, in sequence:

$$\tilde{x}_{i,j} = \tilde{X}_{i,j}^L + \frac{\tilde{X}_{i,j}^U - \tilde{X}_{i,j}^L}{2^m - 1} \sum_{l=1}^{m} a_{m(k-1)+l} \cdot 2^{(m-l)}$$

(17)

where if $j = 1$, the valid region $[\tilde{X}_{i,j}^L, \tilde{X}_{i,j}^U] = [L_{i,j}, U_{i,j}]$; if $j = 2, 3, \cdots, n_i$, the valid region $[\tilde{X}_{i,j}^L, \tilde{X}_{i,j}^U] = [\tilde{x}_{i,j-1} + C_{i,j-1,j}, U_{i,j}]$; The values of $L_{i,j}$, $U_{i,j}$, and $C_{i,j-1,j}$ are specified in the constraints. After all $\tilde{x}_{i,j}$ are evaluated, we can obtain values of the global variables $x$ by the global-local transition equation as indicated in Equation (11).

Now that the above representation scheme has been defined, the procedure of applying the proposed SDP and GA, which is called SDP-GA, to search for the optimal solutions of the mathematical model is shown in Figure 3 and summarized as [26] :

1)  Utilize a random number generator to create an initial population which consists of $z$ individuals, each with a $n_c$ binary string.

2)  Extract the values of variables for each individual by reading $n_b$ sets of $m$ binary digits and decoding into real numbers through the SDP. The fitness values, FMNF, for all individuals are evaluated by solving Equation (8), where the locations of bearings necessitate to be the joint stations of the finite element model, which is coded and tested antecedently.

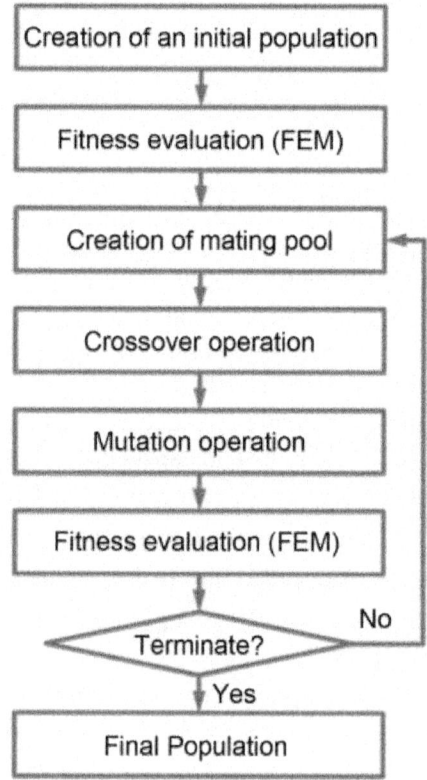

**Figure 3**: Flowchart of the SDP-GA.

3)  Create a mating pool by using the proportional selection procedure, such as a simulated roulette wheel, to choose the members in the initial

population to form a mating pool with a size $Z$, in which the chance of an individual being selected for the mating pool is proportional to its fitness value.

4)   Perform crossover operation on the mating parent pool to produce offspring. The crossover operation adopted in this research involves randomly choosing two chromosomes from the mating pool and an integer $k$ in the range of 1 to $n_c - 1$. The first $k$ positions of the parents are exchanged to produce two children.

5)   Perform mutation operation by visiting every bit of all individuals of the new population and switching the bit from 0 to 1 or 1 to 0 with a mutation probability $P_m$.

6)   Evaluate the population as what we did in step 2. The highest fitness value and the corresponding variable values are stored. A generation is now completed.

7)   If the preset number of generations $n_g$ is complete, the process is stopped; otherwise, go to step 3.

Note that the size of population $Z$, the number of generations $n_g$, and the mutation probability $P_m$ are generally decided with a few experiments to ensure a good convergence. However, to further determine the optimal values of the parameters in this research, we conduct statistical analysis such as design of experiments (DOE), analysis of variance (ANOVA), and linear regression [27].

## AN ILLUSTRATIVE EXAMPLE

To demonstrate the validity of the proposed SDP-GA, a real bearing location optimization problem of a high-speed spindle-bearing system design is exemplified in this section. It is assumed that the concept design stage has already been completed, and the results are summarized as specifications in Table 1 and Table 2 for the spindle shaft and bearings, respectively. The spindle shaft in this illustrative case is made of steel and constitutes 14 cylinders with different diameters. Table 1shows the material properties of the spindle shaft, and the length and diameter of each segment as well as the number of elements that each segment is divided into in the FEM model. There are two different kinds of bearings, types ACB1 and ACB2, used in the example, and their specifications are indicated in Table2 the table also includes the widths of the bearings and the data required in Equation (5) for calculating their stiffnesses with the assumption that preloads applied on bearings are not affected by their locations. The results are also provided in the last column of the table.

If in the system level design stage, four bearings, two of type ACB1 on segment 7 and two of type ACB2 on 12, are planned to be installed in the system, i.e. $n_s = 2$, $n_1 = n_2 = 2$, and $n_b = 4$, we assign the global and

**Table 1**: Dimensions and element numbers for the spindle shaft.

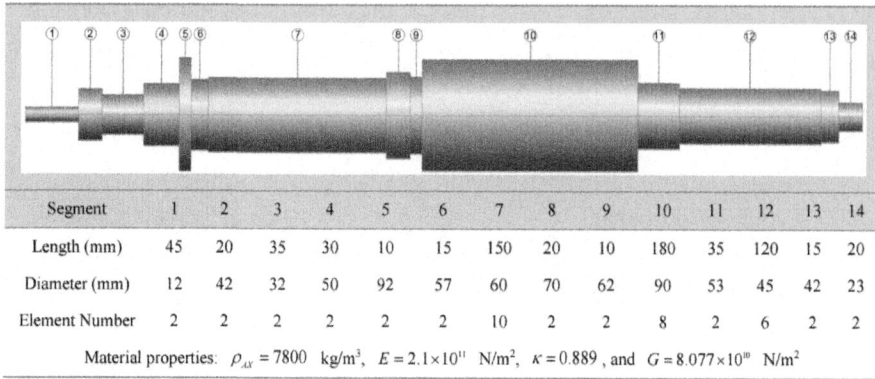

| Segment | 1 | 2 | 3 | 4 | 5 | 6 | 7 | 8 | 9 | 10 | 11 | 12 | 13 | 14 |
|---|---|---|---|---|---|---|---|---|---|---|---|---|---|---|
| Length (mm) | 45 | 20 | 35 | 30 | 10 | 15 | 150 | 20 | 10 | 180 | 35 | 120 | 15 | 20 |
| Diameter (mm) | 12 | 42 | 32 | 50 | 92 | 57 | 60 | 70 | 62 | 90 | 53 | 45 | 42 | 23 |
| Element Number | 2 | 2 | 2 | 2 | 2 | 2 | 10 | 2 | 2 | 8 | 2 | 6 | 2 | 2 |

Material properties: $\rho_{AV} = 7800$ kg/m$^3$, $E = 2.1 \times 10^{11}$ N/m$^2$, $\kappa = 0.889$, and $G = 8.077 \times 10^{10}$ N/m$^2$

**Table 2**: The specifications of the bearings.

| Bearing Type | Width (mm) | $\theta$ (°) | $d_b$ (mm) | $n_a$ | $p_a$ (N) | $c_a$ | $V$ (N/m) |
|---|---|---|---|---|---|---|---|
| ACB1 | 18 | 15 | 23 | 18 | 850 | $1.9 \times 10^6$ | $1.379 \times 10^8$ |
| ACB2 | 16 | 15 | 19 | 16 | 340 | $2.2 \times 10^6$ | $1.021 \times 10^8$ |

$x = [x_1, x_2, x_3, x_4]$ and $[\tilde{x}_1 \ \tilde{x}_2] = [\tilde{x}_{1,1}, \tilde{x}_{1,2}, \tilde{x}_{2,1}, \tilde{x}_{2,2}]$

where the set formed by the two bearings on segment 7 is numbered one and the set on segment 12 is numbered two. Our objective is to maximize $\omega_1$ of the spindle system. The constraints of the design variables can be derived from the data provided in Table 1 and Table2 Assumed that geometry limits are the only sources of Type I constraints. These constraints for the design variables are illustrated as Figure 4. As revealed in Figure 4(a), for example, the lower limit $L_{1,1}$ of $\tilde{x}_{1,1}$ can be obtained by adding the left end coordinate of segment 7 and half of the width of the type ACB1 bearing, i.e. $L_{1,1} = 155 + 0.5 \times 18 = 164$ mm. The upper limit $U_{1,1}$ is equal to subtracting the right end coordinate of segment 7 off 1.5 times of the width of type ACB1 bearing, i.e. $U_{1,1} = 305 - 1.5 \times 18 = 278$ mm, where the origin point is located on the left end of the spindle shaft. The type I constraints for the other three design variables can be obtained in a similar way and the results are shown in Figure 4. Type II constraints specify the least spacing between adjacent bearing pairs. As shown in Figure 5, the least distance between $\tilde{x}_{1,1}$ and $\tilde{x}_{1,2}$ is the width of the bearing ACB1, which is equal to 18 mm, and between $\tilde{x}_{2,1}$ and $\tilde{x}_{2,2}$ is 16 mm, the width of bearing ACB2. Table 3 summarizes all constraints for the four local design variables and also their valid regions utilized in SDP-GA.

Since there are only four variables in this case, we can reach the global optimal solution as $x^* = [189, 296, 558, 574]$ mm and the corresponding $\omega_1^*$ being 794.622 Hz by an exhaustive search with the precision as 0.5 mm. However, even with only four variables, to obtain the results, the FMNF-related eigen-

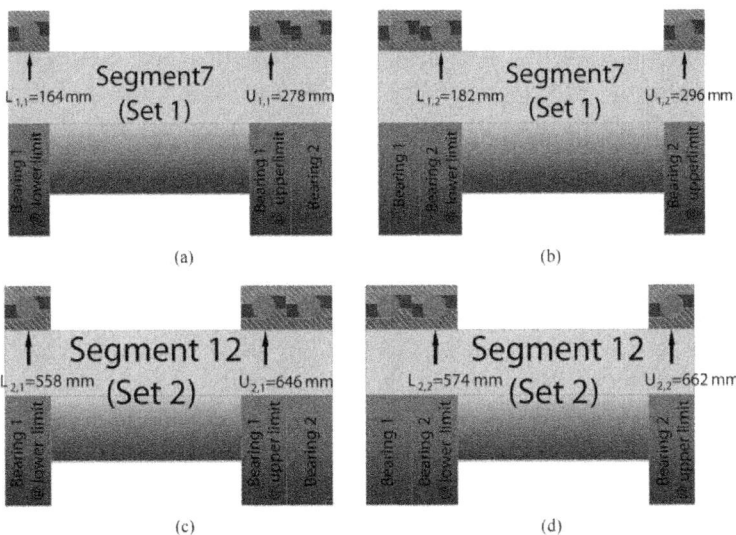

**Figure 4**: Type I constraints for (a) $\tilde{x}_{1,1}$; (b) $\tilde{x}_{1,2}$; (c) $\tilde{x}_{2,1}$; and (d) $\tilde{x}_{2,2}$.

**Table 3**: The constraint equations and valid regions of the design variables (all units are in mm).

| Variable | Type I Constraint | Type II Constraint | Valid Region |
|---|---|---|---|
| $\tilde{x}_{1,1}$ | $164 \leq \tilde{x}_{1,1} \leq 278$ | $\tilde{x}_{1,2} - \tilde{x}_{1,1} \geq 18$ | $[164, 278]$ |
| $\tilde{x}_{1,2}$ | $182 \leq \tilde{x}_{1,2} \leq 296$ | | $[\tilde{x}_{1,1} + 18, 296]$ |
| $\tilde{x}_{2,1}$ | $558 \leq \tilde{x}_{2,1} \leq 646$ | $\tilde{x}_{2,2} - \tilde{x}_{2,1} \geq 16$ | $[558, 646]$ |
| $\tilde{x}_{2,2}$ | $574 \leq \tilde{x}_{2,2} \leq 662$ | | $[\tilde{x}_{2,1} + 16, 662]$ |

**Figure 5**: Type II constraints for (a) $\tilde{x}_{1,1}$ and $\tilde{x}_{1,2}$; and (b) $\tilde{x}_{2,1}$ and $\tilde{x}_{2,2}$.

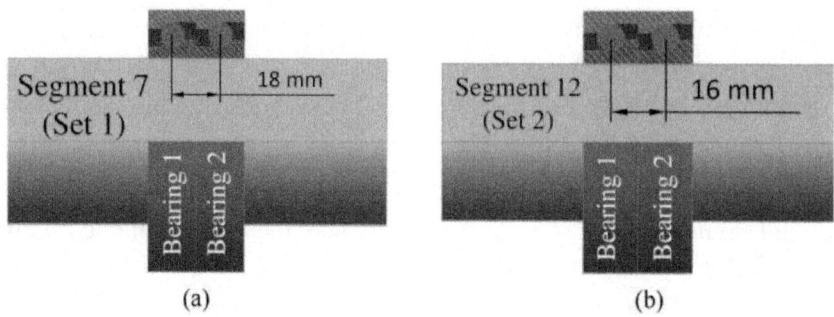

(a)                              (b)

Value-solving subroutine has to be executed $414,855,255^{(=(230\times229/2)\times(178\times177/2))}$ times, which takes about 240 days if one execution takes 0.05 seconds. To a design project, 240 days may be too long to wait for a decision and the schedule is likely to be delayed. To avoid such a tedious, exhaustive search, the alternative GA approach provides a more efficient approach in obtaining an acceptable, near optimal solution.

To code a computer program for searching for the optimal solutions by following the SDP-GA procedures developed in Section 4, we need to decide bit numbers of the design variables first. If the precision is required to be at least 0.5 mm and since the largest interval for the design variables is 114 mm, the binary number $m$, in order to guarantee the required precision, must satisfy $114/(2^m-1)\le0.5$, which leads to $m=8$, and such that the length of chromosome $n_c=32$. Following the procedures indicated in Figure 3, a computer program which constitutes a main script M-file to execute the SDP-GA and a supporting function M-file to calculate the FMNF, is coded in MATLAB. However, we still need to determine the parameters $z$, $n_g$, and $p_m$ before running SDP-GA in searching for the optimal solutions.

In this paper, we use the statistics techniques DOE to identify the optimal values of the parameters under the practical assumption that the allowable time to decide the locations of bearings is one minute, i.e. the number of calculations for finding the eigenvalues must be 1200 at most if each eigenvalue evaluation takes 0.05 seconds.

As the total number of eigenvalue calculations $n_e=z\times(n_g+1)$ in the SDP-GA optimization process, to satisfy the required calculation number 1200, possible values of the parameters $z$ and $n_g$ can be expressed as $(z,n_g)$ combinations that include (16, 74), (20, 59), (30, 39), and (40, 29), where the smallest $z$ is set as 16 since it is the smallest number of chromosomes to prevent a population from ending up with identical chromosomes. Because $z$ and $n_z$ are dependent, we use the levels of $z$ to represent the combinations. The third parameter $P_m$ is typically very small [24], therefore, the possible values considered here are 0.005, 0.01, and 0.02.

To examine convergence, Figure 6 shows the evolution of the optimization process in a certain test run for all four candidates $(z, n_z)$ with $P_m = 0.01$ respectively, where the maximum and average values of FMNF for each of the generations are presented. From the figures, we can find that, as generations go by, not only does the best chromosome of the population improve, but the rest in the entire population also improve. This implies that the proposed SDP-GA approach not only provides a single solution, but also a family of good-quality solutions.

After preliminarily confirming the convergence of the proposed SDP-GA algorithm, to decide the levels for parameters $z$ and $P_m$, a two-factor experiment is performed with FMNF as the response variable and with $z$ and $P_m$ as the main factors. Since $z$ and $P_m$ are with 4 and 3 levels respectively, there are 12 different combinations or cells of the two factors. In a complete balanced experimental design, we obtain 30 observations (replication measurements) from each of the 12 cells. The sums, sums of squares, averages, and standard deviations of the 30 observations for each cell are calculated and shown in Table4 Based on the information provided in Table 4, the ANOVA table for the two-factor problem is shown in Table5 Under the level of significance $\alpha = 0.05$, since the $F$-statistics of $z$ $11.1779 > f_{0.05,3,360} (= 2.6297)$ and of $P_m$ $0.2096 < f_{0.05,2,360} (= 3.0208)$, we conclude that only the main effect $z$ affects FMNF. Furthermore, since the $F$-statistic of interaction $1.3470 < f_{0.05,6,360} (= 2.1238)$, there is no indication of interaction between these two factors. The last column of Table 5 shows that the $P$-value for the test statistic for the main effect $z$ is considerably less than 0.05, while for the main effect $P_m$ and the interaction are greater than 0.05. A graph of the FMNF averages compared to levels of $z$ for each $P_m$ is shown in Figure 7. Notice that of the 12 experimental configurations considered, the optimum configuration employs $z = 16$ at $P_m = 0.01$ in the figure, where the average FMNF is about 790.449 Hz.

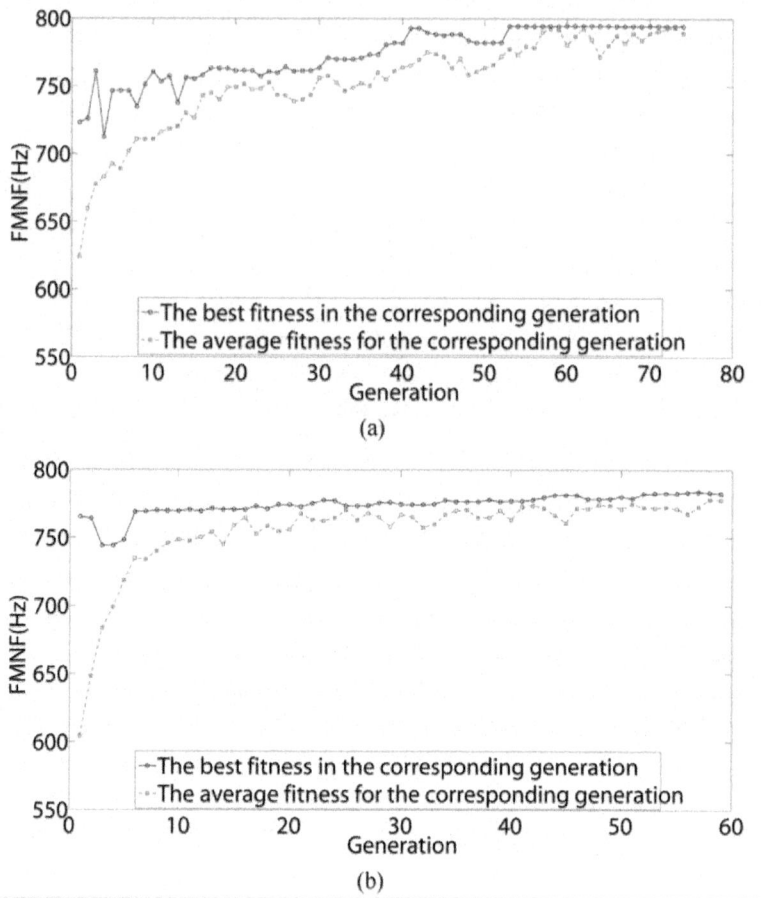

(a)

(b)

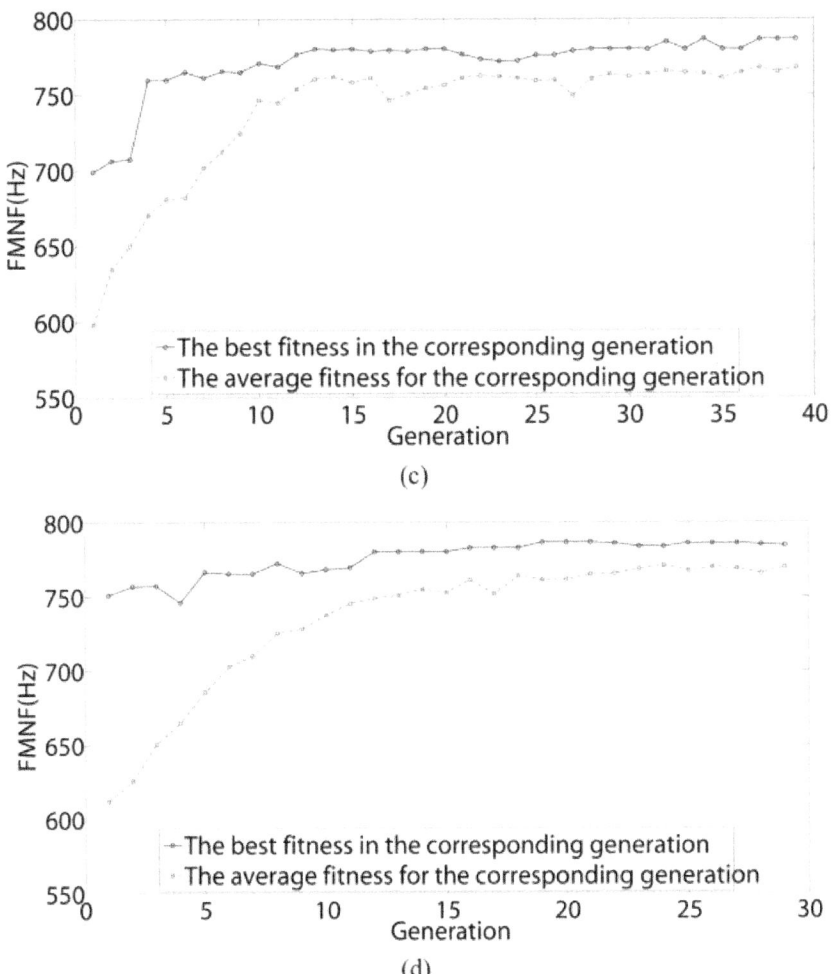

Figure    6.    Results    of    a    certain    test    run    for
$(z, n_g, p_m) = $ (a) $(16,74,0.01)$; (b) $(20,59,0.01)$; (c) $(30,39,0.01)$; and (d) $(40,29,0.01)$

.

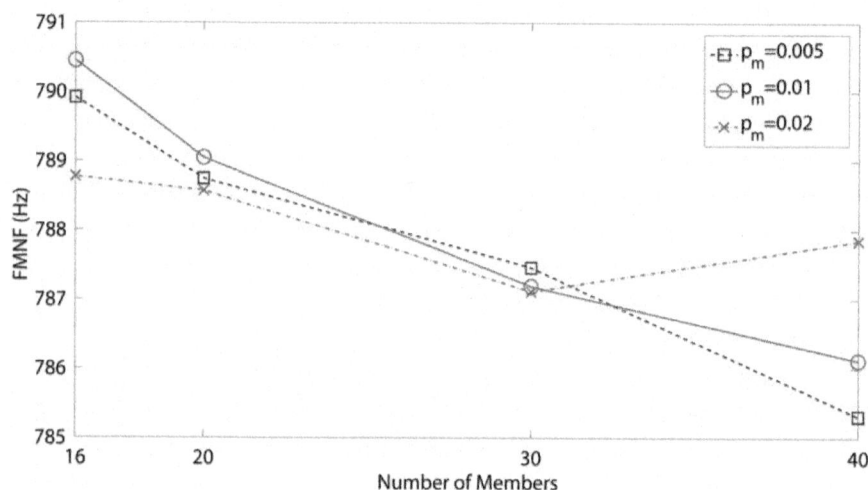

**Figure 7**: Graph of average FMNF versus numbers of members for all values of $P_m$.

**Table 4**: Results of factorial design for parameters of SDP-GA.

| $z$ | $n_g$ | $P_m$ | Sum | Sum of Squares | Average | Standard Deviation |
|------|------|-------|---------|------------|---------|-------------------|
| 16 | 74 | 0.005 | 23697.2 | 18719188.1 | 789.908 | 4.273 |
| 20 | 59 | 0.005 | 23662.2 | 18663884.5 | 788.740 | 4.263 |
| 30 | 39 | 0.005 | 23623.7 | 18603206.6 | 787.455 | 4.621 |
| 40 | 29 | 0.005 | 23559.2 | 18501983.7 | 785.305 | 5.336 |
| **16** | **74** | **0.01** | **23713.5** | **18744704.2** | **790.449** | **3.749** |
| 20 | 59 | 0.01 | 23671.3 | 18677966.8 | 789.044 | 2.967 |
| 30 | 39 | 0.01 | 23615.5 | 18590315.4 | 787.186 | 3.988 |
| 40 | 29 | 0.01 | 23583.5 | 18539994.5 | 786.118 | 4.278 |
| 16 | 74 | 0.02 | 23663.2 | 18665300.5 | 788.774 | 3.512 |
| 20 | 59 | 0.02 | 23657.0 | 18655449.0 | 788.567 | 3.293 |
| 30 | 39 | 0.02 | 23613.3 | 18586860.1 | 787.110 | 4.512 |
| 40 | 29 | 0.02 | 23635.3 | 18621505.7 | 787.844 | 4.253 |

**Table 5**: The analysis of variance table for the FMNF data.

| Source of Variation | Degrees of Freedom | Sum of Squares | Squares Mean | 22 cm F-Statistic | 22 cm p-Value |
|---|---|---|---|---|---|
| $z$ | 3 | 592.4542 | 197.4847 | 11.1779 | $5.0823 \times 10^{-7}$ |
| $p_m$ | 2 | 7.4056 | 3.7028 | 0.2096 | 0.8110 |
| Interaction | 6 | 142.7913 | 23.7985 | 1.3470 | 0.2355 |
| Errors | 348 | 6148.2773 | 17.6675 | | |
| Total | 359 | 6890.9284 | | | |

In summary, under the requirement that the optimal locations of bearings are obtained in one minute, after conducting the two-factor factorial experiments and analysis of variation, the optimal parameters of the GA are decided as $(z, n_g, p_m) = (16, 740.01)$ (highlighted in Table 4), which can be applied to determine the optimal locations of bearings not only for the original design and the engineering changes, but also for the sensitivity studies of the design parameters.

In the 30 outcomes from the experiment of SDP-GA with optimal parameters $(z, n_g, p_m) = (16, 74, 0.01)$, the best and worst ones are 794.625 Hz and 783.589 Hz, which compare to the results obtained by the exhaustive search, the differences are about +0.000% and −388%, respectively. It is noticeable that the best result is slightly better than that of exhaustive search (794.622 Hz) since the precision of the former is finer than the later, and therefore, we replace the value of $\omega_i^*$ with 794.625 Hz. Table 6 summarizes the optimal values of the four design variables, the corresponding FMNF's, and the final spindle-bearing system drawings for the exhaustive search and the best and worst runs. Moreover, since the average and standard deviation of FMNF for this combination are 790.449 Hz and 3.749 Hz respectively, the approximate 95% confidence interval on the mean FMNF $(\mu_{\omega_1})$ is

$$789.107 \leq \mu_{\omega_1} \leq 791.791 \qquad (18)$$

Therefore, if we can tolerate the potential errors of the SDP-GA approach as discussed above and indicated in Table 6, the number of times to run the subroutine for solving the eigenvalues can be reduced to 1200, which is equal to $2.893 \times 10^{-6}$ of that the exhaustive search requires.

Apparently, the quality of the optimal solutions shall be better if we are allowed to perform more calculations in the GA optimization. To investigate the influences of calculating times on the results of SDP-GA by using the optimal parameter values $z = 16$ and $p_m = 0.01$, as obtained in the above analysis, a series of 30-run tests are conducted, with total eigenvalue calculating times

$n_e = 3600$, 6000, 18,000, and 30,000, which correspond to $n_g = 224$, 374, 1124, and 1874 respectively. The results are summarized in Table 7, where the results of $n_e = 1200$ $(n_g = 74)$ are also included. From the table, we can find that as the $n_g$ increases, the average FMNF increases while the standard deviation decreases, which confirms that the qualities of optimal solutions are improved with more iterations by the SDP-GA. The maximum and minimum FMNF obtained in each $n_g$ are also provided in the table, and they also perform better in larger $n_g$. Finally, the last column of the table records the numbers of outcomes hitting $\omega_1^*$, where we can find that the frequencies are greater with larger $n_g$, At $n_g = 1850$, this number even reaches 15, i.e. 50% of the experimental runs hit $\omega_1^*$. By using the extra information of sum and sum of squares provided in Table 7, we can derive the linear regression equation of FMNF vs $n_g$ as [27]

$$\hat{\omega}_1 = 790.4399 + 1.6672 \cdot 10^{-3} \times n_g \qquad (19)$$

Equation (19) can be used to decide the number $n_g$ needed to reach a certain average value of FMNF. For example, if we want the results with an average value 793 Hz, the required $n_g$ can be calculated and rounded to an integer, as 1536, which is equal to 20.5 minutes if each eigenvalue calculation takes 0.05 seconds, since $n_e = 16 \times (1536 + 1) = 24{,}592$ when $z = 16$.

**Table 6.** The results for (A) The exhaustive search; (B) The best run of $(z, n_g, p_m) = (16, 74, 0.01)$; and (C) The worst run of $(z, n_g, p_m) = (16, 74, 0.01)$.

| Case | $x_1$ (mm) | $x_2$ (mm) | $x_3$ (mm) | $x_4$ (mm) | FMNF (Hz) | The spindle system design drawings for the corresponding case |
|------|-----|-----|-----|-----|------|-------------------------------------------------|
| (A) | 189.00 | 296.00 | 558.00 | 574.00 | 794.622 | |
| (B) | 189.04 | 296.00 | 558.00 | 574.00 | 794.625 | |
| (C) | 193.51 | 294.34 | 560.42 | 585.48 | 783.589 | |

**Table 7.** Results of 30 runs for $n_g = 74$, 224, 370, 1124, and 1850.

| $n_g$ | Sum | Sum of Squares | Average | Standard Deviation | Max. FMNF | Min. FMNF | $\omega_1^*$ Occurrence |
|------|------|------|------|------|------|------|------|
| 74 | 23713.466 | 18744704.22 | 790.4489 | 3.7487 | 794.6249 | 783.5886 | 1 |
| 224 | 23717.189 | 18750491.40 | 790.5730 | 3.2823 | 794.5378 | 786.7188 | 0 |
| 370 | 23741.898 | 18789591.44 | 791.3966 | 3.3408 | 794.6249 | 787.1365 | 2 |
| 1124 | 23773.892 | 18840197.42 | 792.4631 | 2.9759 | 794.6249 | 787.2706 | 6 |
| 1850 | 23803.319 | 18886792.39 | 793.4440 | 2.5321 | 794.6249 | 787.5212 | 15 |

# CONCLUSION

This paper has developed a GA-based optimization approach to search for the optimal locations of bearings installed on a spindle shaft to maximize the FMNF. First, an FEM dynamic model of the spindle-bearing system is formulated, and by solving the eigenvalue problem derived from the equations of motion, the natural frequencies of the spindle system can be acquired. Next, the mathematical model is built, which includes the objective function to maximize FMNF and the constraints to limit the locations of the bearings with respect to the geometrical boundaries of the segments they located and the spacings between adjacent bearings. A customized GA optimization procedure, SDP-GA, is designed to accommodate the dependent characteristics of the constraints in the mathematical model. To verify the proposed SDP-GA optimization approach, a four-bearing installation optimization problem of an illustrative spindle system is investigated. The results show that the SDPGA provides good convergence for the optimization searching process. By applying DOE and ANOVA, the optimal values of GA parameters are determined as $z = 16$, $n_g = 74$, and $P_m = 0.01$, under the restriction that the number of executing the eigenvalue calculation subroutine is 1200. The best and worst outcomes for the optimal GA parameter are 794.625 Hz and 783.589 Hz, and for the best one, the locations for the four bearings are $(189.04, 296.00, 558.00, 574.00)$ mm. Compared to the exhaustive search with a maximum FMNF as 794.622 Hz, the 95% confidence interval of mean FMNF obtained by the SDP-GA with optimal parameter values is $[789.107, 791.791]$ Hz; however, the calculation time consumed by SDP-GA is about $2.893 \times 10^{-6}$ of that of the exhaustive search. A linear regression equation is also derived to estimate necessary calculation efforts with respect to the specific quality of the optimization solution. From the results of this illustrative example, we can conclude that the proposed SDP-GA optimization approach is effective and efficient.

## ACKNOWLEDGEMENTS

This study was supported by the National Science Council of Taiwan (grant number NSC 102-2221-E-035-048).

## APPENDIX A

$$M = \sum_{i=1}^{p} A_i^T M_i^e A_i$$

(A.1)

where $p$ is the total element number, and the matrix $A_i$ is defined as:

$$A_i = \left[ a_{jk} \right]_{8 \times 8(p-1)} = \begin{cases} a_{jk} = 1 & \text{if } k = 4i + j - 4, j = 1, 2, \cdots 8, \\ a_{jk} = 0 & \text{otherwise.} \end{cases}$$

(A.2)

and for each element $\overset{\bullet\bullet}{i}$:

$$M^e = M_T^e + M_R^e \quad \text{(A.3)}$$

$$M_T^e = M_{T,1}^e + \Theta M_{T,2}^e + \Theta^2 M_{T,3}^e$$

(A.4)

$$M_{T,1}^e = \frac{\rho_{AX} l}{420(1+\Theta)^2} \cdot \begin{bmatrix} 156 & 0 & 0 & 22l & 54 & 0 & 0 & -13l \\ 0 & 156 & -22l & 0 & 0 & 54 & 13l & 0 \\ 0 & -22l & 4l^2 & 0 & 0 & -13l & -3l^2 & 0 \\ 22l & 0 & 0 & 4l^2 & 13l & 0 & 0 & -3l^2 \\ 54 & 0 & 0 & 13l & 156 & 0 & 0 & -22l \\ 0 & 54 & -13l & 0 & 0 & 156 & 22l & 0 \\ 0 & 13l & -3l^2 & 0 & 0 & 22l & 4l^2 & 0 \\ -13l & 0 & 0 & -3l^2 & -22l & 0 & 0 & 4l^2 \end{bmatrix}$$

(A.5)

$$M_{T,2}^e = \frac{\rho_{AX} l}{420(1+\Theta)^2} \cdot \begin{bmatrix} 294 & 0 & 0 & 38.5l & 126 & 0 & 0 & -31.5l \\ 0 & 294 & -38.5l & 0 & 0 & 126 & 31.5l & 0 \\ 0 & -38.5l & 7l^2 & 0 & 0 & -31.5l & -7l^2 & 0 \\ 38.5l & 0 & 0 & 7l^2 & 31.5l & 0 & 0 & -7l^2 \\ 126 & 0 & 0 & 31.5l & 294 & 0 & 0 & -38.5l \\ 0 & 126 & -31.5l & 0 & 0 & 294 & 38.5l & 0 \\ 0 & 31.5l & -7l^2 & 0 & 0 & 38.5l & 7l^2 & 0 \\ -31.5l & 0 & 0 & -7l^2 & -38.5l & 0 & 0 & 7l^2 \end{bmatrix}$$

(A.6)

$$M_{T,3}^e = \frac{\rho_{AX} l}{420(1+\Theta)^2} \cdot \begin{bmatrix} 140 & 0 & 0 & 17.5l & 70 & 0 & 0 & -17.5l \\ 0 & 140 & -17.5l & 0 & 0 & 70 & 17.5l & 0 \\ 0 & -17.5l & 3.5l^2 & 0 & 0 & -17.5l & -3.5l^2 & 0 \\ 17.5l & 0 & 0 & 3.5l^2 & 17.5l & 0 & 0 & -3.5l^2 \\ 126 & 0 & 0 & 17.5l & 140 & 0 & 0 & -17.5l \\ 0 & 70 & -17.5l & 0 & 0 & 140 & 17.5l & 0 \\ 0 & 17.5l & -3.5l^2 & 0 & 0 & 17.5l & 3.5l^2 & 0 \\ -31.5l & 0 & 0 & -3.5l^2 & -17.5l & 0 & 0 & 3.5l^2 \end{bmatrix}$$

(A.7)

$$M_R^e = M_{R,1}^e + \Theta M_{R,2}^e + \Theta^2 M_{R,3}^e$$

(A.8)

$$M^e_{R,1} = \frac{\rho_{AX} r^2}{120 l (1+\Theta)^2} \cdot \begin{bmatrix} 236 & 0 & 0 & 3l & -36 & 0 & 0 & 3l \\ 0 & 36 & -3l & 0 & 0 & 36 & -3l & 0 \\ 0 & -3l & 4l^2 & 0 & 0 & 3l & -l^2 & 0 \\ 3l & 0 & 0 & 4l^2 & -3l & 0 & 0 & -l^2 \\ -36 & 0 & 0 & 3l & 36 & 0 & 0 & -3l \\ 0 & -36 & 3l & 0 & 0 & 36 & 3l & 0 \\ 0 & 3l & -l^2 & 0 & 0 & 3l & 4l^2 & 0 \\ 3l & 0 & 0 & -l^2 & -3l & 0 & 0 & 4l^2 \end{bmatrix}$$

(A.9)

$$M^e_{R,2} = \frac{\rho_{AX} r^2}{120 l (1+\Theta)^2} \cdot \begin{bmatrix} 0 & 0 & 0 & -15l & 0 & 0 & 0 & -15l \\ 0 & 0 & 15l & 0 & 0 & 0 & 15l & 0 \\ 0 & 15l & 5l^2 & 0 & 0 & -15l & -5l^2 & 0 \\ -15l & 0 & 0 & 5l^2 & 15l & 0 & 0 & -5l^2 \\ 0 & 0 & 0 & 15l & 0 & 0 & 0 & 15l \\ 0 & 0 & -15l & 0 & 0 & 0 & -15l & 0 \\ 0 & 15l & -5l^2 & 0 & 0 & -15l & 5l^2 & 0 \\ -15l & 0 & 0 & -5l^2 & 15l & 0 & 0 & 5l^2 \end{bmatrix}$$

(A.10)

$$M^e_{R,3} = \frac{\rho_{AX} r^2}{120 l (1+\Theta)^2} \begin{bmatrix} 0 & 0 & 0 & 0 & 0 & 0 & 0 & 0 \\ 0 & 0 & 0 & 0 & 0 & 0 & 0 & 0 \\ 0 & 0 & 10l^2 & 0 & 0 & 0 & 5l^2 & 0 \\ 0 & 0 & 0 & 10l^2 & 0 & 0 & 0 & 5l^2 \\ 0 & 0 & 0 & 0 & 0 & 0 & 0 & 0 \\ 0 & 0 & 0 & 0 & 0 & 0 & 0 & 0 \\ 0 & 0 & 5l^2 & 0 & 0 & 0 & 10l^2 & 0 \\ 0 & 0 & 0 & 5l^2 & 0 & 0 & 0 & 10l^2 \end{bmatrix}$$

(A.11)

$$S = \sum_{i=1}^{p} A_i^T S_i^e A_i$$

(A.12)

$$S^e = S_1^e + S_2^e$$

(A.13)

$$S_1^e = \frac{EI}{l^3 (1+\Theta)} \begin{bmatrix} 12 & 0 & 0 & 6l & -12 & 0 & 0 & 6l \\ 0 & 12 & -6l & 0 & 0 & -12 & -6l & 0 \\ 0 & -6l & 4l^2 & 0 & 0 & 6l & 2l^2 & 0 \\ 6l & 0 & 0 & 4l^2 & -6l & 0 & 0 & 2l^2 \\ -12 & 0 & 0 & -6l & 12 & 0 & 0 & -6l \\ 0 & -12 & 6l & 0 & 0 & 12 & 6l & 0 \\ 0 & -6l & 2l^2 & 0 & 0 & 6l & 4l^2 & 0 \\ 6l & 0 & 0 & 2l^2 & -6l & 0 & 0 & 4l^2 \end{bmatrix}$$

(A.14)

$$S_2^e = \frac{EI\Theta}{l^3 (1+\Theta)} \begin{bmatrix} 0 & 0 & 0 & 0 & 0 & 0 & 0 & 0 \\ 0 & 0 & 0 & 0 & 0 & 0 & 0 & 0 \\ 0 & 0 & l^2 & 0 & 0 & 0 & -l^2 & 0 \\ 0 & 0 & 0 & l^2 & 0 & 0 & 0 & -l^2 \\ 0 & 0 & 0 & 0 & 0 & 0 & 0 & 0 \\ 0 & 0 & 0 & 0 & 0 & 0 & 0 & 0 \\ 0 & 0 & -l^2 & 0 & 0 & 0 & l^2 & 0 \\ 0 & 0 & 0 & -l^2 & 0 & 0 & 0 & l^2 \end{bmatrix}$$

(A.15)

$$\Theta = \frac{12EI}{\kappa AGl^2}$$

(A.16)

where $\rho_{AX}$ is the axial mass density, $l$ is the length of the element, $r$ is the radius of the element, $E$ is Young's modulus, $I$ is the moment of inertia, $\kappa$ is the Timoshenko's shear correction factor, $A$ is the cross-section area, and $G$ is the shear modulus of elasticity.

## REFERENCES

1.  Bossmann, B. and Tu, J. (2002) Conceptual Design of Machine Tool Interfaces for High-Speed Machining. Journal of Manufacturing Processes, 4, 16-27.

2.  Lin, C.-W. and Tu, J. (2007) Model-Based Design of Motorized Spindle Systems to Improve Dynamic Performance at High Speeds. Journal of Manufacturing Processes, 9, 94-108.

3.  Al-Shareef, K. and Brandon, J. (1990) On the Effects of Variations in the Design Parameters on the Dynamic Performance of Machine Tool Spindle-Bearing Systems. International Journal of Machine Tools and Manufacture, 30, 431-445.

4.  Kang, Y., Chang, Y.-P., Tsai, J.-W., Chen, S.-C. and Yang, L.-K. (2001) Integrated CAE Strategies for the Design of Machine Tool Spindle-Bearing Systems. Finite Elements in Analysis and Design, 37, 485-511.

5.  Li, H. and Shin, Y. (2004) Analysis of Bearing Configuration Effects on High Speed Spindles Using an Integrated Dynamic Thermomechanical Spindle Model. International Journal of Machine Tools and Manufacture, 44, 347-364.

6.  Maeda, O., Cao, Y. and Altintas, Y. (2005) Expert Spindle Design System. International Journal of Machine Tools and Manufacture, 45, 537-548.

7.  Altintas, Y. and Cao, Y. (2005) Virtual Design and Optimization of Machine Tool Spindles. CIRP Annual, 54, 379-382.

8.  Srinivasan, S., Maslen, E. and Barrett, L. (1997) Optimization of Bearing Locations for Rotor Systems with Magnetic Bearings. Journal of Engineering for Gas Turbines and Power, 119, 464-468.

9.  Nelson, H. and McVaugh, J. (1976) The Dynamics of Rotor-Bearing System Using Finite Elements. Journal of Engineering for Industry, Transactions of the ASME, 93, 593-600.

10. Zorzi, E. and Nelson, H. (1977) Finite Element Simulation of Rotor-Bearing Systems with Internal Damping. Journal of Engineering for Power, Transactions of the ASME, 7, 71-76.

11. Nelson, H. (1980) A Finite Rotating Shaft Element Using Timoshenko Beam Theory. Journal of Mechanical Design, Transactions of the ASME, 102, 793-803.

12. Lantto, E. (1997) Finite Element Model for Elastic Rotating Shaft. ACTA Polytechnica Scandinavica, Electrical Engineering Series, 88, 1-73.

13. Inman, D. (2008) Engineering Vibrations. Pearson Education, Inc., Upper Saddle River.

14. Holland, J.H. (1975) Adaptation in Natural and Artificial Systems. University of Michigan Press, Ann Arbor.

15. Rao, S.S. (2009) Engineering Optimization: Theory and Practice. John Wiley and Sons, Hoboken.

16. Huang, M., Chen, K. and Fung, R. (2010) Comparison between Mathematical Modeling and Experimental Identification of a Spatial Slider-Crank Mechanism. Applied Mathematical Modelling, 34, 2059-2073.

17. Fiandaca, G., Fraga, E. and Brandani, S. (2009) A Multi-Objective Genetic Algorithm for the Design of Pressure Swing Adsorption. Engineering Optimization, 41, 833-854.

18. Gomes, H. and Silva, N. (2008) Some Comparisons for Damage Detection on Structures Using Genetic Algorithms and Modal Sensitivity Method. Applied Mathematical Modelling, 32, 2216-2232.

19. Chakraborty, I., Kumar, V., Nair, S. and Tiwari, R. (2003) Rolling Element Bearing Design through Genetic Algorithms. Engineering Optimization, 35, 649-659.

20. Lin, C.-W. (2001) High Speed Effects and Dynamic Analysis of Motorized Spindles for High Speed End Milling. Ph.D. Thesis, Purdue University, West Lafayette.

21. Wardle, F., Lacey, S. and Poon, S. (1983) Dynamic and Static Characteristics of a Wide Speed Range Machine Tool Spindle. Precision Engineering, 5, 175-183.

22. Zill, D., Wright, W. and Cullen, M. (2011) Advanced Engineering Mathematics. Jones & Bartlett Learning, Burlington.

23. Pham, D. and Karaboga, D. (2000) Intelligent Optimisation Techniques, Genetic Algorithms, Tabu Search, Simulated Annealing and Neural Networks. Springer, New York.

24. Chong, E. and Zak, S. (2008) An Introduction to Optimization. Wiley-Interscience, Hoboken.

25. Zalzala, A. and Fleming, P. (1997) Genetic Algorithms in Engineering Systems. IET, London.

26. Belegundu, A. and Chandrupatla, T. (1999) Optimization Concepts and Applications in Engineering. Prentice Hall, Upper Saddle River.

27. Montgomery, D. and Runger, G. (2011) Applied Statistics and Probability for Engineers. Wiley, Hoboken.

# Chapter 9

# AN EFFICIENT METHOD TO SOLVE THERMAL WAVE EQUATION

Mohamed Salah, R. M. Amer, M. S. Matbuly

Department of Engineering Mathematics and Physics, Faculty of Engineering, Zagazig University, Zagazig, Egypt

## ABSTRACT

In this paper, an efficient technique of differential quadrature method and perturbation method is employed to analyze reaction-diffusion problems. An efficient method is presented to solve thermal wave propagation model in one and two dimensions. The proposed method marches in the time direction block by block and there are several time levels in each block. The global method of differential quadrature is applied in each block to discretize both the spatial and temporal derivatives. Furthermore, the proposed method is validated by comparing the obtained results with the available analytical ones and also compared with the hybrid technique of differential quadrature method and Runge-Kutta fourth order method.

## INTRODUCTION

Thermal wave is the reaction-diffusion equation which plays an ever-increasing role in the study of material parameters. It has been employed in optical investigations of solids, liquids, and gases with photo-acoustic and thermal lens spectroscopy. Thermal waves have also been used to analyze the thermal and thermodynamic properties of materials and for imaging thermal and material features within a solid sample [1].

In the past several decades, there has been greeting activity in developing numerical and analytical methods for the thermal wave equation. Due to the nonlinearity and complexity of such problems, only limited cases can analytically be solved [2-5]. Yan applied the projective Riccati equation method to solve Schrodinger equation in nonlinear optical fibers [2]. Then Mei, Zhang and Jiang employed the same method to get the exact solutions for some reaction-diffusion problems [3]. Abdusalam applied a factorization technique to find exact traveling wave solutions [4]. Chowdhury and Hashim obtain analytical solution for Cauchy reaction-diffusion problems using homotopy perturbation method [5]. Literature on the numerical solution of reaction-diffusion equations is sparse; singular perturbation method has been applied to solve reaction-diffusion equations by Puri et al. in [6]. David, Curtis and John introduced time integration methods to solve thermal wave propagation [7]. Marcus applied Finite difference method to study the dynamics of predator-prey interactions [8], and Chen et al. employed the finite element method to solve advective reaction-diffusion equations [9]. Then Christos et al. applied also the same method to solve the problem with boundary layers [10]. Meral and Sezgin used this method and finite difference method with a relaxation parameter to solve nonlinear reaction-diffusion equation in one and two dimensions [11]. Recently differential quadrature method has been efficiently employed in a variety of engineering problems [12]. Wu and Liu have introduced the generalization of the differential quadrature method to solve linear and non linear differential equations [13]. Moreover, Meral applied differential quadrature method and implicit Euler method to solve density dependent nonlinear reaction-diffusion equation [14]. Kajal applied differential quadrature and Runge-Kutta method to solve thermal wave, a blow-up and a Brusselator chemical dynamics system [15]. Kajal achieved high accuracy, but, there are some difficulties in the previous method that numerical solution is obtained layer by layer in the time direction so it can be expected that the accuracy of numerical solution is decreasing with time due to accumulation of numerical errors, also explicit scheme is used to update the solution using very small step size due to the limitation of stability condition. Also, Salah, Rania and Matbuly solved thermal wave equation in one and two dimensions using Implicit Euler and perturbation method. They compared the numerical solution with the results from Runge-Kutta method. They overcome the limitation of stability and one can use large step size [16]. Shu et al. early presented block-marching technique with DQ discretization to obtain the solution in the time direction block by block to overcome the above difficulty. Moreover, they used successive over-relaxation (SOR) iterative method to complete the solution achieving high accuracy and efficiency [17].

In this research block marching in time and DQ discretization in both the spatial and temporal derivatives are applied to obtain the solution in time direction block by block for thermal wave equation. In each block there are several time levels (layers). So the accumulation error is block by block instead of layer by layer. Perturbation method is used to complete the solution. The obtained results are compared with the previous analytical ones and also compared with the hybrid technique of differential quadrature method and Runge-Kutta fourth order method [15].

## NUMERICAL PROCEDURE OF THERMAL WAVE

The main strategy is to apply perturbation method of second order [18,19] then applying block-marching technique with DQ discretization to reduce the problem to a system of linear algebraic equations.

Propagation of thermal waves through a rectangular plate, is governed by [15]:

$$\frac{\partial U}{\partial t} = \alpha \frac{\partial^2 U}{\partial x^2} + \beta \frac{\partial^2 U}{\partial y^2} + \gamma (U_{max} - U)U^2, t > 0, (0,0) < (x,y) < (a,b)$$

(1)

where:

U is a temperature,

$\alpha$ and $\beta$ are diffusion parameters in direction of x and y, respectively.

$\gamma$ is reaction parameter.

a and b are plate dimensions in direction of x and y, respectively.

$U_{max}$ is a maximum temperature of the system.

Along the external boundaries, the temperatures can be described as:

$$a_1 U(o,y,t) + b_1 \frac{\partial U}{\partial x}\bigg|_{(0,y,t)} = f_1(y,t)$$

(2a)

$$a_2 U(a,y,t) + b_2 \frac{\partial U}{\partial x}\bigg|_{(a,y,t)} = f_2(y,t)$$

(2b)

$$a_3 U(x,0,t) + b_3 \frac{\partial U}{\partial y}\bigg|_{(x,0,t)} = f_3(x,t)$$

(2c)

$$a_4 U(x,b,t) + b_4 \frac{\partial U}{\partial y}\bigg|_{(x,b,t)} = f_4(x,t)$$

(2d)

where $a_i, b_i$ and $f_i, (i = 1,4)$, are known functions.

Then initial temperature may be described as:

$$U(x,y,0) = g(x,y) \qquad (3)$$

where $g(x,y)$ is a known function.

Solution of Equations (1)-(3) can be obtained as follows:

1) In the K$^{th}$ block, the non-uniform distribution is used Chebyshev-Gauss-Lobatto (CGL) in x, y and t directions respectively, such as [12]:

$$x_i = \frac{a}{2}\left[1-\cos\frac{(i-1)\pi}{N-1}\right], i = 1,2,\cdots,N \qquad (4a)$$

$$y_j = \frac{b}{2}\left[1-\cos\frac{(j-1)\pi}{M-1}\right], j = 1,2,\cdots,M \qquad (4b)$$

$$t_k = (K-1)\delta t + \frac{\delta t}{2}\left[1-\cos\frac{(k-1)\pi}{L-1}\right], k = 1,2,\cdots,L \qquad (4c)$$

where N, M is the number of grid points in the x and y direction respectively, L is the number of time levels in the block and $\delta t$ is the length of the block in time direction.

2) We can solve

$$\frac{\partial U}{\partial t} = \alpha\frac{\partial^2 U}{\partial x^2} + \beta\frac{\partial^2 U}{\partial y^2} + \varepsilon\gamma\left(U_{max} - U\right)U^2 \qquad (5)$$

subjected to the prescribed to boundary and initial conditions in Equations (2) and (3), assuming

$$U = U_o + \varepsilon U_1 + \varepsilon^2 U_2 + \cdots + \varepsilon^n U_n \qquad (6)$$

where $U_o, U_1, U_2$ are unknowns functions and $\varepsilon$ is a perturbation parameter.

The following condition is tested to ensure the convergence condition [20] in previous series in Equation (6).

$$\left|\frac{U_{i+1}}{U_i}\right| < 1 \text{ where } i = 0,1,\cdots,n-1 \qquad (7)$$

3) On sustainable substitution from Equation (6) into (5), one can reduce the problem to the following equation.

$$\frac{\partial}{\partial t}\left(U_o + \varepsilon U_1 + \varepsilon^2 U_2\right) = \alpha\frac{\partial^2}{\partial x^2}\left(U_o + \varepsilon U_1 + \varepsilon^2 U_2\right) + \beta\frac{\partial^2}{\partial y^2}\left(U_o + \varepsilon U_1 + \varepsilon^2 U_2\right)$$
$$+ \varepsilon\gamma\left(U_{max} - \left(U_o + \varepsilon U_1 + \varepsilon^2 U_2\right)\right)\left(U_o + \varepsilon U_1 + \varepsilon^2 U_2\right)^2 \qquad (8)$$

4)      Applying zero order perturbation method such that,

$$\frac{\partial U_0}{\partial t} = \alpha \frac{\partial^2 U_0}{\partial x^2} + \beta \frac{\partial^2 U_0}{\partial y^2}$$

(9)

Subjected to boundary and initial conditions in Equations (2), (3), where block-marcing technique and differential quadrature method are used to reduce Equation (9) to a system of linear algebraic equations such that [12],

$$\left. \frac{\partial U}{\partial x} \right|_{(x_i, y_j, t_l)} = \sum_{k=1}^{N} A_{ik}^x U(x_k, y_j, t_l)$$

(10a)

$$\left. \frac{\partial U}{\partial y} \right|_{(x_i, y_j, t_l)} = \sum_{k=1}^{M} A_{jk}^y U(x_k, y_j, t_l)$$

(10b)

$$\left. \frac{\partial U}{\partial t} \right|_{(x_i, y_j, t_l)} = \sum_{k=1}^{L} A_{lk}^t U(x_i, y_j, t_k)$$

(10c)

$$\left. \frac{\partial^2 U}{\partial x^2} \right|_{(x_i, y_j, t_l)} = \sum_{k=1}^{N} B_{ik}^x U(x_k, y_j, t_l)$$

(10d)

$$\left. \frac{\partial^2 U}{\partial y^2} \right|_{(x_i, y_j, t_l)} = \sum_{k=1}^{M} B_{jk}^y U(x_k, y_j, t_l)$$

(10e)

$$\left. \frac{\partial^2 U}{\partial t^2} \right|_{(x_i, y_j, t_l)} = \sum_{k=1}^{L} B_{lk}^t U(x_i, y_j, t_k)$$

(10f)

where A and B are the first and second order weighting coefficients respectively.

By substitution of Equation (10) into Equation (9) result that,

$$\sum_{k=1}^{L} A_{lk}^t U_o(x_i, y_j, t_k) = \alpha \sum_{k=1}^{N} B_{ik}^x U_o(x_k, y_j, t_l) + \beta \sum_{k=1}^{M} B_{jk}^y U_o(x_i, y_k, t_l)$$

(11)

5)      First order perturbation method is applied such that,

$$\frac{\partial U_1}{\partial t} = \alpha \frac{\partial^2 U_1}{\partial x^2} + \beta \frac{\partial^2 U_1}{\partial y^2} + \gamma(U_{max} - U_o)U_o^2$$

(12)

Subjected to the same boundary and initial conditions n Equations (2), (3), reduced to the following algebraic system

$$\sum_{k=1}^{L} A_{lk}^t U_1(x_i, y_j, t_k) = \alpha \sum_{k=1}^{N} B_{ik}^x U_1(x_k, y_j, t_l) + \beta \sum_{k=1}^{M} B_{jk}^y U_1(x_i, y_k, t_l) + \gamma(U_{max} - U_{oi,j,l})U_{oi,j,l}^2$$

(13)

6)      Also second order perturbation method is applied such that,

$$\frac{\partial U_2}{\partial t} = \alpha \frac{\partial^2 U_2}{\partial x^2} + \beta \frac{\partial^2 U_2}{\partial y^2} + \gamma(2U_{max}U_0 U_1 - 3U_0^2 U_1)$$

(14)

Subjected to the same boundary and initial conditions in Equations (2), (3), reduced to the following algebraic system

$$\sum_{k=1}^{L} A_{jk}^t U_2(x_i,y_j,t_k) = \alpha \sum_{k=1}^{N} B_{jk}^x U_2(x_k,y_j,t_i) + \beta \sum_{k=1}^{M} B_{jk}^y U_2(x_i,y_k,t_i) + \gamma \left(2U_{max} U_{0i,j,J} U_{1i,J,J} - 3U_{0i,j,J}^2 U_{1i,J,J}\right)$$ (15)

Finally, the series solution can be written as

$$U_{numerical} = \lim_{\varepsilon \to 1} U = U_o + U_1 + U_2$$ (16)

After obtaining the solution in the K$^{th}$ block, we march to (K + 1)$^{th}$ block where the solution at the last time level at K$^{th}$ block is considered an initial condition for (K + 1)$^{th}$ block. We carry on this process until the specified time is reached.

## RESULTS AND DISCUSSIONS

To ensure the accuracy of the proposed block marching technique, the thermal wave propagating model is solved using presented method and compared with the available analytical solution [15,21] and also compared with the hybrid technique of differential quadrature method and Runge-Kutta fourth order method [15].

### Results for One Dimension Thermal Wave

Consider a one-dimensional problem of thermal wave propagation along x-direction as [16]:

$$\alpha = a_1 = a_2 = U_{max} = 1, \gamma = 1, \beta = a_3 = a_4 = b_1 = b_2 = b_3 = b_4 = f_3 = f_4 = 0 .$$

While

$$f_1(t) = \frac{1}{1+\exp(-t/2)}, f_2(t) = \frac{1}{1+\exp\left((1-t/\sqrt{2})/\sqrt{2}\right)}$$

$$g(x) = \frac{1}{1+\exp(x/\sqrt{2})}, 0 \le x \le 1$$ (17)

The exact solution for such problem can be obtained as [21]:

$$U_{exact}(x,t) = \frac{1}{1+\exp\left((x-t/\sqrt{2})/\sqrt{2}\right)}, t > 0, 0 \le x \le 1$$ (18)

To validate the accuracy of numerical results, the following errors [16] are computed,

$$\text{Root mean square of errors} = R.M.S. \ of \ errors = \sqrt{\left[\sum_{i=1}^{NP}(U_{numerical} - U_{exact})^2\right]/NP}$$ (19a)

$$\text{Root sum of square errors} = R.S.S. \ of \ errors = \sqrt{\left[\sum_{i=1}^{NP}\left(U_{numerical} - U_{exact}\right)^2\right]}$$ (19b)

**Absolute error** $= \left|U_{numerical} - U_{exact}\right|$ (19c)

where P is the number of time steps in the case of using Runge-Kutta method and is considered the number of blocks multiplied by number of time levels in case of block marching technique.

For the numerical computation, the time domain is limited to $0 \le t \le 20$ and N = 7 while $\delta t$ and L are not fixed. The efficiency of block marching technique is tested by CPU time required when the computation reaches to t = 20 s. Also convergence condition in Equation (7) is tested achieving higher accuracy at second order perturbation method as shown in Figure 1.

Table 1 shows R. M. S errors $7 \times 10^{-6}$, R. S. S errors $2 \times 10^{-4}$ and CPU time required is 0.132531 s.

The accuracy are very high at using $\delta t = 0.5$ and L = 4 so the number of blocks used in time interval $0 \le t \le 20$ are 40 blocks.

From Table 2, we increase $\delta t$ to 1.00 and time levels L to 10, the errors are approximately the same but CPU time is reduced slightly to 0.118479 s and we are used 20 blocks only.

From Table 3, we increase $\delta t$ to 2.00 and time levels L to 15, the accuracy of numerical results can be kept the same as the previous cases of $\delta t = 0.5$ and 1.00 but CPU time is increased to 0.129549 s as L is increased so the number of unknowns in each block for this case is much larger than previous cases shown in Tables 1 and 2.

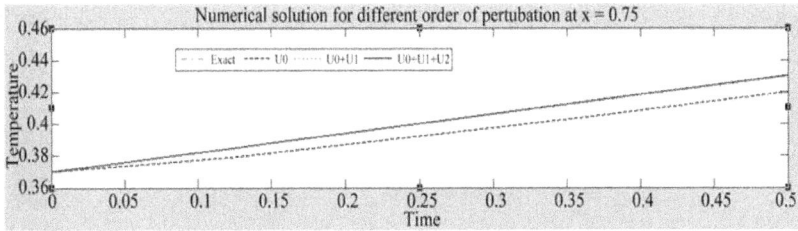

**Figure 1**: Satisfying convergence conditions.

**Table 1**: R.M.S, R.S.S errors, time steps and CPU times by present method for δt = 0.5 and L = 4.

| Time (s) | No. of blocks | R.S.S error | R.M.S error | CPU time (s) |
|----------|---------------|-------------|-------------|--------------|
| 0.5 | 1 | 9.0934E−006 | 1.7185E−006 | 0.081430 |
| 1.0 | 2 | 1.3076E−005 | 1.7473E−006 | 0.085238 |
| 1.5 | 3 | 3.0465E−005 | 3.3240E−006 | 0.085400 |
| 2.00 | 4 | 6.0459E−005 | 5.7128E−006 | 0.085253 |
| 3.00 | 6 | 1.1859E−004 | 9.1493E−006 | 0.085531 |
| 5.00 | 10 | 1.4268E−004 | 8.5270E−006 | 0.097101 |
| 10.00 | 20 | 2.2942E−004 | 9.6950E−006 | 0.107284 |
| 20.00 | 40 | 2.3292E−004 | 6.9599E−006 | 0.132531 |

**Table 2**: R.M.S, R.S.S errors, time steps and CPU times by present method for δt = 1.00 and L = 10.

| Time (s) | No. of blocks | R.S.S error | R.M.S error | CPU time (s) |
|----------|---------------|-------------|-------------|--------------|
| 1.00 | 1 | 1.1547E−005 | 1.3801E−006 | 0.091552 |
| 2.00 | 2 | 7.6912E−005 | 6.5003E−006 | 0.083349 |
| 3.00 | 3 | 1.4346E−004 | 9.8995E−006 | 0.096041 |
| 4.00 | 4 | 1.6410E−004 | 9.8067E−006 | 0.105310 |
| 5.00 | 5 | 1.7305E−004 | 9.2499E−006 | 0.096794 |
| 6.00 | 6 | 2.0736E−004 | 1.0118E−005 | 0.105096 |
| 7.00 | 7 | 2.4614E−004 | 1.1120E−005 | 0.088205 |
| 8.00 | 8 | 2.7221E−004 | 1.1503E−005 | 0.108769 |
| 9.00 | 9 | 2.8597E−004 | 1.1393E−005 | 0.101100 |
| 10.00 | 10 | 2.9231E−004 | 1.1048E−005 | 0.113502 |
| 20.00 | 20 | 2.9676E−004 | 7.9312E−006 | 0.118479 |

**Table 3**: R.M.S, R.S.S errors, time steps and CPU times by present method for δt = 2.00 and L = 15.

| Time (s) | No. of blocks | R.S.S error | R.M.S error | CPU time (s) |
|----------|---------------|-------------|-------------|--------------|
| 2.00 | 1 | 7.82E−05 | 7.63E−06 | 0.088891 |
| 4.00 | 2 | 1.44E−04 | 9.91E−06 | 0.163151 |
| 6.00 | 3 | 1.90E−04 | 1.07E−05 | 0.109113 |
| 8.00 | 4 | 2.48E−04 | 1.21E−05 | 0.125668 |
| 10.00 | 5 | 2.66E−04 | 1.16E−05 | 0.114223 |
| 20.00 | 10 | 2.70E−04 | 8.33E−06 | 0.129549 |

**Table 4**: R.M.S, R.S.S errors, time steps and CPU times by present method for $\delta t = 2.00$ and $L = 10$.

| Time (s) | No. of blocks | R.S.S error | R.M.S error | CPU time (s) |
|---|---|---|---|---|
| 2 | 1 | 6.47E−05 | 7.73E−06 | 0.086238 |
| 4 | 2 | 1.17E−04 | 9.91E−06 | 0.08094 |
| 6 | 3 | 1.56E−04 | 1.07E−05 | 0.095065 |
| 8 | 4 | 2.03E−04 | 1.21E−05 | 0.093627 |
| 10 | 5 | 2.18E−04 | 1.16E−05 | 0.095982 |
| 20 | 10 | 2.21E−04 | 8.36E−06 | 0.10836 |

Furthermore, it can be shown from Table 4, when number of time levels fixed to 10 the accuracy kept the same as the previous cases but the efficiency is more improved as the CPU time reduced to 0.108360 s.

As well as, Figure 2 shows that at x = 0.25 and 0.75 the absolute error $\leq 2 \times 10^{-5}$ at different $\delta t$ and L.

Figure 3 shows that the effects of $\delta t$ and L on the absolute error where $\delta t$ is increasing the accuracy will increase and with proper choice of L will enhance also the efficiency. If δt is too large, the accuracy of numerical results can be greatly reduced if L is not large enough. In all previous cases the convergence condition in Equation (15) is satisfied.

To show the superiority of the block marching method, the RK4 method which is layer marching approach is also applied to solve the same problem. Two time step sizes of 0.001 and 0.005 are used for layer marching in the time direction. The numerical results of these two cases are listed respectively in Tables 5 and 6. From these two tables, it can be seen that RK4 method can achieve high accuracy at very small step size Dt = 0.001 with $R.M.S.\ of\ errors < 2 \times 10^{-7}, R.S.S.\ of\ errors < 5.3 \times 10^{-10}$ and at Dt = 0.005 with $R.M.S.\ of\ errors < 2.7 \times 10^{-7}, R.S.S.\ of\ errors < 1.6 \times 10^{-9}$. Moreover, as shown in Figure 4, as Dt increased slightly to 0.00515 the stability condition will not achieved and the oscillation will occurs in the period $6 \leq t \leq 25$. On the other hand the efficiency is very small as the CPU time required to reach t = 20 s is much larger.

Figure 5 shows that the temperature distribution at different times and locations for numerical solution using block marching technique compared with exact solution and RK4 method.

## Results for Two Dimensions Thermal Wave

Consider also a simple two dimensional problem with $\alpha = \beta = a_1 = a_2 = a_3 = a_4 = 1,$
$\gamma = b_1 = b_2 = b_3 = b_4 = f_1 = f_2 = f_3 = f_4 = 0$
and $g(x,y) = \sin \pi x \sin \pi y, 0 \le x, y \le 1$     Which can be solved exactly as [22,23]:
$U(x,y,t) = e^{-2\pi^2 t} \sin \pi x \sin \pi y$.

The design of the numerical scheme is extended to two dimensions. Table 7 shows that for $\Delta t = 0.01, N = 5, \ M = 4, L = 4, \delta t = 0.01$ the obtained results agree with the analytical ones [21] with $R.M.S.\ of\ errors < 10^{-9}$ and the accuracy of block-marching is greater than the hybrid method.

Figure 6 shows that absolute error for presented method $< 0.005$ while in hybrid method absolute error $< 0.01$, where also ensuring that accuracy of block-marching technique is better than hybrid method.

Figure 7 shows parametric study for the effect on block length and number of blocks, at t = 0.3 s we use one block each has length dt = 0.3, R.M.S error = $9.9012 \times 10^{-6}$ and absolute error $< 3 \times 10^{-4}$ but if we decrease dt = 0.15 using two blocks, R.M.S error = $1.3382 \times 10^{-4}$ and absolute error $< 3 \times 10^{-4}$, number of time levels doesn't effect on the accuracy of solution.

## CONCLUSIONS

The block marching technique with DQ discretization is employed to solve thermal wave propagations in one and two dimensions. The numerical results are obtained block by block in the time direction. Each block has several time levels. The length of the block in time direction $\delta t$ and the number of time levels L can be adjusted to keep high accuracy and efficiency. The validity of the proposed technique is proved by comparing the obtained results with the previous analytical ones. The proposed technique needs a small number of grid points and a little com

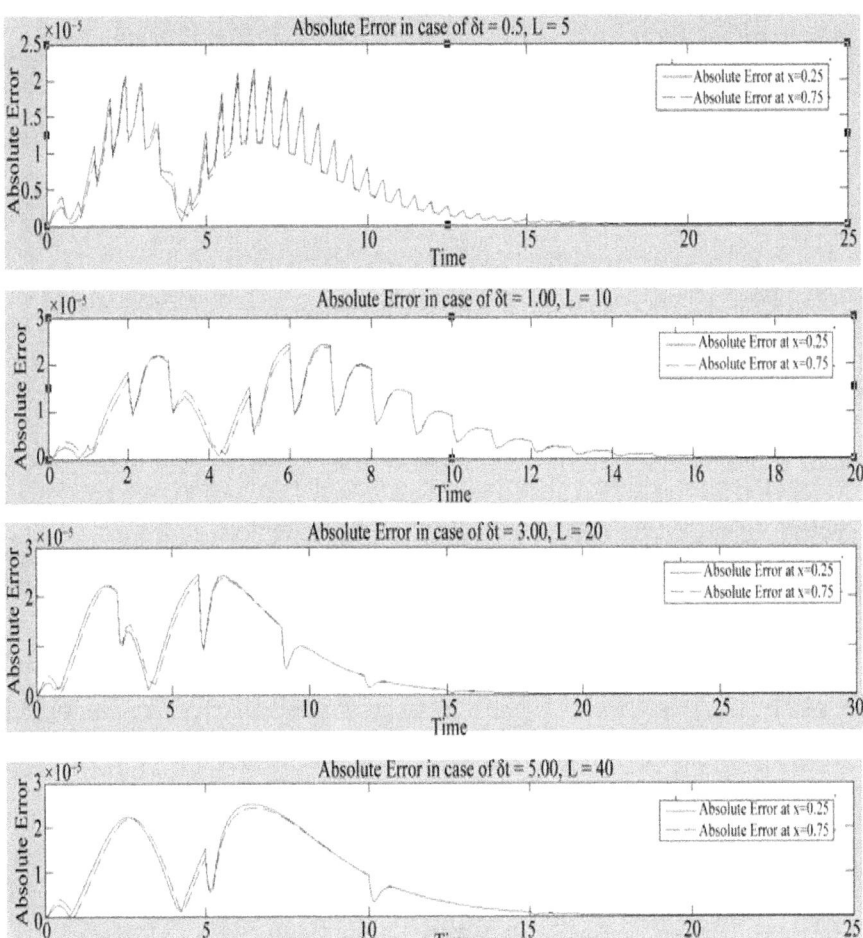

**Figure 2**: Absolute error at different times and locations.

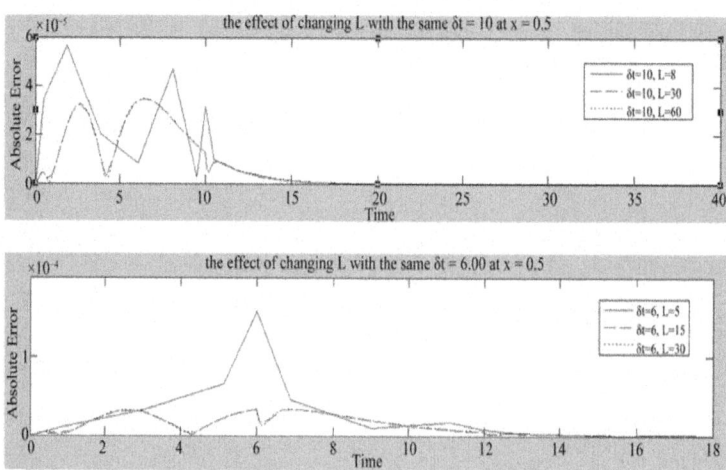

**Figure 3**. Behavior of absolute error with changing δt and L.

**Table 5**. R.M.S, R.S.S errors, time steps and CPU times given by 4-stage Runge-Kutta at Δt = 0.001.

| Time (s) | Time steps | R.S.S error | R.M.S error | CPU time (s) |
|----------|-----------|-------------|-------------|--------------|
| 0.5 | 501 | 9.0852E−008 | 1.5342E−009 | 0.172995 |
| 1.0 | 1001 | 1.4210E−007 | 1.6976E−009 | 0.342234 |
| 1.5 | 1501 | 1.6532E−007 | 1.6128E−009 | 0.582446 |
| 2.0 | 2001 | 1.6865E−007 | 1.4250E−009 | 0.834394 |
| 2.5 | 2501 | 1.7133E−007 | 1.2949E−009 | 1.449289 |
| 5.0 | 5001 | 1.9784E−007 | 1.0574E−009 | 3.656532 |
| 10.0 | 10001 | 1.9874E−007 | 7.5111E−010 | 18.518357 |
| 20.0 | 20001 | 1.9874E−007 | 5.3113E−010 | 79.050595 |

**Table 6**. R.M.S, R.S.S errors, time steps and CPU times given by 4-stage Runge-Kutta at Δt = 0.005

| Time (s) | Time steps | R.S.S error | R.M.S error | CPU time (s) |
|----------|-----------|-------------|-------------|--------------|
| 0.5 | 101 | 1.2586E−007 | 4.7333E−009 | 0.142803 |
| 1.0 | 201 | 1.9692E−007 | 5.2499E−009 | 0.119084 |
| 1.5 | 301 | 2.2902E−007 | 4.9894E−009 | 0.148232 |
| 2.0 | 401 | 2.3356E−007 | 4.4084E−009 | 0.167572 |
| 2.5 | 501 | 2.3732E−007 | 4.0074E−009 | 0.165521 |
| 5.0 | 1001 | 2.7397E−007 | 3.2729E−009 | 0.332336 |
| 10.0 | 2001 | 2.7521E−007 | 2.3254E−009 | 0.788888 |
| 20.0 | 4001 | 2.7521E−007 | 1.6445E−009 | 2.588399 |

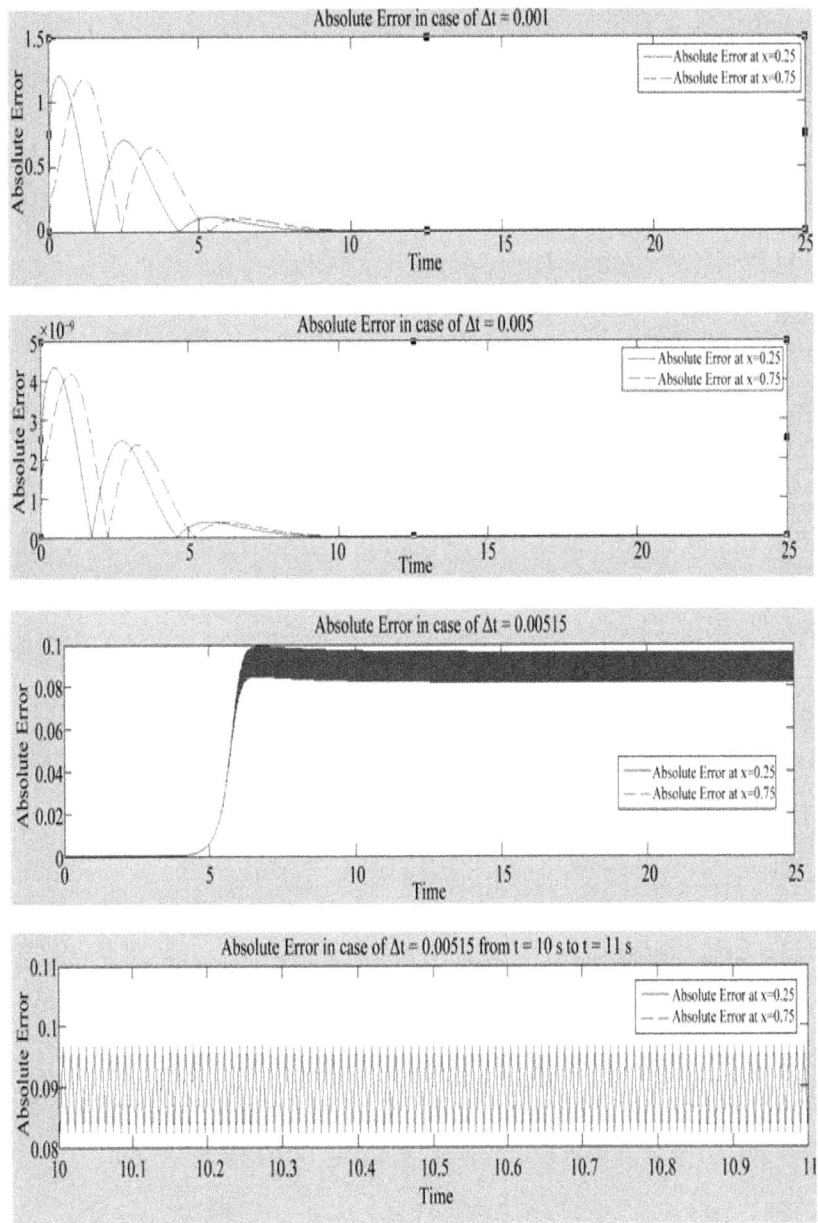

**Figure 4**: Absolute errors at different times and locations.

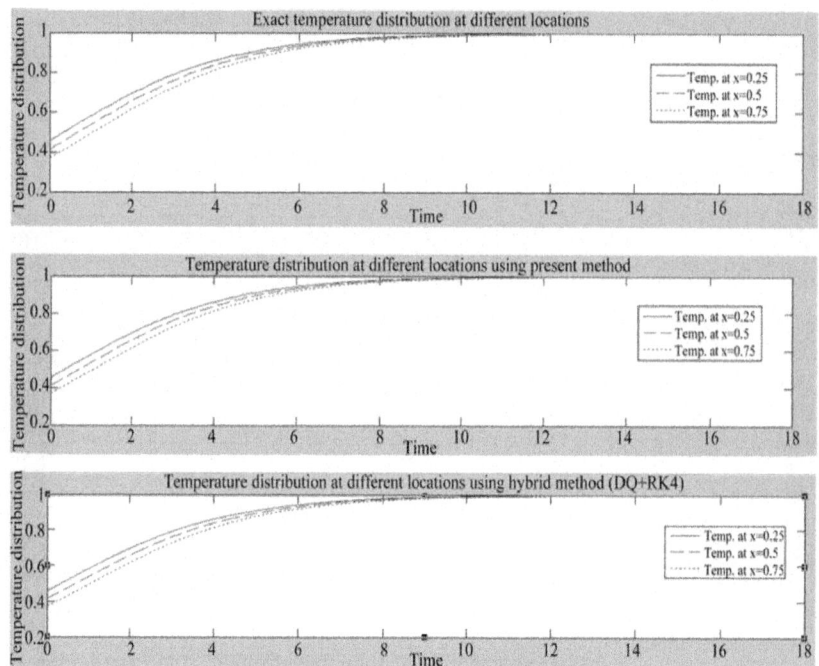

**Figure 5**: Temperature distributions at different times and locations.

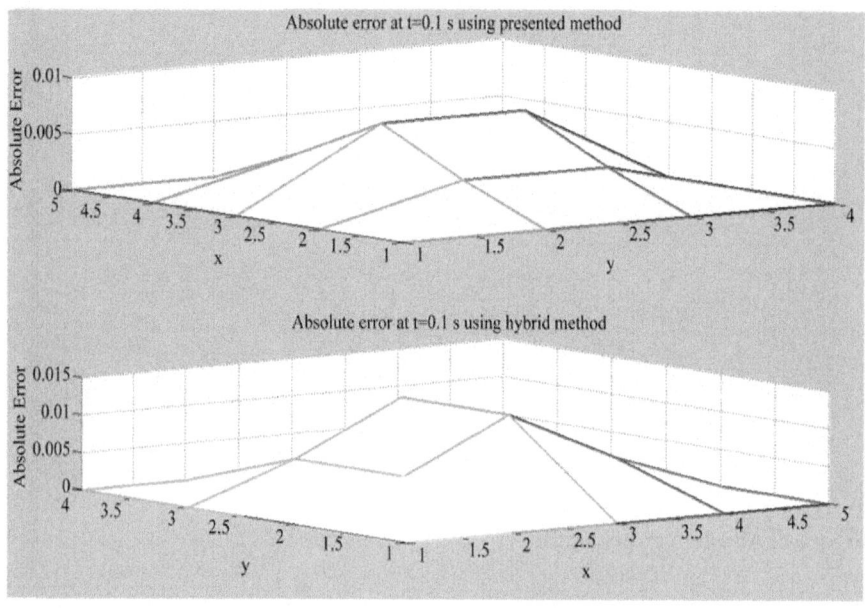

**Figure 6**: Absolute error at t = 0.1 s and different locations of x, y.

**Table 7**: Root mean square of errors for two dimensional thermal wave propagation at $\Delta t = 0.01$, N = 5, M = 4, L = 4, $\delta t = 0.01$

| Time | 0.01 | 0.04 | 0.08 | 0.16 | 0.4 | 1.00 |
|---|---|---|---|---|---|---|
| Root mean square of errors by hybrid method | 0.0027 | 0.0060 | 0.0054 | 0.0021 | 3.9994E−005 | 5.1513E−010 |
| Root mean square of errors by present method | 0.0016 | 0.0036 | 0.0032 | 0.0013 | 2.5528E−005 | 3.7316E−010 |

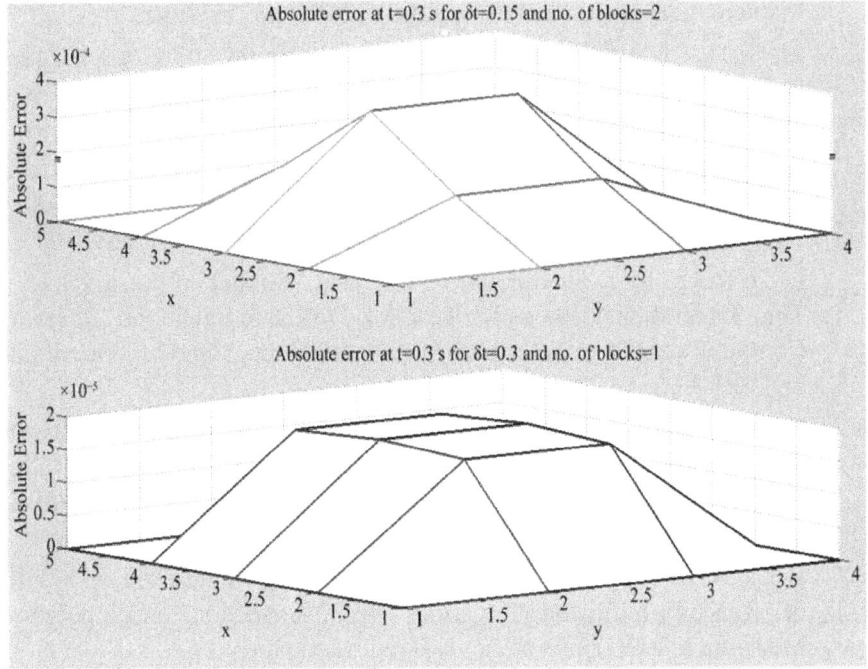

**Figure 7**: The effect of $\delta t$ on absolute error.

putational effort to obtain accurate results with absolute error $\leq 2 \times 10^{-5}$. Furthermore, the obtained results are also compared and agreed with the results of RK4 method.

## REFERENCES

1. J. Opsal, A. Rosencwaig and D. L.Willenborg, "Thermal-Wave Detection and Thin-Film Thickness Measurements with Laser Beam Deflection," Applied Optics, Vol. 22, No. 20, 1983, pp. 3169-3176. http://dx.doi.org/10.1364/AO.22.003169

2. Z. Y. Yan, "Generalized Method and Its Application in the Higher-Order Nonlinear Schrodinger Equation in Nonlinear Optical Fibres," Chaos Solitons and Fractals, Vol. 16, No. 5, 2003, pp. 759-766. http://dx.doi.org/10.1016/S0960-0779(02)00435-6

3. J. Mei, H. Zhang and D. Jiang, "New Exact Solutions for a Reaction-Diffusion Equation and a Quasi-Camassa Holm Equation," Applied Mathematics E-Notes, Vol. 4, 2004, pp. 85-91.

4. E. S. Fahmy and H. A. Abdusalam, "Exact Solutions for Some Reaction Diffusion Systems with Nonlinear Reaction Polynomial Terms," Applied Mathematical Sciences, Vol. 3, No. 11, 2009, pp. 533-540.

5. M. S. H. Chowdhury and I. Hashim, "Analytical Solution for Cauchy Reaction-Diffusion Problems by Homotopy Perturbation Method," Sains Malaysiana, Vol. 39, No. 3, 2010, pp. 495-504.

6. S. Puri and K. Wiese, "Perturbative Linearization of Reaction-Diffusion Equations," Journal of Physics A: Mathematical and General, Vol. 36, 2003, pp. 2043-2054.http://dx.doi.org/10.1088/0305-4470/36/8/303

7. L. D. Ropp, N. J. Shadid and C. C. Ober, "Studies of the Accuracy of Time Integration Methods for Reaction-Diffusion Equations," Journal of Computational Physics, Vol. 194, No. 2, 2004, pp. 544-574. http://dx.doi.org/10.1016/j.jcp.2003.08.033

8. R. G. Marcus, "Finite-Difference Schemes for Reaction-Diffusion Equations Modeling Predator—Prey Interactions in MATLAB," Bulletin of Mathematical Biology, Vol. 69, No. 3, 2007, pp. 931-956. http://dx.doi.org/10.1007/s11538-006-9062-3

9. B. Liu, M. B. Allen, H. Kojouharov and B. Chen, "Finite-Element Solution of Reaction-Diffusion Equations with Advection," Computational Mechanics, 1996, pp. 3-12.

10. X. Christos and L. Oberbroeckling, "On the Finite Element Approximation of Systems of Reaction—Diffusion Equations by p/hp Methods," Global Science, Vol. 28, No. 3, 2010, pp. 386-400.

11. G. Meral and M. S. Tezer, "Solution of Nonlinear Reaction-Diffusion Equation by Using Dual Reciprocity Boundary Element Method with Finite Difference or Least Squares Method," Advances in Boundary Element Techniques, 2008, pp. 317-22.

12. C. Shu, "Differential Quadrature and Its Application in Engineering," Springer Verlag, London, 2000. http://dx.doi.org/10.1007/978-1-4471-0407-0

13. T. Y. Wu and G. R. Liu, "A Differential Quadrature as a Numerical Method

to Solve Differential Equations," Computational Mechanics, Vol. 24, No. 3, 1999, pp. 197-205.http://dx.doi.org/10.1007/s004660050452

14. G. Meral, "Solution of Density Dependent Nonlinear Reaction-Diffusion Equation Using Differential Quadrature Method," World Academy of Science, Engineering and Technology, Vol. 41, 2010, pp. 1178-1183.

15. V. Kajal, "Numerical Solutions of Some Reaction-Diffusion Equations by Differential Quadrature Method," International Journal of Applied Mathematics and Mechanics, Vol. 6, No. 14, 2010, pp. 68-80.

16. M. Salah, R. M. Amer and M. S. Matbuly, "The Differential Quadrature Solution of Reaction-Diffusion Equation Using Explicit and Implicit Numerical Schemes," Applied Mathematics, 2014.

17. C. Shu, Q. Yao and K. S. Yeo, "Block-Marching in Time with DQ Discretization: An Efficient Method for Time-Dependent Problems," Computer Methods in Applied Mechanics and Engineering, Vol. 191, No. 41, 2002, pp. 4587-4579. http://dx.doi.org/10.1016/S0045-7825(02)00387-0

18. J. S. Nadjafi and A. Ghorbani, "He's Homotopy Perturbation Method: An Effective Tool for Solving Nonlinear Integral and Integro-Differential Equations," Computers and Mathematics with Applications, Vol. 58, No. 11-12, 2009, pp. 2379-2390.http://dx.doi.org/10.1016/j.camwa.2009.03.032

19. H. S. Prasad and Y. N. Reddy, "Numerical Solution of Singularly Perturbed Differential-Difference Equations with Small Shifts of Mixed Type by Differential Quadrature Method," American Journal of Computational and Applied Mathematics, Vol. 2, No. 1, 2012, pp. 46-52. http://dx.doi.org/10.5923/j.ajcam.20120201.09

20. J. P. Hambleton and S. W. Sloan, "A Perturbation Method for Optimization of Rigid Block Mechanisms in the Kinematic Method of Limit Analysis," Computers and Geotechnics, Vol. 48, 2013, pp. 260-271. http://dx.doi.org/10.1016/j.compgeo.2012.07.012

21. M. Bastani and D. K. Salkuyeh, "A Highly Accurate Method to Solve Fisher's Equation," Indian Academy of Sciences, Vol. 78, No. 3, 2012, pp. 335-346.

22. W. Y. Yang, W. Cao, T. Chung, J. Morris, et al., "Applied Numerical Methods Using Matlab," John Wiley & Sons, Hoboken, New Jersey, 2005.http://dx.doi.org/10.1002/0471705195

23. E. Kreyszig, "Advanced Engineering Mathematics," John Wiley & Sons, Columbus, 2006

# Chapter 10

## PARAMETER SENSITIVITY ANALYSIS ON DEFORMATION OF COMPOSITE SOIL-NAILED WALL USING ARTIFICIAL NEURAL NETWORKS AND ORTHOGONAL EXPERIMENT

Jianbin Hao and Banqiao Wang

School of Geology Engineering and Geomatics, Chang'an University, Xi'an, Shaanxi 710054, China

## ABSTRACT

Based on the back-propagation algorithm of artificial neural networks (ANNs), this paper establishes an intelligent model, which is used to predict the maximum lateral displacement of composite soil-nailed wall. Some parameters, such as soil cohesive strength, soil friction angle, prestress of anchor cable, soil-nail spacing, soil-nail diameter, soil-nail length, and other factors, are considered in the model. Combined with the in situtest data of composite soil-nail wall reinforcement engineering, the network is trained and the errors are analyzed. Thus it is demonstrated that the method is applicable and feasible in predicting lateral displacement of excavation retained by composite soil-nailed wall. Extended calculations are conducted by using the well-trained intelligent forecast model. Through application of orthogonal table test theory, 25 sets of tests are designed to analyze the sensitivity of factors affecting the maximum lateral displacement of composite soil-nailing wall. The results show that the sensitivity of factors affecting the maximum lateral displacement of composite soil nailing wall, in a descending order, are prestress of anchor cable, soil friction angle, soil cohesion strength, soil-nail spacing, soil-nail length, and soil-nail diameter. The results can provide important reference for the same reinforcement engineering.

## INTRODUCTION

Soil-nailing was first devised in France in the early 1970s and offers advantages of low cost, versatility, and easily negotiating curves and corners. Furthermore, nailing does not require large equipment [1]. Composited with prestressed cables or micropiles, soil-nailed wall's displacement can be reduced [2]. The lateral displacement of soil-nail wall depends on multiple factors such as friction angle, coherence force, and tensile strength of soil mass; diameter, spacing between nails, inclination angle, and length of nail; strength of grout; and prestress of cable and other factors [3]. The same viewpoint was also presented by Yang [2] based on 7 typical case histories of composite soil-nailing walls with prestressed anchors in deep excavations, and he concluded that the lateral displacements were influenced by geology conditions of site, design parameters of soil-nail, and prestressed anchors. These factors will be uncertain for a special case. So, how much the influence of the possible swinging range of these uncertain factors on the lateral displacement of excavation facing is and to which factor the lateral displacement is the most sensitive are often important concerns for a soil-nailing project.

In 1986, Hohenbichler and Rackwitz [4] conducted a study of the sensitivity of reliability on random variables. Then Madsen et al. [5], Bjerager and Krenk [6], Karamchandani and Cornell [7] further studied this issue. There was at present less research on the sensitivity of parameters of composite soil-nail wall in China. The existing research methods for sensitivity have been confined to the finite element method (FEM) [8–13]. The obvious drawback of FEM is that the amount of calculation is very large if too many parameters are required. On the other hand, it is difficult to determine the combination of multiple parameters.

Based on the in situ test trials of a soil-nail wall, under the comprehensive consideration of the factors such as soil cohesive strength, soil friction angle, prestress of anchor cable, soil-nail spacing, soil-nail diameter, soil-nail length, and other factors, an intelligent model for predicting the maximum lateral displacement of soil-nail wall is established by using error back-propagation (BP) algorithm of artificial neural networks (ANNs) and MATLAB ANN toolbox [14–16]. On this basis, the sensitivity of the maximum lateral displacement of soil-nail wall to every parameter is analyzed, by adopting the orthogonal table test theory [16, 17]. It provides reference for the practical application of the same reinforcement engineering.

## OVERVIEW: ARTIFICIAL NEURAL NETWORKS

Artificial neural networks (ANNs) are the result of academic investigations that involve using mathematical formulations to model biological nervous system operations. ANNs are highly parallel computational systems comprising interconnected artificial neurons or processing units. Neural networks use logical parallelism combined with serial operations, as information in one layer is transferred to neurons in another layer [18].

A typical multilayer perceptron consists of an input layer, one or more hidden layers, and one output layer. The structural chart of artificial neural network [19] is shown as Figure 1. A neuron forms the basis of the ANN where it receives, processes, and transmits information (Figure 2). Input layers receive inputs from sources external to the system under study. The output layers send signals out of the system, while the hidden layers are those whose inputs and outputs are within the system [20]. Hidden layers are necessary for the network to learn interdependencies in the model. The units in a neural network are connected by links, where each link has a numeric weight associated with it. Weights are the primary means of long-term storage in neural networks, and learning occurs through changes to the weights.

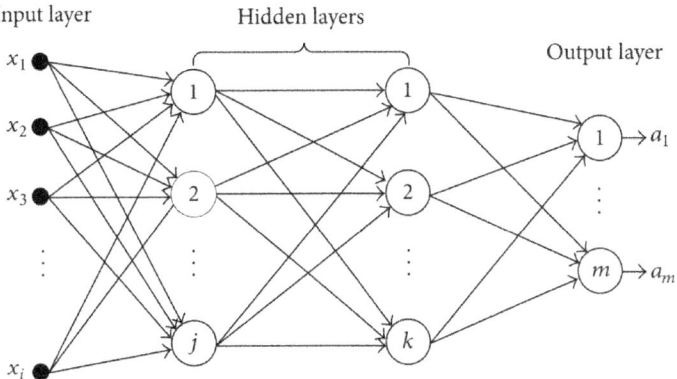

**Figure 1:** Schematic diagram of an ANN architecture showing input, processing center consisting of many neurons (hidden layers), and outputs. 1, 2. . . $j$, $k$ are neurons, 1, 2, . . . , $m$ are outputs, and $x_1$, $x_2$,...,$_i$ are inputs.

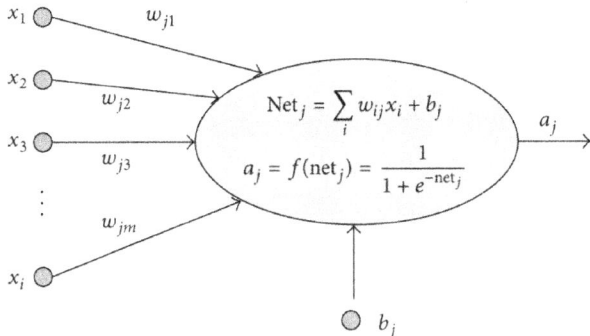

**Figure 2:** Schematic diagram for a simple neuron showing its input (parameters $x$ and weighting $w$), processing center with its task, and an output ($a$) section. $b$ is bias.

Neural networks require some computational method of weight adjustment. Back-propagation, which was proposed by Rumelhart and McClelland in 1985, is a gradient descent algorithm that compares actual outputs with desired outputs. If an error exists, its reduction is accomplished by back-propagating the error through the network and adjusting the weights. It provides a way of dividing the calculation of the gradient among the units, so the change in each weight can be calculated by the unit to which the weight is attached using only local information [21].

Each layer's computation is split into two components. First, a linear component, called the input function, computes the weighted sum of the layer's input values. Second, a nonlinear component, called the activation function (Figure 3), transforms the weighted sum into the layer's final activation value.

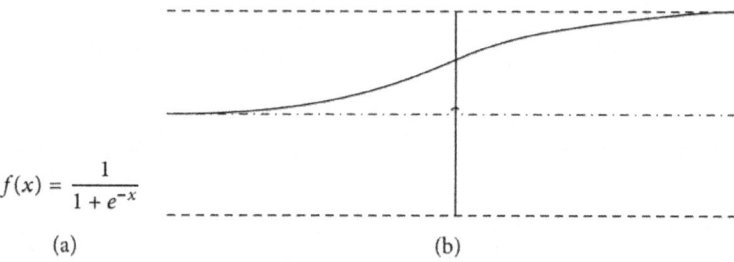

$$f(x) = \frac{1}{1 + e^{-x}}$$

(a)                                                (b)

**Figure 3:** Logistic function: (a) definition and (b) function curve.

The net input to a processing unit j is given by

$$\text{net}_j = \sum_i w_{ij} x_i + b_j,$$

(1)

where the $x_i$ is the output from the previous layer, $w_{ij}$ is the weight of the link connecting unit $i$ to unit $j$, and $b_j$ is the bias of unit $j$, which determines the location of the logsig function on the horizontal axis. The activation value (output) of unit $j$ is given by

$$a_j = f\left(\text{net}_j\right) = \frac{1}{1 + e^{-\text{net}_j}}.$$

(2)

In its simplest form, the weight-update is a scaled step in the opposite direction of the gradient. Hence, the weight-update rule is

$$\Delta_p w_{ij}(t) = -\varepsilon \frac{\partial E_p}{\partial w_{ij}}(t) + \alpha \Delta_p w_{ij}(t - 1),$$

(3)

where $\alpha$ is the momentum term and determines the influence from the previous iteration on the present one. In this equation, the total error is given by

$$E = \sum_{p=1}^{P} \frac{1}{2} E_p = \frac{1}{2} \sum_{p=1}^{Q} \sum_{j=1}^{N} \left( d_{pj} - a_{pj} \right)^2 ,$$

(4)

where $\varepsilon \in (0, 1)$ is a parameter that determines the step size and is called the learning rate and $d_{pj}$ and $a_{pj}$ are the target and the actual response value of output neuron $j$ corresponding to pattern $p$. $Q$ is the number of training patterns and $N$ represents the number of output units. This error information is propagated backwards through the ANNs and the weights are adjusted. After some number of iterations, the training stops when the calculated output values best approximate the desired values.

The learning course is made up of two aspects: forwardand back-propagation. In the course of forward-propagation, the state in every layer only affects the next neuron network. If the outcomes in output layer are not the expected targets, that is, the error between actual outputs and expected targets exceeds the given error, then the course would shift to that of back-propagation. During back-propagation, the error signal would backtrack along the original path. The connected weights in every nerve cell are adjusted and the course gradually propagates to the input layer of neural network.The forward-propagation would restart again. The two repeating courses would not stop until the error between actual outputs and expected targets is less than the given error [22].

# MODELING PROCEDURES

## Modeling

The constraints of this problem are formulated as follows.

### *Determination of Input Layer Parameters*

The node numbers of neurons of the input layer in the neural network model are decided by the numbers of factors that affect the output layer. According to the mechanical characteristics of composite soil-nailing wall, there are six factors considered in this paper: soil cohesive strength, soil friction angle, soil-nail spacing, prestress of anchor cable, soil-nail length, and soil- nail diameter. The input variables ($i$) are 6. The output layer, with 1 output variable ($m$=1), is the predicted value of the maximum lateral displacement of composite soil-nailing wall

### Determination of Parameters of the Hidden Layer

According to Kolmogorov's theorem, with a rational structure and appropriate weights, a three-layer feed-forward network can be trained to produce any continuous function with any desired level of precision. The three-layer BP network established in this paper consists of an input layer, a hidden layer, and an output layer. The network can perform any $n$- dimensional to $m$-dimensional mapping, in which the hidden variables are $2i + 1 = 13$.

### Prediction Model

According to the node numbers of input layers, hidden layers, and output layers and the numbers of the hidden layers, the architecture of the BP forecast model built in this paper is 6-13-1 network (Figure 4).

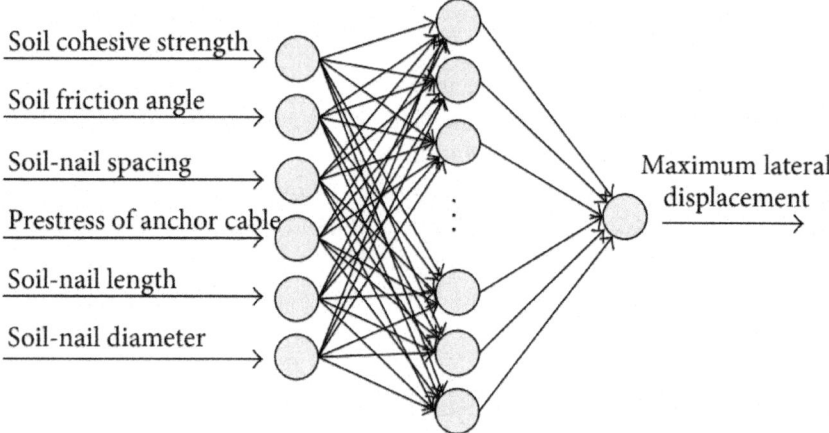

**Figure 4:** Neural network model for predicting the maximum lateral displacement of composite soil-nail wall.

## Model Training and Testing

The constraints of this problem are formulated as follows.

### Training Samples

Ten sets of data obtained from the in situ test of a composite soil-nailing wall were selected as learning samples in this paper, as shown in Table 1. Figure 5 shows the cross section of the composite soil-nailed wall.

**Table 1:** Learning samples of the forecast model

| Number | Soil cohesive strength (kPa) | Soil friction angle (°) | Soil-nail spacing (m) | Prestress of anchor cable (kN) | Soil-nail length (m) | Soil-nail diameter (mm) | Maximum lateral displacement of the soil-nailed wall (mm) |
|---|---|---|---|---|---|---|---|
| 1 | 32.4 | 22.3 | 2.0 | 50 | 12.0 | 20 | 5.06 |
| 2 | 32.4 | 22.3 | 2.0 | 100 | 12.0 | 20 | 4.81 |
| 3 | 32.4 | 22.3 | 2.0 | 180 | 12.0 | 20 | 4.60 |
| 4 | 32.4 | 22.3 | 2.0 | 230 | 12.0 | 20 | 4.47 |
| 5 | 32.4 | 22.3 | 2.0 | 500 | 12.0 | 20 | 4.32 |
| 6 | 32.4 | 22.3 | 1.5 | 180 | 12.0 | 20 | 4.67 |
| 7 | 32.4 | 22.3 | 3.0 | 180 | 12.0 | 20 | 4.83 |
| 8 | 32.4 | 22.3 | 2.0 | 180 | 12.0 | 22 | 4.54 |
| 9 | 32.4 | 22.3 | 2.0 | 180 | 12.0 | 28 | 4.48 |
| 10 | 32.4 | 22.3 | 2.0 | 180 | 15.0 | 20 | 4.55 |

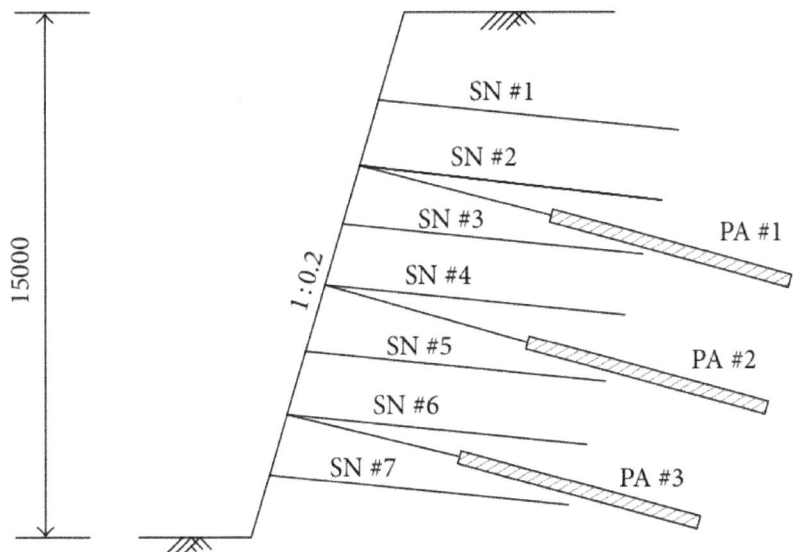

**Figure 5:** Note: SN are soil-nails and PA are prestressed anchors. Cross section of composite soil-nailed wall (mm).

## Training of the Samples

Before training the sample, in order to accelerate the convergence speed of the network and enhance the computational efficiency, using the normalized "premnmx" function offered by MATLAB neural network toolbox, we normalize the input matrix and output matrix to the range [−1, 1].

The constructed model is trained by using the normalized data. The training function is the default "trainlm" of the MATLAB neural network system. In our study, we used an ANN with one hidden layer of 13 neurons. The number of epochs is regulated by net.trainParam.epochs = 1000. The maximum amount of time to stop the training stage and the goal for performance were determined by trainParam.time = inf and net.trainParam.goal = 0.01, respectively. The ANN was trained with resilient propagation for each dataset to produce the lowest errors. Value of 0.01 was used for the learning rate by trainParam.lr = 0.01. Finally, we used 50 for epochs between displays in the training stage by trainParam.show = 50.

As the iteration proceeds, the error decreases from its initial high value (Figure 6). It attains a value of about 0.01 only after 14 iterations.

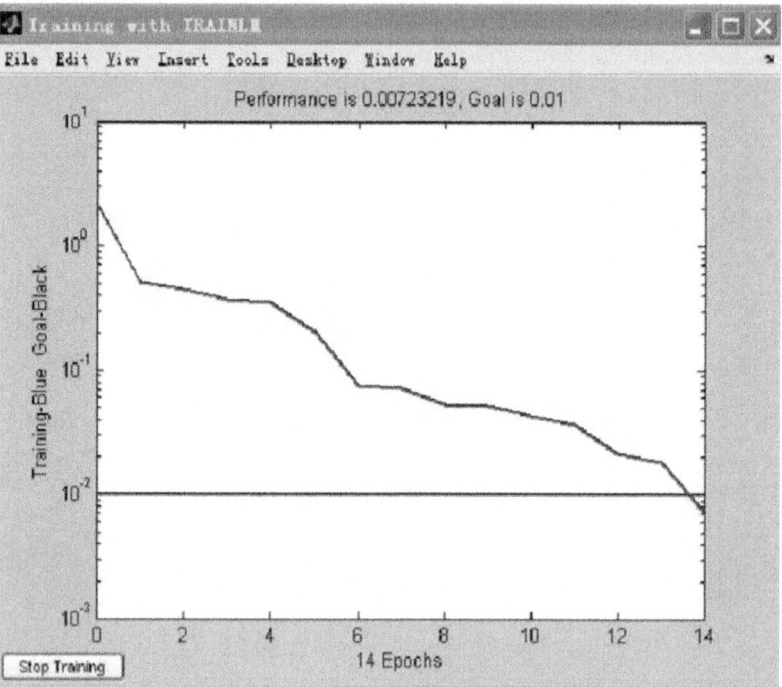

**Figure 6:** Error as a function of the number of iterations (epochs) during training stage.

## Predicted Results and Error Analysis

After training, the network output matches the target. The inherent nonlinear mapping relationship offered by the samples is stored in each layer of the network. Thus it defined the weights of various input and output layers and their thresholds. In order to analyze the accuracy of prediction of the established network model, five groups of in situ test data were firstly selected as the testing samples. The samples were used to get predicted values by using the "sim" function. Then the predicted values were antinormalized to get the final forecasted values by using the "postmnmx" function. Finally, we compared the predicted values and in situ test values, as shown in Table 2. The results show that the average relative error of the predicted results is 2.09%. It can meet the needs in engineering, indicating that the model has certain feasibility and practicality in forecast of the maximum lateral displacement of the pit.

**Table 2:** Predicted results and errors

| Number | Soil cohesive strength (kPa) | Soil friction angle (°) | Soil-nail spacing (m) | Prestress of anchor cable (kN) | Soil-nail length (m) | Soil-nail diameter (mm) | Measured value of the maximum lateral displacement (mm) | Predicted value of the maximum lateral displacement (mm) | Absolute error | Relative error (%) | Average relative error (%) |
|---|---|---|---|---|---|---|---|---|---|---|---|
| 1 | 32.4 | 22.3 | 2.0 | 180 | 12.0 | 20 | 4.60 | 4.7 | 0.1 | 2.1739 | |
| 2 | 32.4 | 22.3 | 2.0 | 230 | 12.0 | 20 | 4.47 | 4.56 | 0.09 | 2.0134 | |
| 3 | 32.4 | 22.3 | 3.0 | 180 | 12.0 | 20 | 4.83 | 4.95 | 0.12 | 2.4845 | 2.09% |
| 4 | 32.4 | 22.3 | 2.0 | 180 | 12.0 | 28 | 4.48 | 4.56 | 0.08 | 1.7857 | |
| 5 | 32.4 | 22.3 | 2.0 | 180 | 15.0 | 20 | 4.55 | 4.46 | -0.09 | 1.9780 | |

# SENSITIVITY ANALYSIS OF FACTORS AFFECTING THE MAXIMUM LATERAL DISPLACEMENT OF EXCAVATION WALL

## Orthogonal Test Method

Orthogonal testing is a designing method of tests that utilizes the orthogonal principle and mathematical statistics, selects typical and representative test points from a large number of test points, arranges reasonably and scientifically the tests by application of the orthogonal tables, and obtains the optimal results with the number of times as less as possible. The conditions that need to be inspected and/or can be controlled in the experiment are called factors. Each status or grade of a factor is called a level of the factor.

## Selecting the Orthogonal Table

According to the principle of orthogonal table selection, the level number of factors determined in our study is 5; there are 6 factors: soil cohesive strength, soil friction angle, soil- nail spacing, prestress of anchor cable, soil-nail length, and soil-nail diameter. Table 3 shows the level of the factors selected. Table 4 shows the $L_{25}$ ($5^6$) orthogonal table selected according to the level numbers and factor numbers.

**Table 3:** The level table of the factors.

| Factor level | Soil cohesive strength | Prestress of anchor cable | Soil-nail spacing | Soil friction angle | Soil-nail diameter | Soil-nail length |
|---|---|---|---|---|---|---|
| 1 | −20% | −20% | −20% | −20% | −20% | −20% |
| 2 | −10% | −10% | −10% | −10% | −10% | −10% |
| 3 | 0% | 0% | 0% | 0% | 0% | 0% |
| 4 | 10% | 10% | 10% | 10% | 10% | 10% |
| 5 | 20% | 20% | 20% | 20% | 20% | 20% |

**Table 4:** Orthogonal Marshall test program.

| Test number | Soil cohesive strength level | Prestress of anchor cable level | Soil-nail spacing level | Soil friction angle level | Soil-nail diameter level | Soil-nail length level |
|---|---|---|---|---|---|---|
| 1 | 1 | 1 | 2 | 4 | 3 | 2 |
| 2 | 2 | 1 | 5 | 5 | 5 | 4 |
| 3 | 3 | 1 | 4 | 1 | 4 | 1 |
| 4 | 1 | 4 | 4 | 4 | | 4 |
| 5 | 5 | 1 | 3 | 2 | 2 | 5 |
| 6 | 2 | 2 | 2 | 2 | 1 | 1 |
| 7 | 2 | 2 | 3 | 4 | | 5 |
| 8 | 3 | 2 | 5 | 4 | 2 | 3 |
| 9 | 4 | 2 | 4 | 5 | 3 | 5 |
| 10 | 5 | 2 | 1 | 1 | 5 | 2 |
| 11 | 1 | 3 | 1 | 5 | 2 | 1 |
| 12 | 2 | 3 | 3 | 1 | 3 | 3 |
| 13 | 3 | 3 | 2 | 3 | 5 | 5 |
| 14 | 4 | 3 | 5 | 2 | 4 | 2 |
| 15 | 5 | 3 | 4 | 4 | 1 | 4 |
| 16 | 1 | 4 | 4 | 2 | 5 | 3 |
| 17 | 2 | 4 | 1 | 4 | 4 | 5 |
| 18 | 3 | 4 | 3 | 5 | 1 | 2 |
| 19 | 4 | 4 | 2 | 1 | 2 | 4 |
| 20 | 5 | 4 | 5 | 3 | 3 | 1 |
| 21 | 1 | 5 | 5 | 1 | 1 | 5 |
| 22 | 2 | 5 | 4 | 3 | 2 | 2 |
| 23 | 3 | 5 | 1 | 2 | 3 | 4 |
| 24 | 4 | 5 | 3 | 4 | 5 | 1 |
| 25 | 5 | 5 | 2 | 5 | 4 | 3 |

## Sensitivity Analysis of the Factors

As mentioned before, the trained network has certain feasibility and practicality in the maximum lateral displacement of excavation side. With the project mentioned before, we predict the maximum lateral displacement of the another side and analyze the sensitivity of those factors. Here, weighted average of the soil cohesive strength is 32.4 kPa, weighted average of the soil friction angle is 22.3°, the nail spacing is 12 m, the prestress of anchor cable is 300 kN, the nail length is 12 m, and the nail diameter is 20 mm.

We predicted, respectively, the above 25 trial samples by using the trained network model. The results are shown in Table 5.

**Table 5:** Prediction results

| Test number | Prediction values | Test number | Prediction values | Test number | Prediction values | Test number | Prediction values | Test number | Prediction values |
|---|---|---|---|---|---|---|---|---|---|
| 1 | 4.11824 | 6 | 4.6932 | 11 | 4.01528 | 16 | 5.70468 | 21 | 6.59516 |
| 2 | 4.42996 | 7 | 4.55132 | 12 | 3.97908 | 17 | 4.8242 | 22 | 7.60396 |
| 3 | 4.46104 | 8 | 4.40032 | 13 | 3.98992 | 18 | 4.69496 | 23 | 4.68976 |
| 4 | 5.50708 | 9 | 4.6068 | 14 | 4.03284 | 19 | 5.66764 | 24 | 4.76712 |
| 5 | 4.61388 | 10 | 4.96008 | 15 | 4.03348 | 20 | 9.29496 | 25 | 6.75368 |

The average value of each index can be obtained for each factor at different levels by averaging the results of five experiments of the same level of each factor. The extreme differences corresponding to the factor variation can be obtained by working out the difference between the maximum and minimum values of indicators of the same factor at different levels. They are shown in Table 6.

**Table 6:** Orthogonal test results

| Level | Soil cohesive strength | Prestress of anchor cable | Soil-nail spacing | Soil friction angle | Soil-nail diameter | Soil-nail length |
|---|---|---|---|---|---|---|
| The first level average | 5.02532 | 4.62604 | 4.79928 | 5.1326 | 5.0764 | 5.41796 |
| The second level average | 5.07772 | 4.64236 | 5.01616 | 4.71848 | 5.2602 | 5.082 |
| The third level average | 4.4472 | 4.01012 | 4.54964 | 6.21784 | 5.33776 | 5.26896 |
| The fourth level average | 4.91628 | 6.03728 | 5.282 | 4.42868 | 4.953 | 4.7028 |
| The fifth level average | 5.9312 | 6.08192 | 5.75064 | 4.90012 | 4.77036 | 4.926 |
| Extreme differences | 1.484 | 2.0718 | 1.201 | 1.78916 | 0.5674 | 0.71516 |
| Sequence of extreme differences | 3 | 1 | 4 | 2 | 6 | 5 |

The experimental results showed the order of the sensitivity of factors affecting the maximum lateral displacement if excavation side is as follows: prestress of anchor cable, soil friction angle, soil cohesive strength, soil-nail spacing, soil-nail length, and soil-nail diameter.

# CONCLUSION

Artificial neural networks (ANNs) have good self-adaptability, self-organization and strong learning, associative memory, fault tolerance and anti-interference ability, and so on. By introducing ANNs into the composite soil-nailed wall system, we can fully consider many factors that affect the lateral displacement of excavation sides and make an intelligent prediction.

The prediction model of the maximum lateral displacement of excavation established in this paper by utilizing ANNs makes a very nonlinear mapping between various factors affecting the lateral displacement and lateral displacement. With the in situ test data of displacement of an excavation side, we have verified the feasibility and applicability of ANNs in excavation engineering, providing a new method reference for predicting the lateral displacement of other similar excavations. In this paper, combining ANNs with the orthogonal trial design, we performed the sensitivity analysis of factors affecting the maximum displacement of excavation side. The analysis results showed that the sensitivity of these factors, in a descending order, is the prestress of anchor cable, soil friction angle, soil cohesive strength, soil-nail spacing, soil-nail length, and soil-nail diameter.

## CONFLICT OF INTERESTS

The authors declare that there is no conflict of interests regarding the publication of this paper.

## ACKNOWLEDGMENTS

This work was partially supported by the National Natural Science Foundation of China through Grants nos. 41102117 and 41272286, the Special Fund for Basic Scientific Research of Central Colleges, Chang'an University, through Grant no. 2013G2261010, and the Key Laboratory of West Mineral Resources and Geotechnical Engineering of Ministry of Education, Chang'an University, China.

## REFERENCES

1.    L. Richard Handy and G. Merlin Spangler, Geotechnical Engineering: Soil and Foundations Principles and Practice, McGraw-Hill, Blacklick, Ohio, USA, 5th edition, 2007.

2.    Y. Yang, "Case studies of composite soil-nailing walls and movement estimate," Chinese Journal of Geotechnical Engineering, vol. 34, no. 4, pp. 734–741, 2012.

3.    L. Cheng, J. Han, and P. Zhang, "Long-term performance and safety assessment of anchorage in geotechnical engineering," Chinese Journal of Rock Mechanics and Engineering, vol. 27, no. 5, pp. 865–872, 2008.

4.    M. Hohenbichler and R. Rackwitz, "Sensitivities and importance measures in structural reliability," Civil Engineering Systems, vol. 3, no. 4, pp. 203–209, 1986.

5.    H. O. Madsen, S. Krenk, and N. C. Lind, Methods of Structural Safety, Springer, New York, NY, USA, 1986.

6.    P. Bjerager and S. Krenk, "Parametric sensitivity in first order reliability theory," Journal of Engineering Mechanics, vol. 115, no. 7, pp. 1577–1582, 1989.

7.   A. Karamchandani and C. A. Cornell, "Sensitivity estimation within first and second order reliability methods," Structural Safety, vol. 11, no. 2, pp. 95–107, 1992.

8.   E. Song and Y. Qiu, "Deformation analysis of composite soil nailing by FEM," Building Construction, vol. 23, no. 6, pp. 370–374, 2001.

9.   Z. C. Wu, L. S. Tang, Z. Q. Liao, X. G. Liu, and B. Yan, "FLAC-3D simulation of deep excavation with compound soil nailing support," Chinese Journal of Geotechnical Engineering, vol. 28, pp. 1460–1465, 2006.

10.  L. Tang, M. Song, H. Liao, Z. Wu, and T. Xu, "Analysis of stress and deformation of prestressed anchor cable composite soil nailing," Chinese Journal of Rock Mechanics and Engineering, vol. 27, no. 2, pp. 410–417, 2008.

11.  B. S. Xu, R. C. Liu, L. X. Li, M. Gong, and Y. K. Wang, "Study of slope deformation and parameters sensitivity in supporting design of composite soil nailing wall," Rock and Soil Mechanics, vol. 32, no. 2, pp. 393–400, 2011.

12.  T. Lai, "The analysis of sensitivity of composite soil nailing designed parameter," Journal of Lanzhou Railway University (Natural Sciences), vol. 22, no. 1, pp. 27–29, 2003.

13.  Z. Wu, L. Tang, X. Liu, Q. Zhang, and Y. Liu, "Study on large-scale in-situ test on compound soil nailing wall and deformation property analysis," Chinese Journal of Rock Mechanics and Engineering, vol. 26, no. 1, pp. 2974–2980, 2007.

14.  L. Jiao, Neural Network Computing, Electronic and Science University Press, Xi'an, Shaanxi, China, 1995.

15.  K. Zhou and Y. Kang, Neural Network Model and MATLAB Simulation Program Design, Tsinghua University Press, Beijing, China, 2005.

16.  S. F. Yao, "Design of neural networks based on orthogonal experiments," Mathematics in Practice and Theory. Shuxue de Shijian yu Renshi, vol. 38, no. 13, pp. 116–122, 2008.

17.  Machinery Industry Ministry of the People's Republic of China, Process parameter optimization method of orthogonal test method (JB/T, 7510-1994), China Academy of Machinery Science & Technology, Beijing, China, 1994.

18.  M. B. Perry, J. K. Spoerre, and T. Velasco, "Control chart pattern recognition using back propagation artificial neural networks," International Journal of Production Research, vol. 39, no. 15, pp. 3399–3418, 2001.

19. A. Neyamadpour, S. Taib, and W. A. T. Wan Abdullah, "Using artificial neural networks to invert 2D DC resistivity imaging data for high resistivity contrast regions: a MATLAB application," Computers and Geosciences, vol. 35, no. 11, pp. 2268–2274, 2009.

20. B. D. Ripley, Pattern Recognition and Neural Networks, Cambridge University Press, New York, NY, USA, 1996.

21. S. Russell and P. Norvig, Artificial Intelligence: A Modern Approach, Prentice-Hall, Upper Saddle River, NJ, USA, 1995.

22. G. Li, Intelligent Control and MATLAB Application, Publishing House of Electronics Industry, Beijing, China, 2005.

# Chapter 11

# MATHEMATICAL MODEL AND MATLAB SIMULATION OF STRAPDOWN INERTIAL NAVIGATION SYSTEM

Wen Zhang, [1] Mounir Ghogho,[2,3] and Baolun Yuan[1]

[1]College of Opto-Electronic Science and Technology, National University of Defense Technology, Changsha 410073, China

[2]School of Electronic and Electrical Engineering, University of Leeds, Leeds LS2 9JT, UK

[3]International University of Rabat, Rabat 11 100, Morocco

## ABSTRACT

Basic principles of the strapdown inertial navigation system (SINS) using the outputs of strapdown gyros and accelerometers are explained, and the main equations are described. A mathematical model of SINS is established, and its Matlab implementation is developed. The theory is illustrated by six examples which are static status, straight line movement, circle movement, s-shape movement, and two sets of real static data.

## INTRODUCTION

Many navigation books and papers on inertial navigation system (INS) provide readers with the basic principle of INS. Some also superficially describe simulation methods and rarely provide the free code which can be used by new INS users to help them understand the theory and develop INS applications. Commercial simulation software is available but is not free. The objective of this paper is to develop an easy-to-understand step-by-step development method for simulating INS. Here we consider the most popular INS which is the strapdown inertial navigation system (SINS). The mathematical operations required in our work are mostly matrix manipulations and more generally basic linear algebra [1]. In this paper, Matlab [2] is chosen as the simulation environment. It is a popular computing environment to perform complex matrix calculations and to produce sophisticated graphics in a relatively easy manner. A large collection of Matlab scripts are now available for a wide variety of applications and are often used for university courses. Matlab is also becoming more and more popular in industrial research centers in the design and simulation stages.

The main purposes of this paper are to establish a mathematical model and to develop a comprehensive Matlab implementation for SINS. The structure of the proposed mathematical model and Matlab simulation of SINS is shown in Figure 1. In Section 2, the INS-related orthogonal coordinates (the body frame, the inertial frame, the Earth frame, the navigation frame, the ENU-frame, and the wander azimuth navigation frame) are described and figures to illustrate the relationship between the frames are provided. The basic principle of SINS is described in the wander azimuth navigation frame ($p$-frame). In Section 3, two important direction cosine matrices (DCMs), the vehicle attitude DCM and the position DCM, and the related important attitude and position angles are defined. In Section 4, the simulation for data generation of gyros and accelerometers is described in ENU-frame. Instead of $p$-frame, ENU-frame is chosen because the outputs of gyros and accelerometers are easier to obtain in this frame. The Matlab implementation is given and described step by step. Four kinds of scenarios (static, straight, circle, and s-shape) are set as examples of different kinds of vehicle trajectories. In Section 5, the mathematical model of SINS is set up and the calculation steps in $p$-frame are provided. In Section 6, the required initial parameters and other initial data calculation for the SINS model are given for the different simulation scenarios. In Section 7, Matlab implementation code functions are listed and described. Further, simulation results for the four above-mentioned scenarios are presented; two examples from real SINS experiment data are also provided to verify the validity of the developed codes. Finally, conclusions are drawn. Mathematical details are given in Appendices A–D.

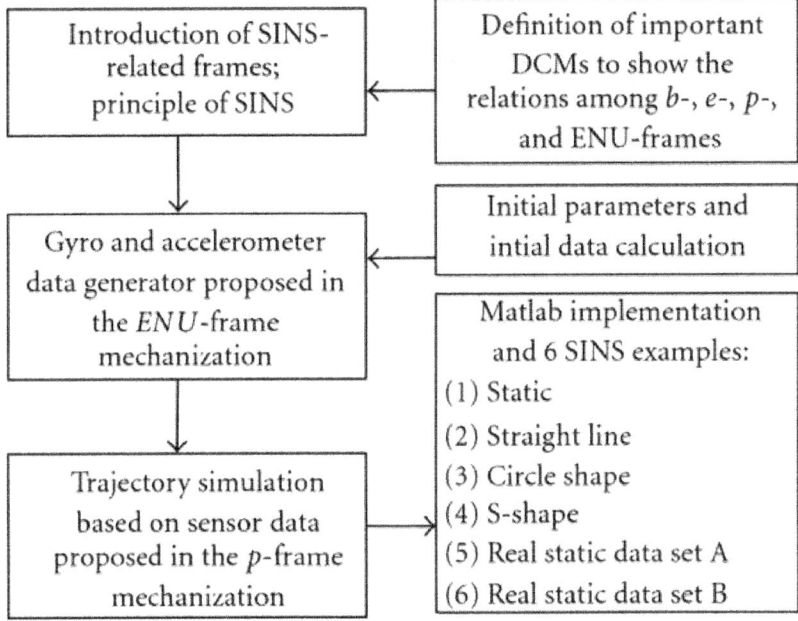

**Figure 1:** The schema of the proposed mathematical model and Matlab simulation of SINS.

## PRINCIPLES

A fundamental aspect of inertial navigation is the precise definition of a number of Cartesian coordinate reference frames. Each frame is an orthogonal, right-handed, coordinate frame or axis set. For all the coordinate frames used in this paper, a positive rotation about each axis is taken to be in a clockwise direction looking along the axis from the origin, as indicated in Figure 2. A negative rotation corresponds to an anti-clockwise direction. This convention is used throughout this paper. It is also worth pointing out that a change in attitude of a body, which is subjected to a series of rotations about different axes, is not only a function of the rotation angles, but also on the order in which the rotations occur. In this paper, the following coordinate frames are used [3].

(1)    The body frame (*b*-frame): the *b*-frame, depicted in Figure 2, is an orthogonal axis set which has its origin at the center of the vehicle, point *P*, and is aligned with the pitch *Pxb* axis, roll *Pyb* axis, and yaw *Pzb* axis of the vehicle in which the navigation system is installed.

(2)     The inertial frame ($i$-frame): the $i$-frame, depicted in Figure 3, has its origin at the center of the Earth and its axes nonrotating with respect to fixed stars; these axes are denoted by $Oxi$, $Oyi$, and $Ozi$, with $Ozi$ being coincident with the Earth polar axis.

(3)     The Earth frame ($e$-frame): the $e$-frame, depicted in Figure 3, has its origin at the center of the Earth and axes nonrotating with respect to the Earth; these axes are denoted by $Oxe$, $Oye$, and $Oze$. The axis $Oze$ is the Earth polar axis. The axis $Oxe$ is along the intersection of the plane of the Greenwich meridian and the Earth equatorial plane. The Earth frame rotates with respect to the inertial frame at a rate $\omega ie$ about the axis $Ozi$.

(4)     The navigation frame ($n$-frame): the $n$-frame, depicted in Figure 3, is a local geographic navigation frame which has its origin at the location of the navigation system, point $P$ (the navigation system is fixed inside the vehicle and we assume that the navigation system is located exactly at the center of the vehicle), and axes aligned with the directions of east $PE$, north $PN$ and the local vertical up $PU$. When the $n$-frame is defined in this way, it is called the "ENU-frame." The turn rate of the navigation frame with respect to the Earth-fixed frame, $\omega en$, is governed by the motion of the point $P$ with respect to the Earth. This is often referred to as the transport rate.

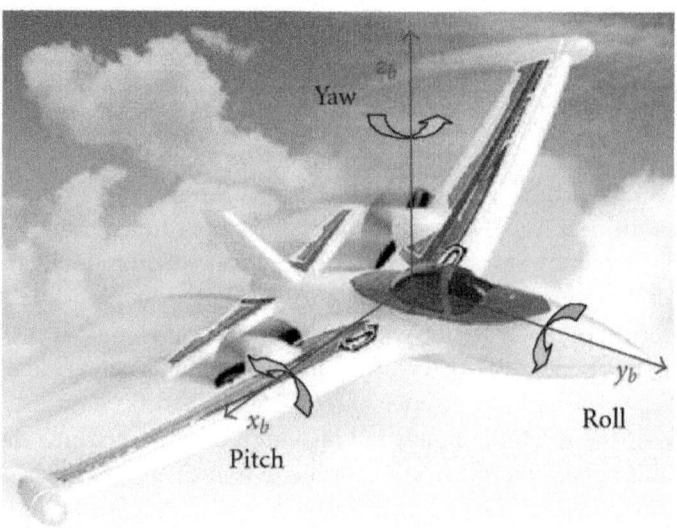

**Figure 2:** The $b$-frame illustration and the definition of axis rotations.

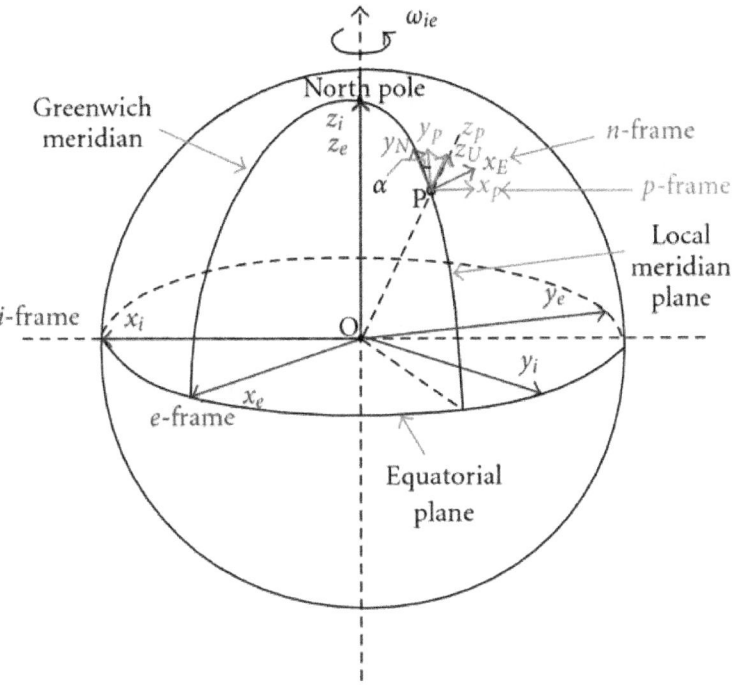

**Figure 3:** The Reference Frames.

(5) The wander azimuth navigation frame (p-frame): the p-frame, depicted in Figure 3, may be used to avoid the singularities in the computation which occur at the poles of the n-frame. Like the n-frame, it is of a local level but rotates through the wander angle about the local vertical. Here we do not call this frame w-frame (w for wander) for notation clarity since w and $\omega$ may look similar when printed. Letter p in p-frame stands for platform; indeed the wander azimuth navigation frame is of a local level and thus forms a horizontal platform.

In this paper, we choose the p-frame as the navigation frame for vehicle trajectory calculation, for the following reason. In the local geographic navigation frame mechanization, the n-frame is required to rotate continuously as the system moves over the surface of the Earth in order to keep its $Py_N$ axis pointing to the true north. In order to achieve this condition worldwide, the n-frame must rotate at much greater rates about its $Pz_U$ axis as the navigation system moves over the surface of the Earth in the polar regions, compared to the rates required at lower latitudes. It is clear that near the polar areas the local geographic navigation frame must rotate about its $Pz_U$ axis rapidly in order to maintain the $Py_N$ axis pointing to the pole. The heading direction will abruptly change by 180° when moving past the pole. In the most extreme case, the turn rate becomes infinite when passing over the pole. One way of avoiding the singularity, and also providing a navigation system with worldwide capability, is to adopt a wander azimuth mechanization in which the z-component of $\omega_{ep}^p$ is always set to zero, that is, $\omega_{epz}^p = 0$. A wander axis system is a local level frame which moves over the Earth surface with the moving vehicle, as depicted in Figure 3. However, as the name implies, the azimuth angle $\alpha$ between $Py_N$ axis and $Py_p$ axis varies with the vehicle position on Earth. This variation is chosen in order to avoid discontinuities in the orientation of the wander frame with respect to Earth as the vehicle passes over either the north or South Pole.

In the remainder of this section, the main principle of SINS in the p-frame is described.

Along the same lines as in [3], a navigation equation for a wander azimuth system can be constructed as follows:

$$\dot{v}_e^p = C_b^p f^b - \left(2C_e^p \omega_{ie}^e + \omega_{ep}^p\right)v_e^p + g^p, \tag{1}$$

where $C_b^p$ is the direction cosine matrix used to transform the measured specific force vector in b-frame into p-frame. This matrix propagates in accordance with the following equation:

$$\dot{C}_b^p = C_b^p \Omega_{pb}^b, \tag{2}$$

where $\Omega_{pb}^b$ is the skew symmetric form of $\omega_{pb}^b$, the b-frame angular rate with respect to the p-frame.

Equation (1) is integrated to generate estimates of the vehicle speed in the wander azimuth frame, $v_e^p$. This is then used to generate the turn rate of the wander frame with respect to the Earth frame, $\omega_{ep}^p$. The direction cosine matrix which relates the wander frame to the Earth frame, $C_e^p$, may be updated using the following equation:

$$\dot{C}_p^e = C_p^e \Omega_{ep}^p, \tag{3}$$

$$\left(\dot{C}_p^e\right)^T = \left(\Omega_{ep}^p\right)^T \left(C_p^e\right)^T = -\Omega_{ep}^p \left(C_p^e\right)^T, \tag{4}$$

where the superscript $T$ means matrix transposition.

Since $(\dot{C}_p^e)^T = \dot{C}_e^p$, $(C_p^e)^T = C_e^p$ and skew symmetric matrix is $(\Omega_{ep}^p)^T = -\Omega_{ep}^p$, (4) can be rewritten as

$$\dot{C}_e^p = -\Omega_{ep}^p C_e^p, \qquad (5)$$

where $\Omega_{ep}^p$ is a skew symmetric matrix formed from the elements of the angular rate vector $\omega_{ep}^p$; we could have $\omega_{ep}^p = -\omega_{pe}^e$ when the rotation angles are reciprocal. Because the $z$-component of $\omega_{ep}^p$ is set to zero, $\omega_{epz}^p = 0$, the matrix expression of $\omega_{ep}^p$ is $\omega_{ep}^p = [\omega_{epx}^p \ \omega_{epy}^p \ 0]^T$. This process is implemented iteratively and enables any singularities to be avoided.

In the next section, the two important DCMs, the vehicle attitude DCM and vehicle position DCM, are defined, as well as the vehicle-attitude-related attitude angles and vehicle-position-related position angles.

# DIRECTION COSINE MATRICES (DCMS)

In this section, the vehicle attitude DCM with the corresponding attitude angles and the vehicle position DCM with the corresponding position angles are described separately.

## Vehicle Attitude DCM $C_b^p$

The definition of the rotation sequence from $p$-frame to $b$-frame is (see Figure 4)

$$x_p y_p z_p \xrightarrow{z_p, \ \psi_G} x_e' y_e' z_e' \xrightarrow{y_p'', \ \theta} x_e'' y_e'' z_e'' \xrightarrow{y_p''', \ \gamma} x_b y_b z_b, \qquad (6)$$

where $\psi_G$ is the gird azimuth angle ($0$–$360°$), $\theta$ is the pitch angle ($-90°$–$90°$), and $\gamma$ is the roll angle ($-180°$–$180°$). The above rotation can be written in the following matrix form:

$$C_p^b = C_3 C_2 C_1$$

$$= \begin{bmatrix} \cos\gamma & 0 & -\sin\gamma \\ 0 & 1 & 0 \\ \sin\gamma & 0 & \cos\gamma \end{bmatrix} \begin{bmatrix} 1 & 0 & 0 \\ 0 & \cos\theta & \sin\theta \\ 0 & -\sin\theta & \cos\theta \end{bmatrix}$$

$$\times \begin{bmatrix} \cos\psi_G & \sin\psi_G & 0 \\ -\sin\psi_G & \cos\psi_G & 0 \\ 0 & 0 & 1 \end{bmatrix}. \qquad (7)$$

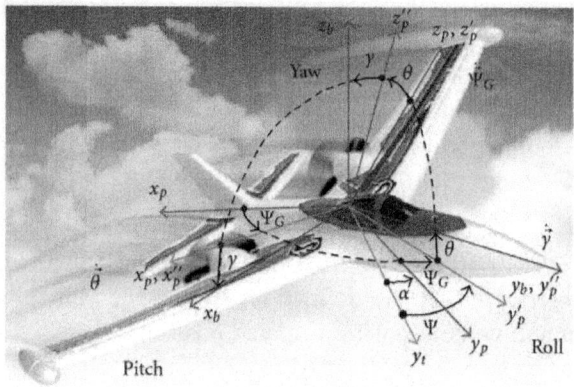

**Figure 4:** The Relation between $b$-Frame and $p$-Frame.

The vehicle attitude DCM $T_b^p$ is then obtained as

$$\mathbf{C}_b^p = \left(\mathbf{C}_p^b\right)^T = \begin{bmatrix} \cos\gamma\cos\psi_G - \sin\gamma\sin\theta\sin\psi_G & -\cos\theta\sin\psi_G & \sin\gamma\cos\psi_G + \cos\gamma\sin\theta\sin\psi_G \\ \cos\gamma\sin\psi_G + \sin\gamma\sin\theta\cos\psi_G & \cos\theta\cos\psi_G & \sin\gamma\sin\psi_G - \cos\gamma\sin\theta\cos\psi_G \\ -\sin\gamma\cos\theta & \sin\theta & \cos\gamma\cos\theta \end{bmatrix}.$$

$$(8)$$

For the $p$-frame system, the angle between the grid north $y_p$ and the true north $y_N$ is the wander azimuth angle $\alpha$. So the angle between the horizontal projection along $y_p'$ axis of the vehicle's vertical axis $z_b$ and the real north $y_N$ is the heading angle $\psi$. We have that

$$\psi = \psi_G + \alpha. \qquad (9)$$

So the direction cosine matrix $C_b^n$ from $b$-frame to $n$-frame is

$$\mathbf{C}_b^n = \begin{bmatrix} \cos\gamma\cos\psi - \sin\gamma\sin\theta\sin\psi & -\cos\theta\sin\psi & \sin\gamma\cos\psi + \cos\gamma\sin\theta\sin\psi \\ \cos\gamma\sin\psi + \sin\gamma\sin\theta\cos\psi & \cos\theta\cos\psi & \sin\gamma\sin\psi - \cos\gamma\sin\theta\cos\psi \\ -\sin\gamma\cos\theta & \sin\theta & \cos\gamma\cos\theta \end{bmatrix}.$$

$$(10)$$

The gimbal angles $\psi$, $\theta$, and $\gamma$ and the gimbal rates $\dot\psi$, $\dot\theta$, and $\dot\gamma$ are related to the body rate $\omega_{nb}^b$, which is the turn rate of the $b$-frame with respect to $n$-frame and measured in $b$-frame as follows:

$$\begin{bmatrix} \omega^b_{nbx} \\ \omega^b_{nby} \\ \omega^b_{nbz} \end{bmatrix} = \begin{bmatrix} 0 \\ \dot{\gamma} \\ 0 \end{bmatrix} + C_3 \begin{bmatrix} \dot{\theta} \\ 0 \\ 0 \end{bmatrix} + C_3 C_2 \begin{bmatrix} 0 \\ 0 \\ \dot{\psi} \end{bmatrix}$$

$$= \begin{bmatrix} \cos\gamma\dot{\theta} - \sin\gamma\cos\theta\dot{\psi} \\ \dot{\gamma} + \sin\theta\dot{\psi} \\ \sin\gamma\dot{\theta} + \cos\gamma\cos\theta\dot{\psi} \end{bmatrix}. \tag{11}$$

## Vehicle Position DCM $C^p_e$

Position matrix $C^p_e$ is the DCM from $e$-frame to $p$-frame. It has the following rotating sequence (see Figure 5)

$$x_e y_e z_e \xrightarrow{z_e, \lambda} x'_e y'_e z'_e \xrightarrow{y'_p, 90°-\varphi} x''_e y''_e z''_e$$

$$\xrightarrow{z''_p, 90°} x_E y_N z_U \xrightarrow{z_U, \alpha} x_p y_p z_p, \tag{12}$$

where $\lambda$ is the longitude angle $(-180°-180°)$, $\varphi$ is the latitude angle $(-90°-90°)$, and $\alpha$ is the wander azimuth angle $(0-360°)$. The above rotation can be written in the following matrix form:

$$C^p_e = \begin{bmatrix} -\sin\alpha\sin\varphi\cos\lambda - \cos\alpha\sin\lambda & -\sin\alpha\sin\varphi\sin\lambda + \cos\alpha\cos\lambda & \sin\alpha\cos\varphi \\ -\cos\alpha\sin\varphi\cos\lambda + \sin\alpha\sin\lambda & -\cos\alpha\sin\varphi\sin\lambda - \sin\alpha\cos\lambda & \cos\alpha\cos\psi \\ \cos\varphi\cos\lambda & \cos\varphi\sin\lambda & \sin\varphi \end{bmatrix}. \tag{13}$$

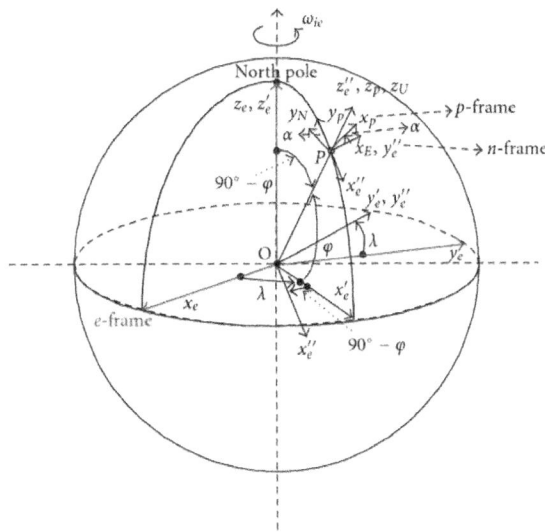

**Figure 5:** The Relation between $e$-Frame and $p$-Frame.

In Section 4, a trajectory simulation method in the ENU-frame is described step by step to generate sensor data. In Section 5, a trajectory and attitude simulator method in the $p$-frame is described step by step to derive the desired trajectory and attitude from the simulated sensor data or real sensor data; Section 6 provides the initial parameters and initial data calculation.

## SENSOR DATA GENERATOR

The Purpose of Trajectory Simulation is to Generate Data of the 3 Orthogonal Gyros and the 3 Orthogonal Accelerometers According to the Designed Trajectory. It is mentioned in Section 2 That $p$-Frame is set up to avoid the Singularities When the Vehicle Passes over Either the North or South Pole. But in Most Applications, The SINS Systems Are Seldom Operated under This Extreme Environment. The ENU-Frame Can be Implemented Easier than $p$-Frame, so it is Chosen as the Navigation Frame. Figure 6 Shows the Whole Process of the SINS Principal in the ENU-Frame Mechanization. First, The Vehicle Trajectory in the ENU-Frame is Set. Then, The Sensor Ideal Output is derived using the Inverse Principle of INS. The Sensor Simulation Data Can be obtained by Adding Noise to the Ideal Data. Then, we Use the Simulated Sensor Data to Derive the Noise-Corrupted Simulated Trajectory. Besides, The Difference between the Ideal and Simulated State Vectors Can be set as the Input for the Observed Measurements in the Kalman Filter.

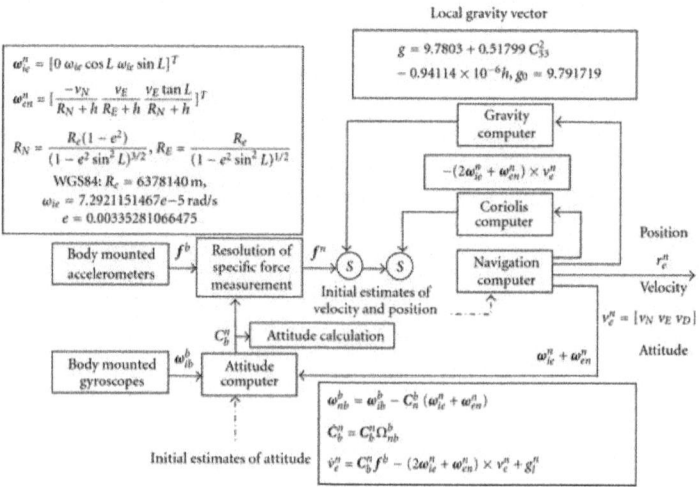

**Figure 6:** SINS ENU-Frame Mechanization.

## The Initial Parameters

For the designed trajectory, the initial parameters are

(1)    Initial position, latitude $\varphi_0$, longitude $\lambda_0$, height $h_0$;

(2)    Initial velocity $\mathbf{v} = [v_{E0}, v_{N0}, v_{U0}]$;

(3)    The designed variation of acceleration $\mathbf{a}$, which varies with time according to the designed trajectory;

(4)    the designed variations of the attitude angles, pitch $\theta$, roll $\gamma$, and heading $\psi$, and attitude angle rates, $\dot{\theta}, \dot{\gamma}$, and $\dot{\psi}$, which vary with time according to the designed trajectory.

The Update of Velocity

The velocity is updated as

$$\mathbf{v} \leftarrow \mathbf{v} + \mathbf{a}\Delta t \qquad (14)$$

where $\Delta t$ is the time step.

## The Update of Position

The position is updated as

latitude: $L \leftarrow L + \dfrac{v_N \Delta t}{R_N}$,

longitude: $\lambda \leftarrow \lambda + \dfrac{v_E \Delta t \sec L}{R_E}$,

altitude: $h \leftarrow h + v_U \Delta t$. $\qquad (15)$

## The Update of Attitude

The attitude angles are updated as

pitch: $\theta \leftarrow \theta + \Delta\theta$,

roll: $\gamma \leftarrow \gamma + \Delta\gamma$,

heading: $\psi \leftarrow \psi + \Delta\psi$. $\qquad (16)$

The attitude rates are updated as

pitch: $\dot{\theta} \leftarrow \dot{\theta} + \Delta\dot{\theta}$,

roll: $\dot{\gamma} \leftarrow \dot{\gamma} + \Delta\dot{\gamma}$,

heading: $\dot{\psi} \leftarrow \dot{\psi} + \Delta\dot{\psi}$. $\qquad (17)$

The expressions for $\Delta\theta, \Delta\gamma, \Delta\psi, \Delta\dot{\theta}, \Delta\dot{\gamma}$, and $\Delta\dot{\psi}$ depend on the designed trajectory.

The direction cosine matrix $C_b^n$ can be calculated using matrix expression (10). We have that $C_n^b = (C_b^n)^T$.

## Gyro Data Generator

The output of the gyros is

$$\omega_{ib}^b = \left(I - S_g\right)\left(C_n^b(\omega_{ie}^n + \omega_{en}^n) + \omega_{nb}^b\right) + \varepsilon^b, \tag{18}$$

where $\omega_{ib}^b$ is the simulated actual output, $I$ is the 3×3 unit matrix, $S_g$ is the 3×3 diagonal matrix whose diagonal elements correspond to the 3 gyros' scale factor errors, and $\varepsilon^b$ is the gyro's drift and can be simulated as the sum of a constant noise and a random white noise:

$\varepsilon^b = \varepsilon_{const}^b + \varepsilon_{random}^b$:

$$\omega_{ie}^n = \begin{bmatrix} 0 \\ \omega_{ie} \cos L \\ \omega_{ie} \sin L \end{bmatrix}. \tag{19}$$

In a static base, $\omega_{nb}^b$ is equal to zero, whereas, in a moving base it is obtained as

$$\omega_{nb}^b = \begin{bmatrix} \cos y \dot\theta - \sin y \cos \theta \dot\psi \\ \dot y + \sin \theta \dot\psi \\ \sin y \dot\theta + \cos y \cos \theta \dot\psi \end{bmatrix}. \tag{20}$$

$\omega_{en}^n$ is related to velocity $\mathbf{v}=[v_E, v_N, v_U]^T$ and can be expressed as

$$\omega_{en}^n = \begin{bmatrix} \dfrac{-v_N}{R_N} \\ \dfrac{v_E}{R_E} \\ \dfrac{v_E \tan L}{R_N} \end{bmatrix}. \tag{21}$$

## Accelerometer Data Generator

The measurement of the accelerometer is the specific force:

$$\mathbf{f}^b = (I - S_a)C_n^b \mathbf{f}^n + \eta^b,$$

$$\mathbf{f}^n = \mathbf{a} + (2\omega_{ie}^n + \omega_{en}^n) \times \mathbf{v} - \mathbf{g}, \tag{22}$$

where $\mathbf{f}^b$ is the simulated actual output, $I$ is the 3×3 unit matrix. $S_a$ is the 3×3 diagonal matrix whose diagonal elements correspond to the 3 accelerometers' scale factor errors, $\eta b$ is the bias considered as the sum of a constant noise and a random white noise $\eta^b = \eta_{const}^b + \eta_{random}^b$. $\mathbf{g} = [0\ 0\ g]^T$, and

$g = 9.7803 + 0.051799C_{33}^2 - 0.94114 \times 10^{-6}h \,(\text{m/s}^2)$, where $C_{33}$ is the 9th element of $C_e^p$ and $h$ is the vehicle altitude.

## Examples

For four examples of static, straight line, circle, and s-shape situations, details will be given next under the conditions that the vehicle is moving on the surface of the Earth with no attitude change except for the heading angle, which means that the pitch angle, roll angle, and altitude are constants during the simulation process:

$$\Delta\theta = 0,$$
$$\Delta\gamma = 0,$$
$$\Delta\dot{\theta} = 0,$$
$$\Delta\dot{\gamma} = 0. \qquad (23)$$

The calculation method for the other parameters for the four situations is described as follows.

(1)   Static:

$$L = \text{constant},$$
$$\lambda = \text{constant},$$
$$v_E = \text{constant},$$
$$v_N = \text{constant},$$
$$\Delta\psi = 0,$$
$$\Delta\dot{\psi} = 0. \qquad (24)$$

(2)   Straight

$$a_E = \text{constant},$$
$$a_N = \text{constant},$$
$$v_E = v_E + a_E\Delta t,$$
$$v_N = v_N + a_N\Delta t,$$
$$\psi = \psi_0 + \arctan\left(\frac{v_E}{v_N}\right),$$
$$\dot{\psi} = \frac{a_N v_E - a_E v_N}{v_E^2 + v_N^2}. \qquad (25)$$

(3)    Circle:

$$v_g = \text{constant},$$

$$\Delta\psi = \text{mod}\left(\frac{2\pi\Delta t}{T_{\text{circle}}}, 2\pi\right),$$

$$\Delta\dot{\psi} = \frac{2\pi}{T_{\text{circle}}},$$

$$a_E = -\frac{2\pi v_g \cos\psi}{T_{\text{circle}}},$$

$$a_N = -\frac{2\pi v_g \sin\psi}{T_{\text{circle}}}.$$

(26)

(4)    S-shape:

$$v_g = \text{constant},$$

$$a_E = -\frac{v_g \cos\left(\psi_0 + A_{\text{sshape}} \sin\left(2\pi t/T_{\text{sshape}}\right)\right)}{T_{\text{sshape}}} \cdot$$

$$\cdot \frac{2\pi A_{\text{sshape}} \cos\left(2\pi t/T_{\text{sshape}}\right)}{T_{\text{sshape}}},$$

$$a_N = -\frac{v_g \sin\left(\psi_0 + A_{\text{sshape}} \sin\left(2\pi t/T_{\text{sshape}}\right)\right)}{T_{\text{sshape}}} \cdot$$

$$\cdot \frac{2\pi A_{\text{sshape}} \cos\left(2\pi t/T_{\text{sshape}}\right)}{T_{\text{sshape}}},$$

$$\psi = \psi_0 + A_{\text{sshape}} \sin\left(\frac{2\pi t}{T_{\text{sshape}}}\right),$$

$$\dot{\psi} = \frac{2\pi A_{\text{sshape}} \cos\left(2\pi t/T_{\text{sshape}}\right)}{T_{\text{sshape}}}.$$

(27)

# MATHEMATICAL MODEL AND TRAJECTORY CALCULATION STEPS

After the Gyro and Accelerometer Data Are Simulated using the Method Described in the Previous Section under the Designed Scenario, The Next Step we have to do is to Figure out the Mathematical Model of SINS and the Calculation Steps to Process the Sensor Data to Get the Calculated Trajectories. Based on the Basic Principles of Strapdown Inertial Navigation System [4], we draw the Mathematical Model in the $p$-Frame Mechanization in Figure 7. The Calculation Steps Are Described Below. Although the Situation That the Vehicle Passes Over Either the North or South Pole Seldom Happens, The Universal $p$-Frame is Still Chosen Instead of the Simpler ENU-Frame to Give a Navigation Illustration in a Different Frame.

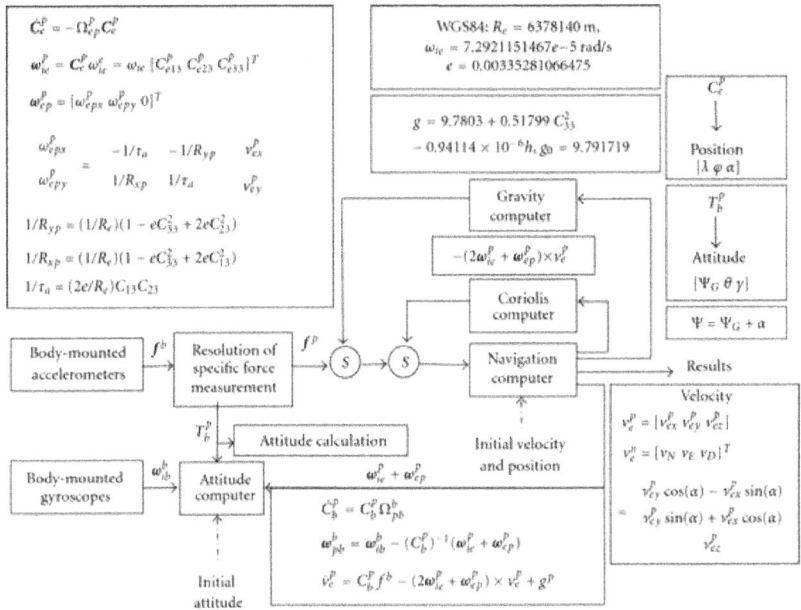

**Figure 7:** SINS $p$-Frame Mechanization.

## Quaternion $Q$ Update and Optimal Normalization

There are three kinds of strapdown attitude representations: DCM, Euler angle, and quaternion. In this paper, we choose quaternion. The reason why quaternion is chosen is explained in [3].

The quaternion formed by a rotating body frame around the platform frame is

$$Q = q_0 + q_1 i_b + q_2 j_b + q_3 k_b. \tag{28}$$

The update for the quaternion can be obtained by solving the following quaternion differential equation:

$$\begin{bmatrix} \dot{q}_0 \\ \dot{q}_1 \\ \dot{q}_2 \\ \dot{q}_3 \end{bmatrix} = \frac{1}{2} \begin{bmatrix} 0 & -\omega_{pbx}^b & -\omega_{pby}^b & -\omega_{pbz}^b \\ \omega_{pbx}^b & 0 & \omega_{pbz}^b & -\omega_{pby}^b \\ \omega_{pby}^b & -\omega_{pbz}^b & 0 & \omega_{pbx}^b \\ \omega_{pbz}^b & \omega_{pby}^b & -\omega_{pbx}^b & 0 \end{bmatrix} \begin{bmatrix} q_0 \\ q_1 \\ q_2 \\ q_3 \end{bmatrix}. \tag{29}$$

Based on the Euclide norm minimized indicator [4], the optimal normalization for the quaternion is

$$Q \longleftarrow \frac{Q}{\sqrt{q_0^2 + q_1^2 + q_2^2 + q_3^2}}. \tag{30}$$

$C_b^{p}$ Calculation

$C_b^{p}$ is vehicle attitude DCM which transforms the measured angle in the b-frame to the p-frame, with its 9 components $T_{ij}, i, j = 1, 2, 3$. (Here we use $T_{ij}$ to distinguish it from the components $C_{ij}$, $i, = 1, 2, 3$ of $C_b^{p}$ which is used below.)

After obtaining $q_0$, $q_1$, $q_2$, and $q_3$ using (29), $C_b^{p}$ can be calculated as

$$C_b^{p}$$

$$= \begin{bmatrix} T_{11} & T_{12} & T_{13} \\ T_{21} & T_{22} & T_{23} \\ T_{31} & T_{32} & T_{33} \end{bmatrix}$$

$$= \begin{bmatrix} q_0^2 + q_1^2 - q_2^2 - q_3^2 & 2(q_1 q_2 - q_0 q_3) & 2(q_1 q_3 + q_0 q_2) \\ 2(q_1 q_2 + q_0 q_3) & q_0^2 - q_1^2 + q_2^2 - q_3^2 & 2(q_2 q_3 - q_0 q_1) \\ 2(q_1 q_3 - q_0 q_2) & 2(q_2 q_3 + q_0 q_1) & q_0^2 - q_1^2 - q_2^2 + q_3^2 \end{bmatrix}. \tag{31}$$

Specific Force Transformation from $f^b$ in b-Frame to $f^p$ in p-Frame

The specific force $\mathbf{f}^b$ in the b-frame can be transformed to $\mathbf{f}^p$ in the p-frame by multiplication with DCM $C_b^{p}$

$$\mathbf{f}^p = C_b^{p} \mathbf{f}^b,$$

$$\begin{bmatrix} f_x^p \\ f_y^p \\ f_z^p \end{bmatrix} = C_b^{p} \begin{bmatrix} f_x^b \\ f_y^b \\ f_z^b \end{bmatrix}, \tag{32}$$

## Velocity $v_e^p$ Calculation

The velocity $v_e^p$ update can be obtained by solving the following differential equation:

$$\dot{v}_e^p = \mathbf{f}^p - \left(2\omega_{ie}^p + \omega_{ep}^p\right) v_e^p + \mathbf{g}^p,$$

$$\begin{bmatrix} \dot{v}_x \\ \dot{v}_y \\ \dot{v}_z \end{bmatrix} = \begin{bmatrix} f_x^p \\ f_y^p \\ f_z^p \end{bmatrix} - \begin{bmatrix} 0 \\ 0 \\ g \end{bmatrix}$$

$$+ \begin{bmatrix} 0 & 2\omega_{iez}^p & -\left(2\omega_{iey}^p + \omega_{epy}^p\right) \\ -2\omega_{iez}^p & 0 & 2\omega_{iex}^p + \omega_{epx}^p \\ 2\omega_{iey}^p + \omega_{epy}^p & -\left(2\omega_{iex}^p + \omega_{epx}^p\right) & 0 \end{bmatrix}$$

$$\times \begin{bmatrix} v_x \\ v_y \\ v_z \end{bmatrix}. \tag{34}$$

## Position Matrix $C_e^p$ Update

The update for the position matrix $C_e^p$ can be obtained by solving the following differential equation, noticing that $\omega_{epz}^p = 0$:

$$\dot{C}_e^p = -\Omega_{ep}^p C_e^p,$$

$$C_e^p = \begin{bmatrix} C_{11} & C_{12} & C_{13} \\ C_{21} & C_{22} & C_{23} \\ C_{31} & C_{32} & C_{33} \end{bmatrix},$$

$$\begin{bmatrix} \dot{C}_{11} & \dot{C}_{12} & \dot{C}_{13} \\ \dot{C}_{21} & \dot{C}_{22} & \dot{C}_{23} \\ \dot{C}_{31} & \dot{C}_{32} & \dot{C}_{33} \end{bmatrix} = \begin{bmatrix} 0 & 0 & -\omega_{epy}^p \\ 0 & 0 & \omega_{epx}^p \\ \omega_{epy}^p & -\omega_{epx}^p & 0 \end{bmatrix} \begin{bmatrix} C_{11} & C_{12} & C_{13} \\ C_{21} & C_{22} & C_{23} \\ C_{31} & C_{32} & C_{33} \end{bmatrix}. \tag{35}$$

## Position Angular Velocity $\omega_{ep}^p$ Update

In the chosen wander azimuth navigation frame, we have $\omega_{epz}^p = 0$, and

$$\begin{bmatrix} \omega_{epx}^p \\ \omega_{epy}^p \end{bmatrix} = \begin{bmatrix} -\dfrac{1}{\tau_a} & -\dfrac{1}{R_{yp}} \\ \dfrac{1}{R_{xp}} & \dfrac{1}{\tau_a} \end{bmatrix} \begin{bmatrix} v_{ex}^p \\ v_{ey}^p \end{bmatrix}, \tag{36}$$

Where

$$\frac{1}{R_{yp}} = \frac{1}{R_e}\left(1 - eC_{33}^2 + 2eC_{23}^2\right),$$

$$\frac{1}{R_{xp}} = \frac{1}{R_e}\left(1 - eC_{33}^2 + 2eC_{13}^2\right).$$

$$\frac{1}{\tau_a} = \frac{2e}{R_e}C_{13}C_{23}. \tag{37}$$

where the elements of position matrix $C_e^p$ can be obtained using (35).

Earth Angular Velocity $\omega_{ie}^p$ and Attitude Angular Velocity $\omega_{pb}^b$ Calculation
We Have That

$$\omega_{ie}^p = C_e^p \omega_{ie}^e = \begin{bmatrix} C_{11} & C_{12} & C_{13} \\ C_{21} & C_{22} & C_{23} \\ C_{31} & C_{32} & C_{33} \end{bmatrix} \begin{bmatrix} 0 \\ 0 \\ \omega_{ie} \end{bmatrix} = \begin{bmatrix} \omega_{ie}C_{13} \\ \omega_{ie}C_{23} \\ \omega_{ie}C_{33} \end{bmatrix}, \tag{38}$$

$$\omega_{pb}^b = \omega_{ib}^b - \omega_{ip}^b = \omega_{ib}^b - \left(C_b^p\right)^{-1}\left(\omega_{ie}^p + \omega_{ep}^p\right), \tag{39}$$

## Attitude Angle Calculation

The relation between attitude matrix $C_b^p$ and the three attitude angles, grid azimuth angle $\psi G$, pitch angle $\theta$, and roll angle $\gamma$, is

$$C_b^p = \begin{bmatrix} \cos\gamma\cos\psi_G - \sin\gamma\sin\theta\sin\psi_G & -\cos\theta\sin\psi_G & \sin\gamma\cos\psi_G + \cos\gamma\sin\theta\sin\psi_G \\ \cos\gamma\sin\psi_G + \sin\gamma\sin\theta\cos\psi_G & \cos\theta\cos\psi_G & \sin\gamma\sin\psi_G - \cos\gamma\sin\theta\cos\psi_G \\ -\sin\gamma\cos\theta & \sin\theta & \cos\gamma\cos\theta \end{bmatrix} \tag{40}$$

Thus, the principal values of $\psi_G$, $\theta$, and

$$\theta_{\text{principal}} = \sin^{-1} T_{32},$$

$$\gamma_{\text{principal}} = \tan^{-1} \frac{-T_{31}}{T_{33}},$$

$$\varphi_{G\text{principal}} = \tan^{-1} \frac{-T_{12}}{T_{22}}. \tag{41}$$

Considering the defined range of the angles, the expressions of the real values of $\psi_G$, $\theta$, $\gamma$, and are

$$\gamma \longleftarrow \begin{cases} \gamma_{\text{principal}}, & \text{if } T_{33} > 0, \\ \gamma_{\text{principal}} + 180°, & \text{if } T_{33} < 0, \gamma_{\text{principal}} < 0, \\ \gamma_{\text{principal}} - 180°, & \text{if } T_{33} < 0, \gamma_{\text{principal}} > 0, \end{cases}$$

$$\psi_G \longleftarrow \begin{cases} \psi_{G\text{principal}}, & \text{if } T_{22} > 0, \psi_{G\text{principal}} > 0, \\ \psi_{G\text{principal}} + 360°, & \text{if } T_{22} > 0, \psi_{G\text{principal}} < 0, \\ \psi_{G\text{principal}} + 180°, & \text{if } T_{22} < 0. \end{cases} \tag{42}$$

## Position Angle Calculation

The relation between position matrix $C_e^p$ and the 3 position angles, longitude $\lambda$, latitude $\varphi$, and wander azimuth angle $\alpha$, is

$$C_e^p = \begin{bmatrix} -\sin\alpha\sin\varphi\cos\lambda - \cos\alpha\sin\lambda & -\sin\alpha\sin\varphi\sin\lambda + \cos\alpha\cos\lambda & \sin\alpha\cos\varphi \\ -\cos\alpha\sin\varphi\cos\lambda + \sin\alpha\sin\lambda & -\cos\alpha\sin\varphi\sin\lambda - \sin\alpha\cos\lambda & \cos\alpha\cos\psi \\ \cos\varphi\cos\lambda & \cos\varphi\sin\lambda & \sin\varphi \end{bmatrix}. \tag{43}$$

Thus, the principal values of $\varphi$, $\lambda$, and $\alpha$ are

$$\varphi_{\text{principal}} = \sin^{-1} C_{33},$$

$$\lambda_{\text{principal}} = \tan^{-1} \frac{C_{32}}{C_{31}},$$

$$\alpha_{\text{principal}} = \tan^{-1} \frac{C_{13}}{C_{23}}. \tag{44}$$

Considering the defined range of the angles, the expressions of the real values of $\varphi$, $\lambda$, and $\alpha$ are

$$\varphi \longleftarrow \varphi_{\text{principal}},$$

$$\lambda \longleftarrow \begin{cases} \lambda_{\text{principal}}, & \text{if } C_{31} > 0, \\ \lambda_{\text{principal}} + 180°, & \text{if } C_{31} < 0, \lambda_{\text{principal}} < 0, \\ \lambda_{\text{principal}} - 180°, & \text{if } C_{31} < 0, \lambda_{\text{principal}} > 0, \end{cases}$$

$$\alpha \longleftarrow \begin{cases} \alpha_{\text{principal}}, & \text{if } C_{23} > 0, \alpha_{\text{principal}} > 0, \\ \alpha_{\text{principal}} + 360°, & \text{if } C_{23} > 0, \alpha_{\text{principal}} < 0, \\ \alpha_{\text{principal}} + 180°, & \text{if } C_{23} < 0. \end{cases} \tag{45}$$

## Heading Angle Calculation

The heading angle $\psi$ is calculated as

$$\psi = \psi_G + \alpha \qquad (46)$$

To make sure that $\psi$ will not be out of range, we should determine it according to

$$\psi \longleftarrow \begin{cases} \psi, & \text{if } \psi < 360°, \\ \psi - 360°, & \text{if } \psi \geq 360°. \end{cases} \qquad (47)$$

## Velocity $v_e^n$ in $n$-Frame Calculation

We Have That

$$v_e^n = \begin{bmatrix} v_E \\ v_N \\ v_U \end{bmatrix} = \begin{bmatrix} v_{ey}^p \cos\alpha - v_{ex}^p \sin\alpha \\ v_{ey}^p \sin\alpha + v_{ex}^p \cos\alpha \\ v_{ez}^p \end{bmatrix}. \qquad (48)$$

## Altitude Calculation

For the calculation of the altitude, damped methods should be used because it diverges with time. To simplify problems, in our simulations, we set the altitude to zero, that is, surface of the Earth.

## Local Gravity g Calculation

The local gravity $g$ is calculated as [5]

$$g = 9.7803 + 0.051799C_{33}^2 - 0.94114 \times 10^{-6}h \ (\text{m/s}^2), \qquad (49)$$

where $C_{33} = \sin\varphi$, $\varphi$ is the latitude and $h$ is the altitude above sea level.

Before we carry out the implementation of the above described mathematical model of SINS, we have to know the initial parameters of the system, which will be described in the following Section.

# INITIAL PARAMETERS AND INITIAL DATA CALCULATION

For the calculations in Section 5, we first need to know the given initial parameters and the corresponding initial data.

## Initial Parameters

(1)   Initial position, latitude $\varphi_0$, longitude $\lambda 0$, height $h_0$. The values of these parameters should be the same as the corresponding ones in Section 4.1.

(2)    Initial wander azimuth angle $\alpha_0$. We could choose $\alpha_0 = 0$ at the very beginning. The value should be the same as the corresponding ones in Section 4.1.

(3)    Initial velocity $v_{E0}$, $v_{N0}$, $v_{U0}$.

(4)    If barometric altimeter applied, initial external reference height $h_{ref0}$ can be supplied.

## Initial Alignment Data

(1)    Initial attitude matrix is determined by initial alignment process $C_{b0}^p$. $C_{b0}^p = C_{b0}^n$ when $\alpha_0 = 0$.

(2)    Initial position matrix is determined by initial alignment process $C_{e0}^p$. $C_{e0}^p = C_{e0}^n$ when $\alpha_0 = 0$.

## Initial Data Calculation

(1)    Initial attitude angles $\varphi_0$, $\lambda 0$, and $\alpha_0$ determination: The initial attitude angles $\psi_{G0}$, $\theta_0$, and $\gamma_0$ can be calculated using (41) and (42). Because $\alpha_0 = 0$, heading angle $\psi_0 = \psi_{G0}$.

(2)    Initial quaternion calculation: From the diagonal elements in (31) and the quaternion constraint equation, we have that

$$q_0^2 + q_1^2 - q_2^2 - q_3^2 = T_{11},$$

$$q_0^2 - q_1^2 + q_2^2 - q_3^2 = T_{22},$$

$$q_0^2 - q_1^2 - q_2^2 + q_3^2 = T_{33},$$

$$q_0^2 + q_1^2 + q_2^2 + q_3^2 = 1, \qquad (50)$$

The solution to (50)

$$|q_1| = \frac{1}{2}\sqrt{1 + T_{11} - T_{22} - T_{33}},$$

$$|q_2| = \frac{1}{2}\sqrt{1 - T_{11} + T_{22} - T_{33}},$$

$$|q_3| = \frac{1}{2}\sqrt{1 - T_{11} - T_{22} + T_{33}},$$

$$|q_0| = \sqrt{1 - q_1^2 - q_2^2 - q_3^2}. \qquad (51)$$

Assuming $q_0$ to be positive, according to (31), we have that

$$\text{sign}(q_0) = \text{sign}(1),$$
$$\text{sign}(q_1) = \text{sign}(T_{32} - T_{23}),$$
$$\text{sign}(q_2) = \text{sign}(T_{13} - T_{31}),$$
$$\text{sign}(q_3) = \text{sign}(T_{21} - T_{12}). \qquad (52)$$

(3)     Initial position matrix $C_{e0}^p$: Substituting initial position, latitude $\varphi_0$, longitude $\lambda_0$ and initial wander azimuth $\alpha_0 = 0$ into (43), we can obtain the initial position matrix: $C_{e0}^p$.

(4)     Initial Earth angular velocity $\omega_{ie0}^p$ and initial attitude angular velocity $\omega_{pb0}^p$ calculations: use (38) and (39).

(5)     Initial position angular velocity $\omega_{ep0}^p$ calculation: use (36) and (37).

(6)     Initial gravity $g_0$ calculation: use (49) and element $C_{33}$ in $C_{e0}^p$.

(7)     Initial ground velocity $v_{g0}$ calculation: use (34).

At this point, the whole SINS model, including sensor data generator and initial parameters, is fully described. The following Section will provide a Matlab implementation of the SINS theory.

# MATLAB IMPLEMENTATION AND SIMULATION EXAMPLES

First, the Matlab program structure and the main codes are given. The Matlab implementation is illustrated using six examples: static, straight, circle, s-shape, and the other two from real SINS experimental data.

## Matlab Implementation and Codes

The program structure is given in Figure 8. The program starts from "Begin" and ends at "Stop." The gyro and accelerometer data are obtained either from a sensor data generator described in Section 4 or from the real SINS experiment logged files. Processing the sensor data with the initial parameters, using the method described in Section 5, we get the attitude, velocity and position values of the system at specific times. After all data are processed, the program will stop and the results will be provided.

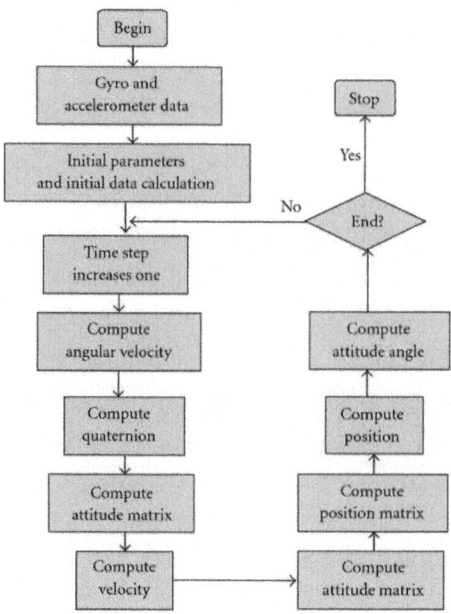

**Figure 8:** SINS Program Structure.

The main Matlab codes are presented next.

(1)    Initial settings:

(a)    initSettings.m contains initial parameters and constants used in the simulation project.

(2)    Trajectory part:

(a)    initialCalculation_static.m gives the initial calculation for the static situation;

(b)    trajectorySimulater_static.m simulates gyro and accelerometer data for the static situation;

(c)    initialCalculation_straight.m gives the initial calculation for the straight line situation;

(d)    trajectorySimulater_straight.m simulates gyro and accelerometer data for the straight line situation;

(e)    initialCalculation_cirlce.m gives the initial calculation for the circle situation;

(f)    trajectorySimulater_circle.m simulates gyro and accelerometer data for the circle situation;

(g)     initialCalculation_Sshape.m gives the initial calculation for the s-shape situation;

(h)     trajectorySimulater_Sshape.m simulates gyro and accelerometer data for the s-shape situation;

(3)     Simulation part:

(a)     INSmain.m is the main program; the simulation starts from here;

(b)     AltitudeParamete.m calculates the four damping parameters to damp the altitude error according to the input parameters $k4$ and $\tau$, to be used with the external reference altitude;

(c)     InitializePosition.m gives the initial position initLong, initLat, initAlt, the external reference altitude extAlt, and the wander azimuth angle wanderAzimuth; it calculates the initial position matrix and then orthogonalizes the matrix;

(d)     InitializeAttitude.m gives the initial alignment error and calculates the attitude matrix (strapdown matrix);

(e)     InitializeQuaternion.m calculates the quaternion according to the input attitude matrix;

(f)     ComputeAngularVelocity.m calculates the position angular velocity, earth angular velocity, and position angle increment in the $p$-frame and resets the gyroscopes and accelerometers;

(g)     ComputeQuaternionRungeKutta.m computes the quaternion using Runge-Kutta method [6]; see Appendices B and C;

(h)     ComputeAttitudeMatrix.m computes the attitude matrix and transfers the raw data of the accelerometers to the $p$-frame ;(i)ComputeVelocity.m computes the velocity, in the wander azimuth frame ($p$-frame) and ENU-frame, the ground velocity and altitude;

(j)     ComputePositionMatrix.m computes the position matrix.

(k)     ComputePosition.m computes latitude, longitude and wanderAzimuth;

(l)     ComputeAttitudeAngle.m computes the attitude angle of pitchAngle, tiltAngle, gridAzimuth andcourseAngle;

(m)     OrthogonalizeMatrix.m computes matrix orthogonalization;

(n)     QuaCofMatrix.m is called by ComputeQuaternionRungeKutta.m;

(o)     PlotResult.m plots the results of the simulation project.

## Simulation Examples

In this subsection, there are 6 SINS simulation examples. Example 1 is the static situation simulation, where the vehicle trajectory in the $n$-frame is a fixed point. Example 2 is the straight line situation simulation, where the vehicle trajectory in the $n$-frame is a straight line. Example 3 is the circle situation simulation, where the vehicle trajectory in the $n$-frame is a circle. Example 4 is the s-shape situation simulation, where the vehicle trajectory in the $n$-frame is an s-shape line. Here, high-accuracy SINS simulation is applied to the four situations. The initial latitude and longitude errors are set to be 1 minute. The simulation time is set to 3600 seconds. The initial positions are dependent on the designed trajectories.

In order to verity the validity of the Matlab codes further, two sets of real static data are used, and we refer to these as Examples 5 and 6. The two sets of real data, set A and set B, are collected from the same SINS in the same place but at different times. The 2 data sets are 24 hours long.

All the errors (the angle error, the velocity error, and the position error) will contribute to the distance error in the INS trajectory calculation. Thus, the distance error is a key index of an INS system. The distance error will increase with time, so it is always associated with a time stamp.

Example 1 (Static situation simulation). The static situation is the most basic and simple situation where the output of the gyro is the Earth rotating angular velocity and the output of the accelerometer is the gravity. Figure 9 shows the designed true trajectory. Figure 10 shows the difference between the calculated angle and the true angle. Figure 11 shows the differences between the calculated PV (position and velocity) and the true PV. The maximum value of the distance error in 1 hour is 3.5 nm (nautical mile).

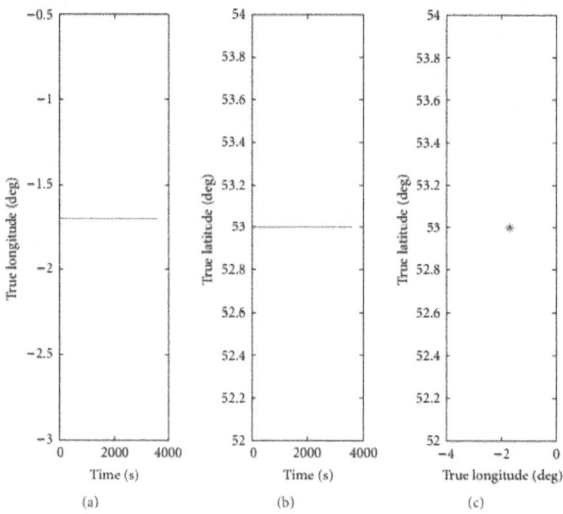

**Figure 9:** The designed trajectory of static simulation.

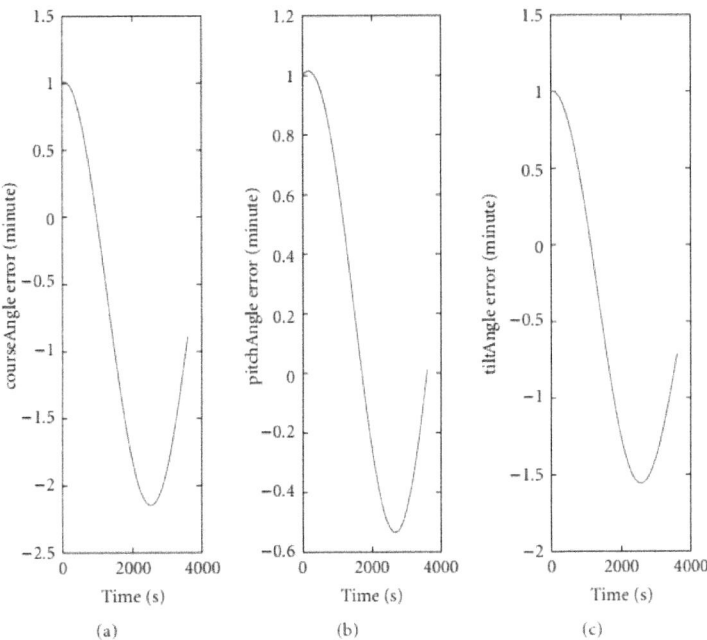

**Figure 10:** Angle error of static simulation.

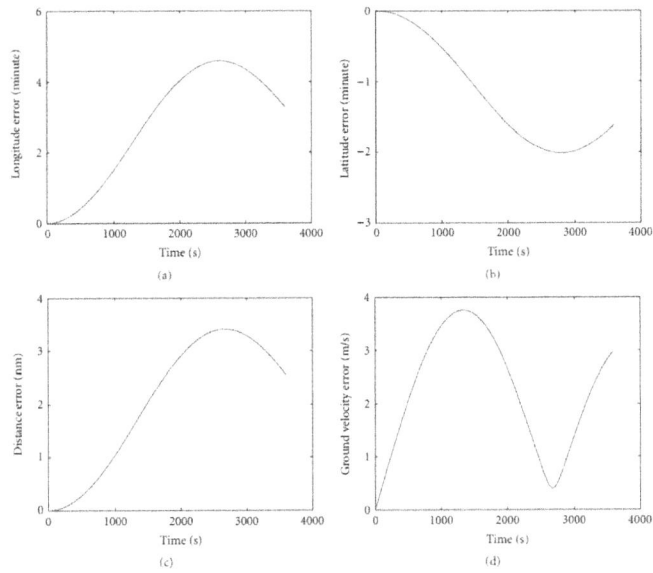

**Figure 11:** Position and velocity error of static simulation.

Example 2 (Straight line situation simulation). The straight line situation corresponds to a vehicle moving along the northwest direction. Figure 12 shows the designed true trajectory. Figure 13 shows the difference between the calculated angle and the true angle. Figure 14 shows the differences between the calculated PV and the true PV. The maximum value of the distance error in 1 hour is 3.7 nm.

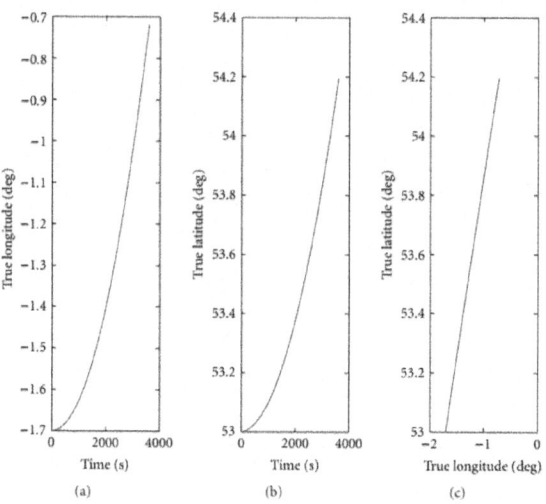

**Figure 12:** The designed trajectory of straight line simulation.

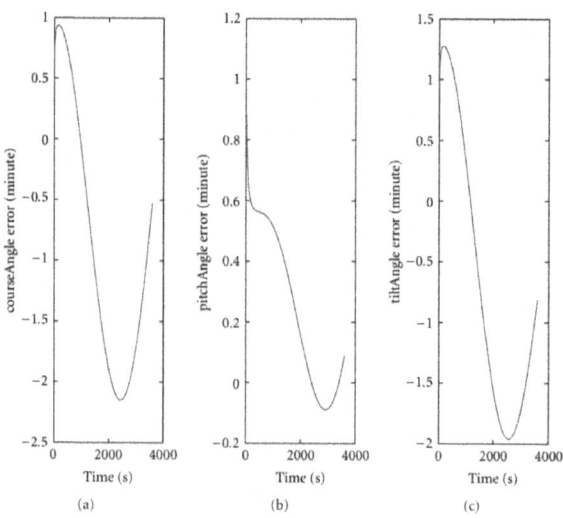

**Figure 13:** Angle error of straight line simulation.

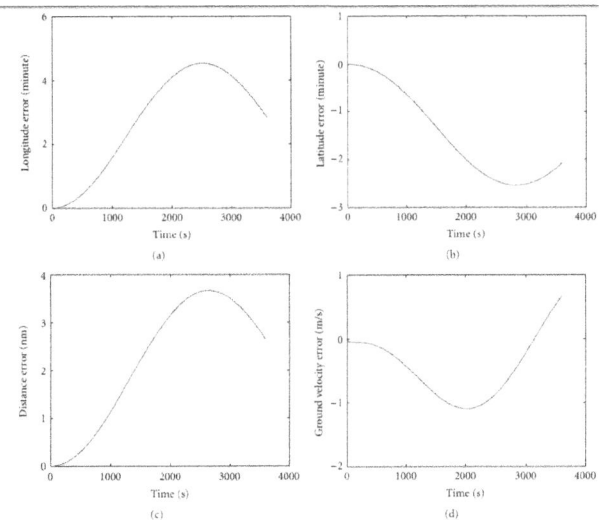

**Figure 14:** Position and velocity error of straight line simulation.

Example 3 (Circle situation simulation). The circle situation corresponds to a vehicle moving along a circle. Figure 15 shows the designed true trajectory. Figure 16 shows the difference between the calculated angle and the true angle. Figure 17 shows the difference between the calculated PV and the true PV. The maximum value of the distance error in 1 hour is 3.0 nm.

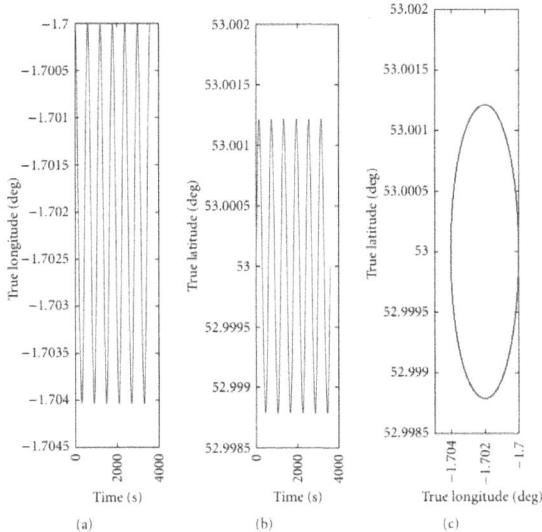

**Figure 15:** The designed trajectory of circle simulation.

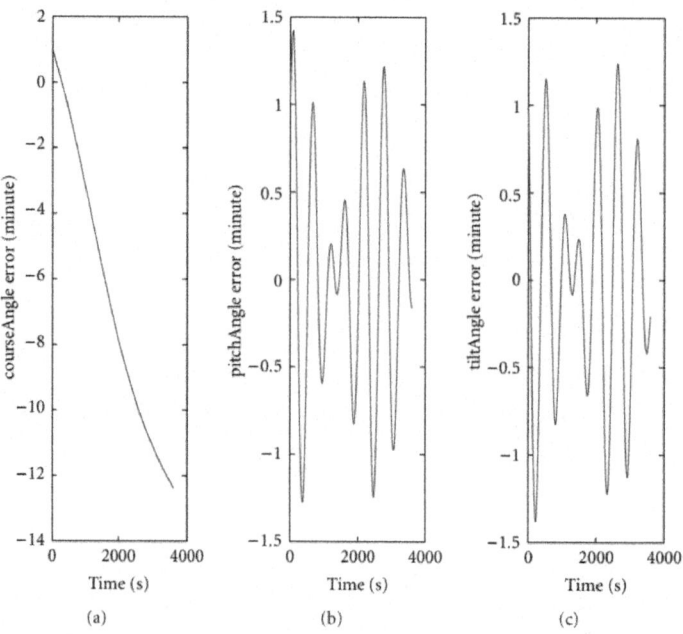

**Figure 16:** Angle error of circle simulation.

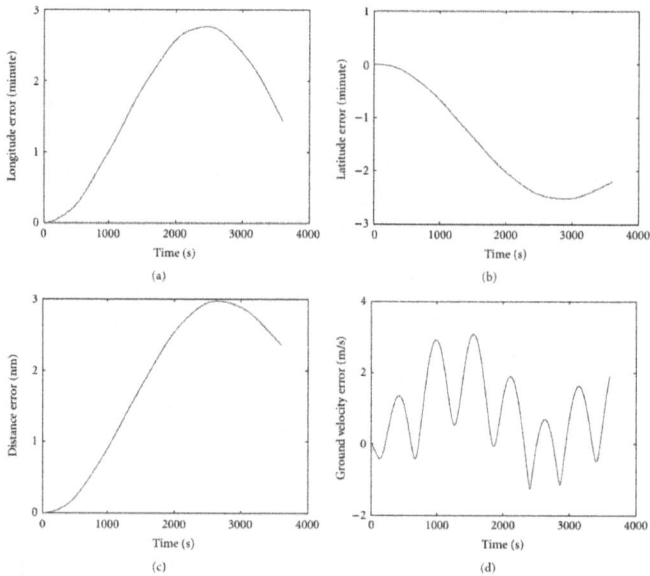

**Figure 17:** Position and velocity error of circle simulation.

Example 4 (S-shape situation simulation). The s-shape situation corresponds to a vehicle moving along an s-shaped line. Figure 18 shows the designed true trajectory of s-shape situation simulation. Figure 19 shows the difference between the calculated angle and the true angle. Figure 20 shows the differences between the calculated PV and the true PV. The maximum value of the distance error in 1 hour is 3.3 nm.

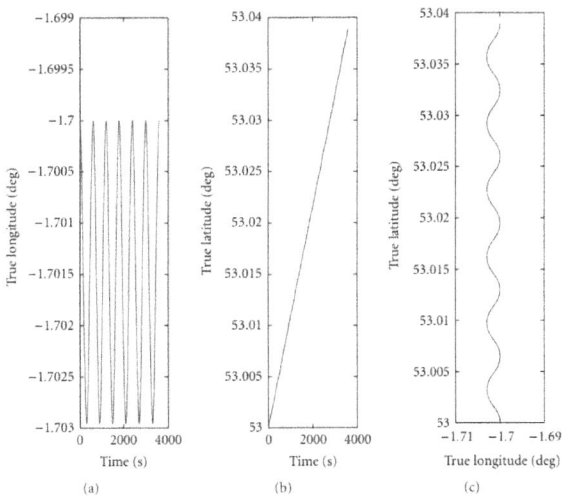

**Figure 18:** The designed trajectory of s-shape simulation.

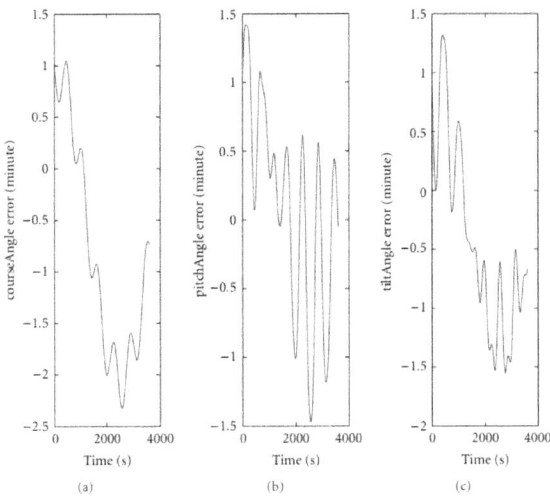

**Figure 19:** Angle error of s-shape simulation.

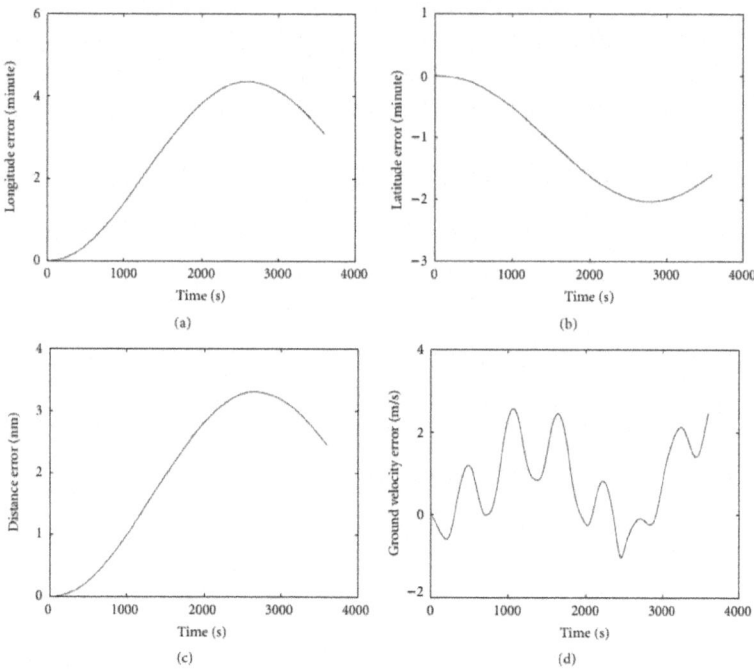

**Figure 20:** Position and velocity error of s-shape simulation.

Example 5 (Real static data set A simulation). First, we process data set A [7]. Figure 21 shows the trajectory for the real data set A; from the figure we can conclude that the system is static. In Figure 22, the red line corresponds to the three attitude angle errors of the real system, while the blue line corresponds to the three attitude angle errors processed by the Matlab code. We can also show that the difference between the red and blue lines is negligible. In Figure 23, the red line corresponds to the position and velocity errors of the real system, while the blue line corresponds to the position and velocity errors processed by the Matlab code. We can also see that the difference between the red and blue lines is negligible and this validates the correctness of the Matlab code. The error described by the red lines (output from the real system) is slightly smaller than that described by the blue lines (simulation). This is due to the fact that the real system is processed in a much higher rate and thus its input is more accurate than the simulated system.

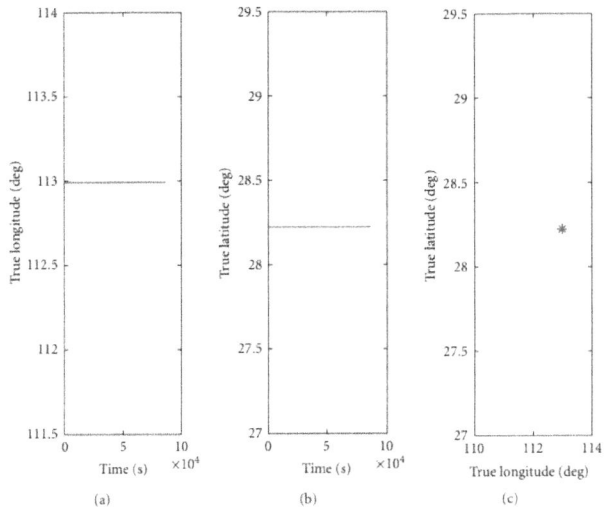

**Figure 21:** The trajectory of real data set A.

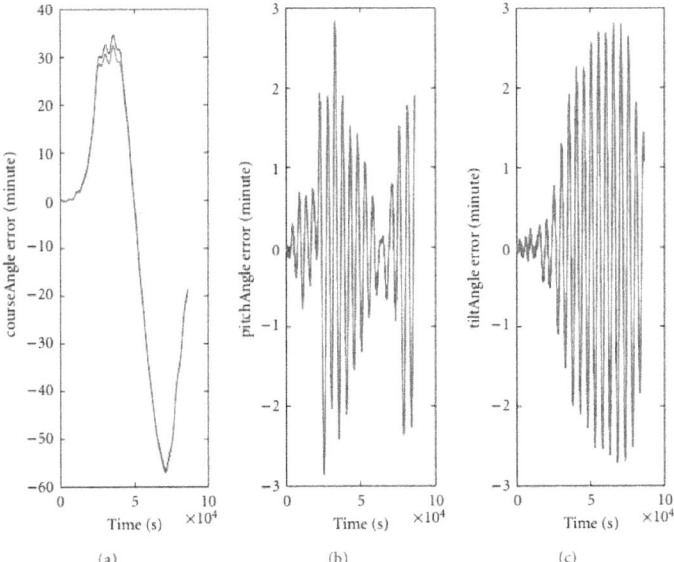

**Figure 22:** Angle error of real data set A.

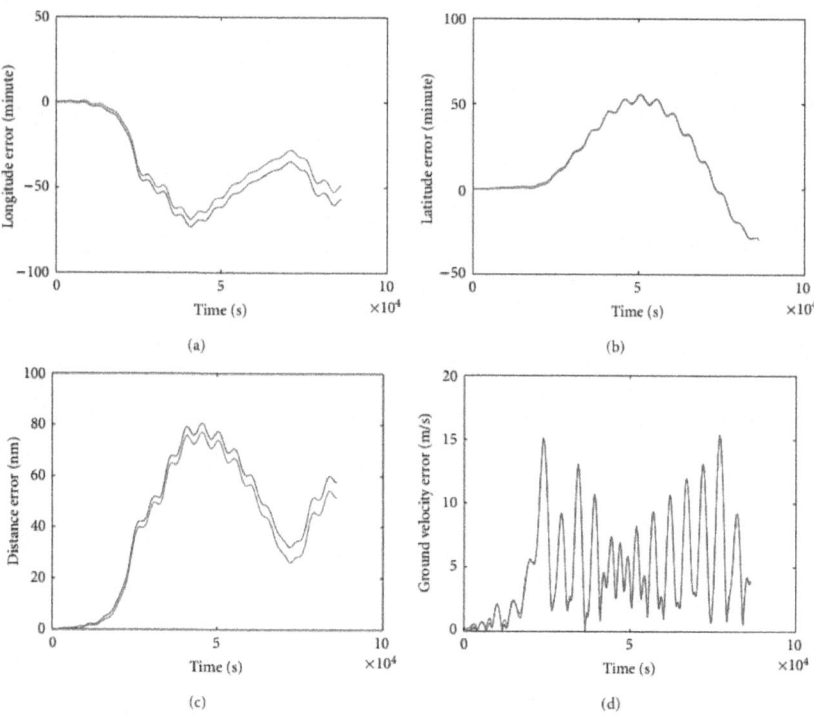

**Figure 23:** Position and velocity error of real data set A.

Example 6 (Real static data set B simulation). Figure 24 shows the trajectory of the real data set B; from the figure we can conclude that the system is static too. In Figure 25, the red line corresponds to the three attitude angle errors of the real system, while the blue line corresponds to the three attitude angle errors obtained by the Matlab code when applied to the real raw sensor data set B. We can also see that the difference between the red and blue lines is negligible. In Figure 26, the red line corresponds to the position and velocity errors of the real system, while the blue line corresponds to the position and velocity errors obtained by the Matlab code when applied to the real raw sensor data set B. We can also see that the difference between the red and blue lines is negligible, and this further validates the correctness of the Matlab code.

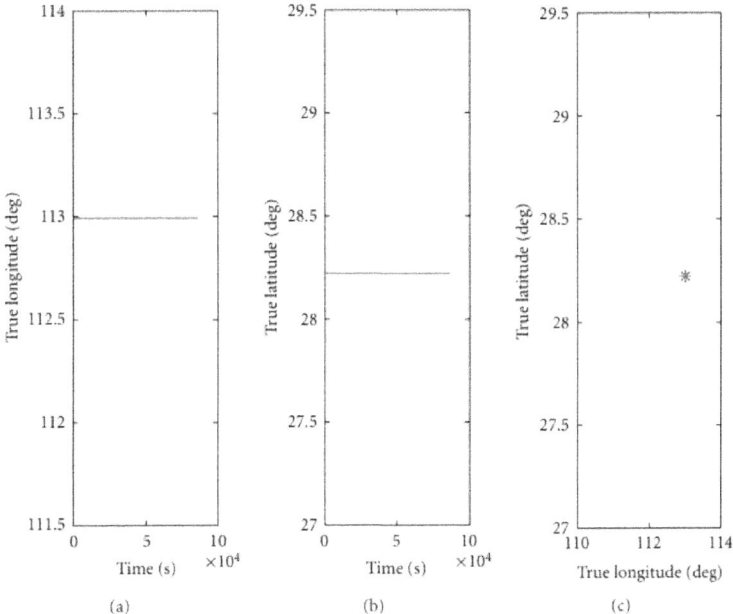

**Figure 24:** The trajectory of real data set B.

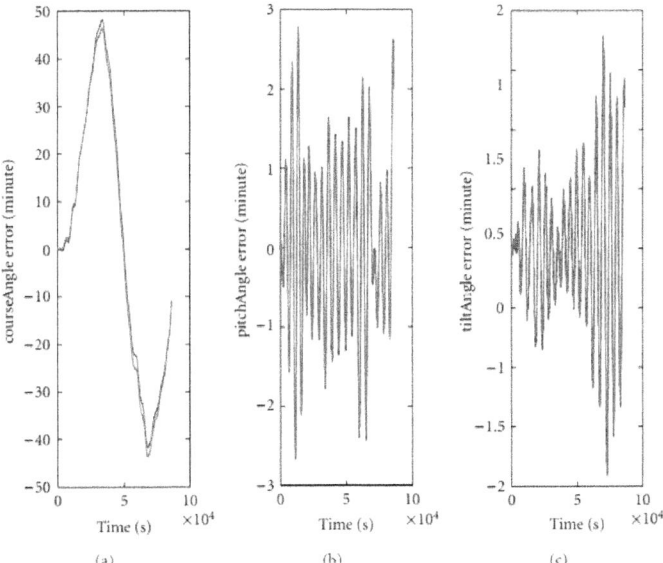

**Figure 25:** Angle error of real data set B.

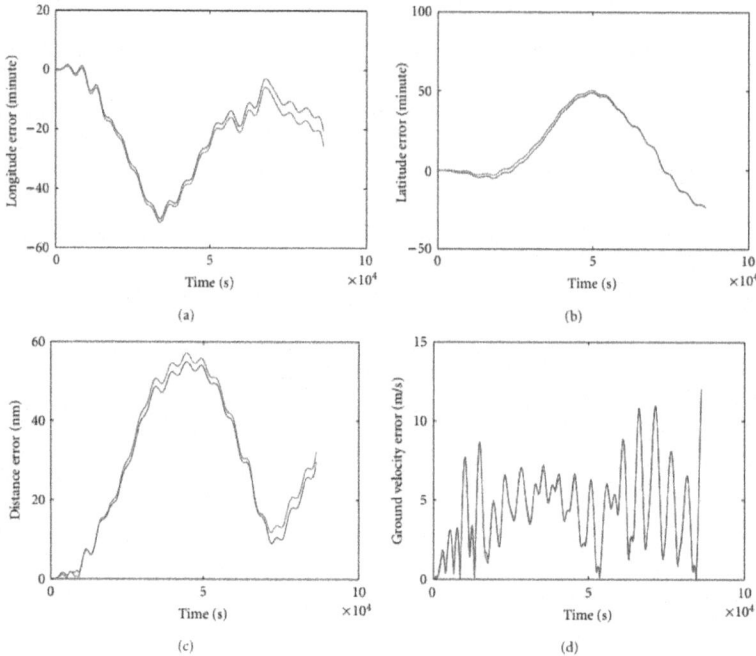

**Figure 26:** Position and Velocity Error of Real Data Set B.

## CONCLUSIONS

In this paper, a mathematical model for the strapdown inertial navigation system (SINS) is built and its Matlab implementation is developed. First, a number of Cartesian coordinate reference frames that relate to SINS are introduced, the basic principle of SINS in the wander azimuth navigation frame ($p$-frame) is explained, and the main equations are described. Second, the important attitude direction cosine matrix and position direction cosine matrix in the $p$-frame are defined in detail. Third, the mathematical model for SINS simulation is described in detail. Fourth, a trajectory simulator model is set up to generate data from three orthogonal gyros and three orthogonal accelerometers. The initial parameters and initial data calculations for the mathematical model are also carried out. Finally, a Matlab implementation of SINS is developed. The proposed simulation method is illustrated with four examples, static, straight line, circle, and s-shape trajectories; details are given under the condition that the pitch angle, roll angle, and altitude are constant during the simulation process. Further, two sets of real experimental data are processed to verify the validity of the Matlab code.

# REFERENCES

1.  H. Schneider and N. E. George Philip Barker, Matrices and Linear Algebra, Dover Publications, New York, NY, USA, 1989.

2.  A. Gilat, Matlab: An Introduction with Applications, John Wiley & Sons, New York, NY, USA, 3rd edition, 2008.

3.  D. H. Titterton and J. L. Weston, Strapdown Inertial Navigation Technology, Institution of Engineering and Technology, Stevenage, UK, 2004.

4.  Z. Chen, Strapdown Inertial Navigation System Principles, China Astronautic Publishing House, Beijng, China, 1986.

5.  P. S. Maybeck, "Wander azimuth implimentation algorithm for a strapdown inertial system," Air Force Flight Dynamics Laboratory AFFDL-TR-73-80, Tech. Rep., Ohio, USA, 1973.

6.  J. C. Butcher, Numerical Methods for Ordinary Differencial Equations, John Wiley & Sons, New York, NY, USA, 2003.

7.  B. Yuan, Research on Rotating Inertial Navigation System with Four-Frequency Differential Laser Gyroscope, Graduate School of National University of Defense Technology, Changsha, China, 2007.

8.  I. Y. Bar-Itzhack, "Iterative optimal orthogonalization of the strapdowm matrix," IEEE Transactions on Aerospace and Electronic Systems, vol. 11, no. 1, pp. 30–37, 1975.

# Chapter 12

## INTERVAL ARITHMETIC FOR NONLINEAR PROBLEM SOLVING

Benito A. Stradi-Granados

Department of Materials Science and Engineering, Institute of Technology of Costa Rica, Cartago 07050, Costa Rica

## ABSTRACT

Implementation of interval arithmetic in complex problems has been hampered by the tedious programming exercise needed to develop a particular implementation. In order to improve productivity, the use of interval mathematics is demonstrated using the computing platform INTLAB that allows for the development of interval-arithmetic-based programs more efficiently than with previous interval-arithmetic libraries. An interval-Newton Generalized-Bisection (IN/GB) method is developed in this platform and applied to determine the solutions of selected nonlinear problems. Cases 1 and 2 demonstrate the effectiveness of the implementation applied to traditional polynomial problems. Case 3 demonstrates the robustness of the implementation in the case of multiple specific volume solutions. Case 4 exemplifies the robustness and effectiveness of the implementation in the determination of multiple critical points for a mixture of methane and hydrogen sulfide. The examples demonstrate the effectiveness of the method by finding all existing roots with mathematical certainty.

## INTRODUCTION

There are a large number of problems that require the computation of stationary points and roots of equations. These are found in the fields of optimization [1], economics and finance [2, 3], thermodynamics [4, 5], applied mathematics [6], and similar contributions over the last thirty years. The application of interval arithmetic involves from algebraic equations with known solutions to more complicated systems representing physical phenomena. Among the earlier problems, second- and third-order polynomial problems serve to illustrate the effectiveness of the implementation. Similarly, more complicated multidimensional problems serve to illustrate stability and robustness of the implementation. In particular, the determination of critical points is of interest. Critical point computations are a well-known highly nonlinear problem that has been studied for a long time. The literature is endowed with some worthy analyses of the problem. Michelsen and Heidemann [7] discuss the calculation of critical points from cubic two-constant equations of state. They numerically determine the critical points of mixtures solving the highly nonlinear critical point equations. Hoteit et al. [8] claim an efficient algorithm based on bisection, secant, and inverse quadratic programming methods where their method is apparently faster but incapable of handling the presence of multiple critical points without restarting the program after each critical point determination. Nichita and Gomez [9] discuss the use of the tunneling global optimization method to determine all the of global minima where they concentrate on the determination of the critical points of mixtures. They use temperature and molar volume as primary variables. Their method lacks the strength of being able to stop when all critical points have been found, and this follows from the fact that at the beginning of the process the number of critical points to be determined needs to be specified. Michelsen [10] develops some earlier ideas that deal with the computation of phase envelopes with use of finite differences. Michelsen [11] discusses the isothermal flash separation and critical point computations with the use of tangent plane analysis. Michelsen and Kistenmacher [12] describe the disadvantage of using composition-dependent binary interaction coefficients in computations. Michelsen and Heidemann [13] present the calculation of tricritical points using tangent-plane distance analysis with some extra algebra in the mathematical development. The results are generated using the Peng-Robinson and Soave-Redlich-Kwong (SRK) equations of state. Michelsen [14] illustrates the determination of the critical points and phase boundaries using the real-arithmetic Newton-Raphson method with temperature and pressure as variables; in his analysis rather close initial estimates to the solution are needed for convergence.

There are also other more general contributions such as that of Cai et al. [15] who developed expressions for the spinodal criterion, critical criterion, and various stability tests for systems containing one discrete component and one polydisperse polymer. Lindvig et al. [16] propose the EFV (Entropic-FV) model for predicting the miscibility behavior of paints. Cismondi and Michelsen [17] use a Newton procedure for the calculation of critical lines, critical endpoints, and three-phase lines for binary mixtures. Carstensen and Petković [18] propose the use of a hybrid method combining normal (Nourein's method) and interval arithmetic to arrive to the solution of polynomial equations efficiently. Maranas and Floudas [19] propose an approach based on convex lower bounding and domain partitioning to achieve convergence to all solutions within a domain. Stradi et al. [4] discuss the computation of critical points using the INTBIS library [20].

This paper presents the solution of nonlinear problems using interval arithmetic in INTLAB (INTerval LABoratory) [21] by using the interval-Newton Generalized-Bisection method (IN/GB). The advantage of this approach is that INTLAB is developed over MATLAB [22], a well-known and powerful computational package widely used in mathematics and engineering applications [23–25]. Although, with the cost of interpretation overhead [26], INTLAB facilitates interval computations by using the MATLAB user interface and debugging tools, this makes production times smaller and results generation faster [27]. Section 2 of the paper provides a description of the problems to be solved in the paper, Section 3 proposes the methodology to solve the problems using the IN/GB method, Section 4 presents the results and discussion comparing the IN/GB method with the simple bisection method, and Section 5 highlights the conclusions derived from this study.

## PROBLEM DESCRIPTION

This section starts with two examples to demonstrate the use and performance of the IN/GB method. The first deals with two second-order equations where the solution is to be found over a relatively large domain. The second illustrates a particularly important case of a repeated root that causes the first derivative to be zero, and consequently it would traditionally create problems during the process of solution for the real-arithmetic Newton-Raphson method [23]. The third and fourth examples are related to the behavior of fluids in the two phase and critical regions.

Consider the general case where all of the roots to the nonlinear problem $(x) = 0$ have to be found within a given domain in the absence of an initial guess.

Case 1. The first problem is a set of two second-order equations as follows:

$$(xy)^2 + 4x - 4 = 0,$$

$$y^2 + 4xy - 4 = 0.$$
$$(1)$$

The domain for the search is the same for both variables: $x \in [-100, 100]$, $y \in [-100, 100]$. Three solutions are found for these two equations with mathematical certainty in a single run of the IN/GB method implemented over the INTLAB program.

Case 2. The second problem is a third-order polynomial with three real roots, where one of the roots is repeated:

$$x^3 - 5x^2 + 7x - 3 = 0. \tag{2}$$

This is an important example because the derivative of the function is zero at the repeated root. The traditional real-arithmetic Newton-Raphson method [23] would fail if we needed to apply the method near the root where the derivative is close to zero. The domain of the search is $x \in [-100, 100]$

The interval Newton is expected to have problems because of its reliance on the interval derivative of the function to find a solution, but the combination with the bisection method should solve the problem.

Case 3. The third problem deals with the determination of volume solutions for the Peng-Robinson equation of state [28] for carbon dioxide at a given temperature and pressure, which is given by the following expression that related pressure, temperature, and volume:

$$P - \frac{RT}{(v-b)} + \frac{a}{(v*(v+b)+b*(v-b))} = 0. \tag{3}$$

This can also be written in the form

$$P(v-b)(v*(v+b)+b*(v-b))$$
$$- RT(v*(v+b)+b*(v-b)) + a*(v-b) = 0,$$

$$a = \frac{0.45724R^2T_C^2}{P_C\left(1 + (0.3764 + 1.54226w - 0.2699w^2)\left(1 - T_r^{1/2}\right)\right)^2},$$

$$b = \frac{0.07780RT_C}{P_C}. \tag{4}$$

The necessary conditions for this problem are the following

$T_C = 304.21$ K,

$P_C = 73.83$ bar,

$w = 0.224$,

$T = 301$ K,

$P = 68.18 * 10^5$ Pa,

$R = 83.14$ (Pa cm$^3$/gmol K),

and the interval for the volume search is given by $V \in [1.1b, 16b]$..

This is a classic chemical engineering problem where, for a given equation of state, there may be more than one molar volume that satisfies the equation of state at a given temperature and pressure. In particular, if the conditions are such that two phases coexist, like in this case, then there will be three solutions. The traditional interpretation is that the solutions represent the liquid molar volume (smallest volume solution) and the vapor molar volume (largest volume solution). The middle volume solution is an unstable solution, and thus not observable.

Case 4. This is a highly nonlinear problem subject to analysis by different research groups for a number of years [4, 7, 14, 17]. In this case, there is a mixture of methane and hydrogen sulfide where the problem is to determine all of the critical points for a given composition and domain ranges for temperature and molar volume using the Redlich-Kwong equation of state [28]. The problem was solved in a previous development with the use of the INTBIS library and some extensive programming.

The domains of the variables under study are indicated as follows:

$T \in [180, 600]K,$

$v \in [1.1, 4]b,$

$\Delta x_1 \in [0, 1],$

$\Delta x_2 \in [-1, 1].$  (5)

The criticality conditions are written as follows:

$$\sum_{j=1}^{C} A_{ij}\Delta n_j$$
$$= \frac{RT}{n}\left(\frac{\Delta n_i}{y_i} + F_1\left(\beta_i\bar{N} + \bar{\beta}\right) + \beta_i F_1^2\bar{\beta}\right)$$
$$+ \frac{a}{bn}\left(\beta_i\bar{\beta}F_3 - \frac{F_5}{a}\sum_{j=1}^{C}a_{ij}\Delta n_j + F_6\left(\beta_i\bar{\beta} - \alpha_i\bar{\beta} - \bar{\alpha}\beta_i\right)\right),$$
$$i = 1, \ldots, C,$$  (6)

The mass balance for a mixture of constant composition is given by

$$\sum_{i=1}^{C}\Delta n_i^2 - 1 = 0.$$  (7)

Meanings are given in the appendix.

# METHODOLOGY

We employed the interval mathematics platform INTLAB to implement the interval-Newton Generalized-Bisection (IN/GB) method for the solution of nonlinear problems. The advantage of this procedure is that INTLAB is user friendly and runs over MATLAB; this makes INTLAB a very promising platform by putting the tools of MATLAB at the disposal of the INTLAB programmer. In the past, one of the major problems with the use of interval mathematics was the use of libraries that required extensive programming and lacked utilities for fast debugging. As a result, large time investments needed to be made in programming and debugging prior to the implementation and generation of first results, a situation that obviously derives in making interval arithmetic less attractive to a larger scientific audience. However, with INTLAB, the situation is different: program sizes are much smaller, debugging is more efficient, and results are generated more rapidly.

## The Interval-Newton Generalized-Bisection Method

An interval-Newton Generalized-Bisection method (IN/GB) is presented to demonstrate the use of the INTLAB platform for nonlinear problems of different levels of complexity.

The interval IN/GB method is a powerful computational approach that allows for the computation of all of the roots of a nonlinear problem without the need for initial guesses for the variables [1, 29, 30].

The IN/GB method requires only that we specify the domain of the variables of interest. The program tests a sequence of enclosures determined through the IN/GB method. If there is a root, then the IN/GB method determines a very thin slice in which the root lies. On the other hand, if there is no root, then the algorithm also determines with mathematical certainty the absence of solutions. The most important component part of the IN/GB method is the Newton method in interval arithmetic. This is written as follows:

$$J(X)(N - x) = -f(x). \tag{8}$$

This equation is of the form $AX = b$, and consequently suitable for solution with simultaneous linear equations solvers.

In this equation, capital letters are intervals and small case letters are real numbers. $(X)$ is the interval extension of the Jacobian matrix, which is determined by evaluating the Jacobian matrix, $J$, using the interval domain $X$, rather than the real variable $x$, where $x$ is generally the midpoint of $X$. This system of equations is solved using the Gauss-Seidel method [1], or a similar algorithm to determine $N$, where $N$ is the image $X$ generated by the application of the IN/GB method. If $N$ and $X$ intersect, then the process is repeated, noting that in successive applications it is the intersection of $N$ and $X$ ($N \cap X$) that serves as initial interval. The process continues until a sufficiently thin interval around the solution is derived. Other implementations, when sufficiently closed to the solution, switch to the real-arithmetic NewtonRaphson method to accelerate convergence [20].

If $N$ is a subset of $X$, then there is a single root in the domain of interest [29]. Similarly, if $N$ does not intersect $X$, then there is no root in the domain, and a new subdomain is tested for the presence of roots

It is important to indicate at this point that there are several problems not generally mentioned in the literature that may occur when using the IN/GB method. If the interval $X$ is too large, then the image $N$ will contain and be larger than the original interval $X$, generating the problem where the image is larger than the original interval and no convergence is achieved with the interval-Newton portion of the method alone. The other problem is where the image $N$ is equal to $X$; in this case the search will not advance, and no solution will be achieved within the permissible execution time. In both cases, the action taken is to proceed with the partition of the box, as shown in Table 1, and restart the search for each subdomain generated. The partition is done with the Bisection method [23] over each dimension of the domain followed by a range test applied to all subdomains generated.

**Table 1:** Interval-Newton Generalized-Bisection method results and actions

| Operation | Result | Comment | Action |
|-----------|--------|---------|--------|
| $N \cap X$ | $I$ | $I$ smaller in size | Continue with interval Newton |
| $N \cap X$ | $X$, other | No reduction in the original domain size | Bisect |

Application of the Bisection method generates two intervals from each interval that is bisected. Consequently, in order to explore all of the possibilities, the combinations of these intervals are needed to determine the existence and values of all possible solutions.

The existence of a root is ascertained by evaluating the original function over the interval, or subdomain, of interest, using interval arithmetic [30]. If the resulting interval contains zero, $0 \subset (X)$, then there may exist at least one root in that domain. This process of testing for the existence of a root in the domain of interest is called range test. If the evaluation of the function over the interval of interest, $X$, results in an interval image that does not contain zero, then there is a mathematical guarantee that there is no root within the interval $X$

For example, if three intervals are bisected, then six subintervals are generated, and eight combinations of them would need to be tested first in the range test to ascertain whether there may be a root in the subdomain, and second in the IN/GB method search for the solution.

Generally, multiple applications of the Bisection method, part of the IN/GB method, are needed when large variable domains are used. This situation is particularly true when zeros are contained in diagonal elements of the interval Jacobian expression, $J$. The presence of these zeros generates interval images, $N$, in the IN/GB method with infinite spans for some or all variables. Consequently, bisection is needed to divide and eliminate those subdomains where infinite spans are generated. These computational problems are managed efficiently in INTLAB. In INTLAB, the programming environment allows for processing warnings identified by the flags infinite (inf) and not-a-number (NaN) without an error that would stop the program execution. The flag inf would identify quantities that exceed real number representation such as $e\ 1^{000}$ and also results generated by the division of a number by zero: 1.0/0.0. The flag NaN would identify quantities that are mathematically undefined such as 0.0/0.0 or intervals that do not intersect. Consequently, if these flags occur, then bisection is applied because they are indicative of problems with the interval-Newton portion of the algorithm. This capacity to handle numerically undefined quantities saves extensive programming time and debugging effort invested on the solution of a particular problem

The search for a solution can only end with one of two outcomes, either a root is found within an interval or no root is found with mathematical certainty. The main steps of the algorithm are represented in Figure 8.

# RESULTS AND DISCUSSION

## Case 1

Table 2 presents the results obtained using the IN/GB and the Bisection methods applied to Case 1. The IN/GB method determined all of the roots and searched the entire domain with 685 subdivisions. The Bisection method with interval arithmetic was applied, and convergence to all roots was achieved with 820 subdivisions of the original interval. It is important to mention that, before applying either the IN/GB method or the Bisection method, a range test is applied to the subdomain under consideration to determine whether a root may be found within the subdomain. Three solutions are found for these two equations. The algorithm developed found three solutions without the need for initial guesses over a fairly large domain. This case is a baseline case where the solutions can be calculated also with the real-arithmetic Newton method, but then at least three different initial guesses would need to be provided.

**Table 2:** Results for Case 1

| Domain | Equations | Roots | Subset | Method |
|---|---|---|---|---|
| $[-100, 100]$ | $(x_1 x_2)^2 + 4x_1 - 4 = 0$ | $x_1 = -0.68443924$ | 647 | N |
| | $x_2^2 + 4x_1 x_2 - 4 = 0$ | $x_2 = 3.79247516$ | 809 | B |
| | | $x_1 = 0.465822965$ | 654 | N |
| | | $x_2 = -3.13799232$ | 814 | B |
| | | $x_1 = 0.84600538$ | 677 | N |
| | | $x_2 = 0.92770308$ | 819 | B |
| | | Total | | |
| | | | 685 | N |
| | | | 820 | B |

N: IN/GB method, B: Bisection method.

Figure 1 shows the evolution of the error with the subinterval number for the Bisection method; Figure 2 shows the evolution of the error with the subinterval number for the IN/GB method. The error limit is represented with a horizontal line at $1 * 10^{-6}$, and the trends are approximated with a third-order polynomial in a semilog plot for the Bisection and the IN/GB methods. The interval-Newton part of the IN/GB method may generate some very large errors particularly when the derivative, or the Jacobian matrix, contains a zero in the diagonal-element intervals. Consequently, in order to keep the scale to a legible size, large errors described as inf and nondefined quantities described as NaN appear with an error value of 10 on the graph's $y$- -axis. For both the bisection method and the IN/GB method, the solutions are found toward the end of the search of subdomains.

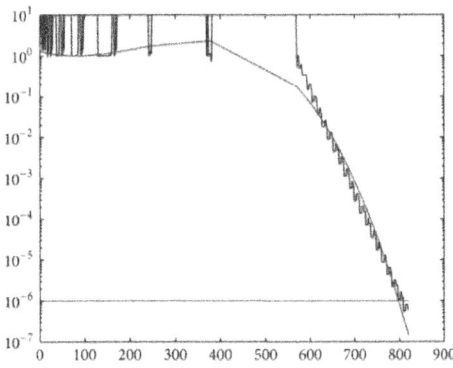

**Figure 1:** Evolution of the relative error (-axis) with the subinterval number for the Bisection method for Case 1.

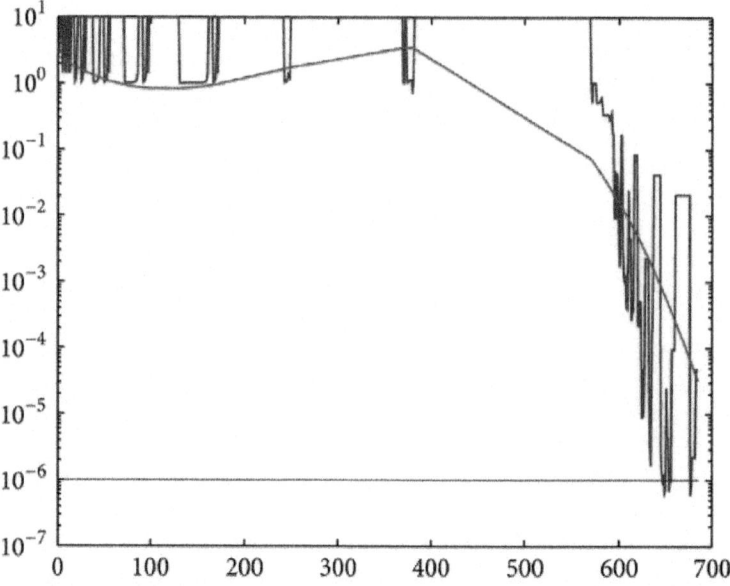

**Figure 2:** Evolution of the relative error (-axis) with the subinterval number for the IN/GB method for Case 1.

## Case 2

Table 3 presents the results for Case 2, where the IN/GB method determines all of the roots to this third-order equation using 17723 subdivisions to complete the search over the entire domain. The bisection method requires 17981 subdivisions to search the entire domain and find all of the roots. It is important to notice that the first root is found much faster by the IN/GB method by a factor of more than a hundred times over the bisection method. However, the convergence speed is smaller at the second root where both algorithms need closely the same number of subdivisions. For this simple example, the roots are evident and can be determined by inspection. However, if using the real-arithmetic Newton-Raphson method, there would be no immediate way to determine that one of the roots is repeated, and this situation would require additional time to test additional initial guesses or execute a polynomial deflation procedure with the roots already found.

**Table 3:** Results for Case 2

| Domain | Equation | Roots | Subset | Method |
|---|---|---|---|---|
| $[-100, 100]$ | $x^3 - 5x^2 + 7x - 3 = 0$ | $x = 3.00000000$ | 46 | N |
| | | $x = 2.99997777$ | 9121 | B |
| | | $x = 0.99807270$ | 12543 | N |
| | | $x = 0.99807270$ | 12801 | B |
| | | (double root) | | |
| | | | Total | |
| | | | 17723 | N |
| | | | 17981 | B |

Figure 3 shows the evolution of the error with the subinterval number for the Bisection method. Figure 4 shows the evolution of the error with the subinterval number for the IN/GB method.

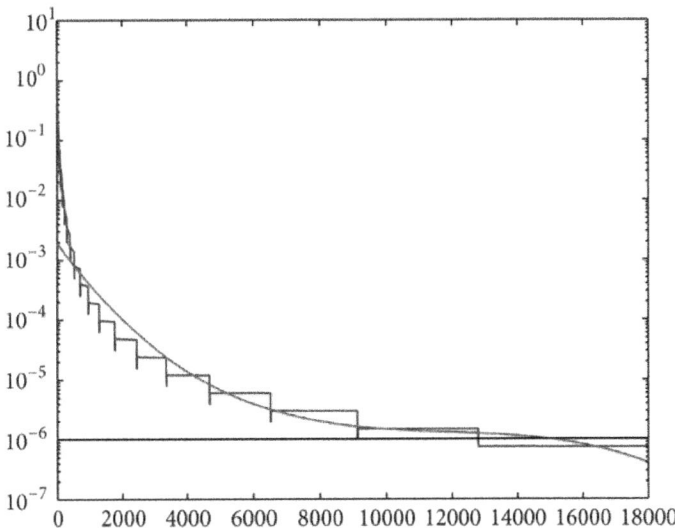

**Figure 3:** Evolution of the relative error (*y*-axis) with the subinterval number for the Bisection method for Case 2.

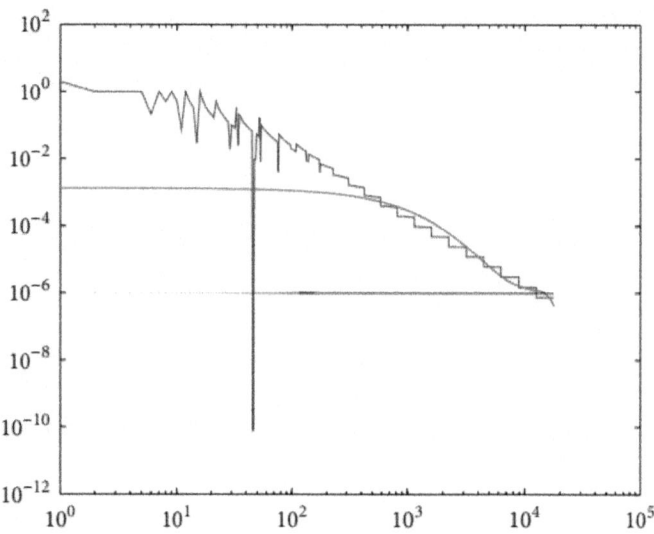

**Figure 4:** Evolution of the relative error (*y*-axis) with the subinterval number for the IN/GB method for Case 2.

In the case of the Bisection method, it is particularly different the convergence profile, compared to Case 1; both cases have been approximated with a third-order polynomial determined with a least-squares fit. The IN/GB method presents some very large errors at the beginning of the search, and a log-log plot is used to better depict this behavior. Large error values are constrained to a value of 10 for graphical clarity, and they are indicative of inf and NaN results; in these cases, the algorithm switches from the interval-Newton to the Bisection method. In this case, the roots are found by IN/GB method both at the beginning and at the end of the process of testing all generated subintervals as indicated by the values below $1 * 10^{-6}$.

## Case 3

Table 4 presents the results for Case 3. This is a particularly interesting example because it exposes some of the difficulties in interval computations. In this example, the interval-Newton method does not result in an interval, N, with the desired characteristics: (1) that N be of smaller size than the original interval, X, and (2) that N and X intersect. During the search process with the interval-Newton, this portion of the IN/GB method generates NaN or inf messages, and the program switches to the Bisection method. This point is very important because it clearly demonstrates that if for some reason the interval-Newton portion of the algorithm is incapable of dealing with the problem then the other half of the algorithm (i.e., the Bisection method) takes over and provides a solution, making the IN/GB a very efficient method.

**Table 4:** Results for Case 3

| Domain | Equations | Roots | Subset | Method |
|---|---|---|---|---|
| [29.31, 426.43] | alpha = $(v * (v + b) + b * (v - b))$ | $v = 100.0632$ | 4031 | N |
| | beta = $(v - b)$ | $v = 100.0632$ | 4031 | B |
| | $P * \text{alpha} * \text{beta} - RT * \text{alpha} + a * \text{beta} = 0$ | $v = 160.6770$ | 4181 | N |
| | $T = 301K$ | $v = 160.6770$ | 4181 | B |
| | $P = 68.18 \, \text{bar}$ | $v = 79.6437$ | 4238 | N |
| | $b = 26.6520264 \, \text{cm}^3/\text{gmol}$ | $v = 79.6437$ | 4238 | B |
| | | Total | | |
| | | | 4332 | N |
| | | | 4332 | B |

Once an interval is bisected, a range test on each subdomain created is performed, and if a root is detected (i.e., $0 \in (X)$), then that interval is stacked for further processing. This process will continue until either each subdomain is eliminated for not containing zero (i.e., $0 \notin (X)$) or a root is found within the subdomain with a prescribed error tolerance. In this example, the IN/GB method finds the solutions and scans the entire domain by dividing the domain in 4332 subsets. Consequently, the graph in Figure 5 for the Bisection and the IN/GB methods is the same.

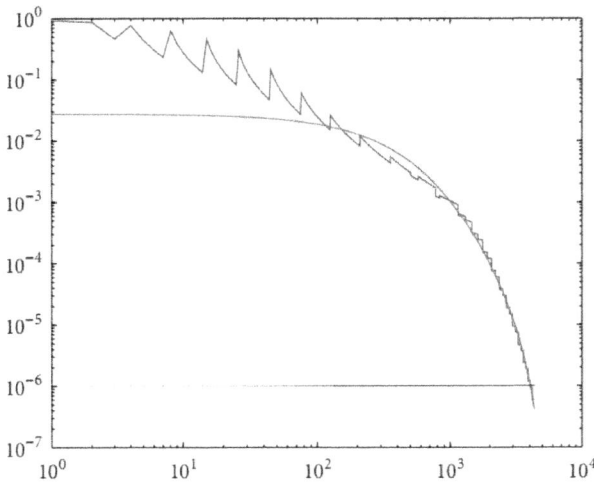

**Figure 5:** Evolution of the relative error ($y$-axis) with the subinterval number for the Bisection and IN/GB methods for Case 3.

## Case 4

Table 5 presents the results for Case 4. This is a more elaborate example because for a multicomponent mixture there is no procedure to determine a priori whether there will be none, one, or more than one critical points. The IN/GB method is much faster (Figure 6) than the Bisection method (Figure 7), and the order for finding the solutions differs between the methods. The IN/GB finds the solutions and scans the entire domain by dividing the domain in 345733 subsets, while the Bisection method finds the solutions and scans the entire domain by dividing the domain in 865970 subsets.

**Table 5:** Results for Case 4.

| Domain | Equations | Roots | Subset | Method |
|---|---|---|---|---|
| $[0.30, 1]$ $T$ | Redlich-Kwong-Soave equation of state | $T = 0.48143566$ | 168781 | N |
| $[0.275, 1]$ $v$ | Mixture of methane and hydrogen sulfide | $v = 0.56426799$ | 829620 | B |
| $[0, 1]$ $\Delta n_1$ | Find all of the critical points in the domain | $\Delta n_1 = 0.08460040$ | | |
| $[-1, 1]$ $\Delta n_2$ | $T$ scaled by $600K$ | $\Delta n_2 = -0.99641496$ | | |
| $x_1 = 0.49$ CH$_4$ | $v$ scaled by $4b$ | | | |
| $x_2 = 0.51$ H$_2$S | | | | |
| | | $T = 0.34691664$ | 237185 | N |
| | | $v = 0.29124484$ | 824978 | B |
| | | $\Delta n_1 = 0.6814699$ | | |
| | | $\Delta n_2 = -0.73189541$ | | |
| | | $T = 0.38612396$ | 285239 | N |
| | | $v = 0.36953418$ | 779696 | B |
| | | $\Delta n_1 = 0.52293204$ | | |
| | | $\Delta n_2 = -0.85237438$ | | |
| | | Total | | |
| | | | 345733 | N |
| | | | 856970 | B |

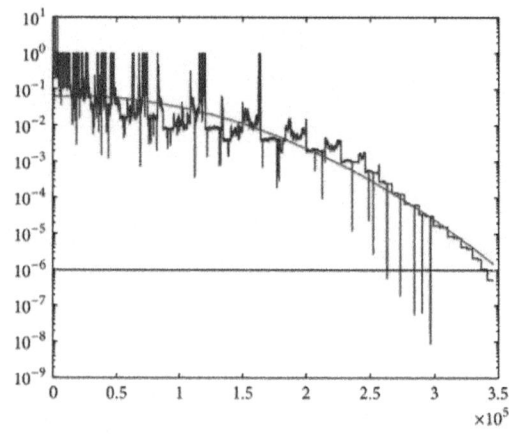

**Figure 6**: Evolution of the relative error ($y$-axis) with the subinterval number for the IN/GB method for Case 4.

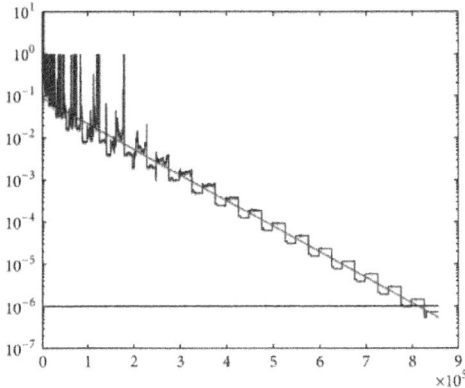

**Figure 7:** Evolution of the relative error (y-axis) with the subinterval number for the Bisection method for Case 4.

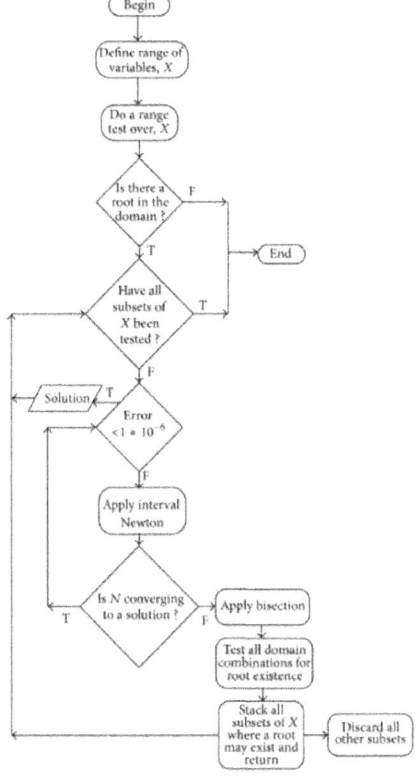

Figure 8

It is illustrative to plot the error in the IN/GB method versus the number of subdivisions of the initial domain. There are some very large errors, particularly at the beginning, which are cut to a value of 10 units, in order to preserve a reasonable scaling in the figures. As it was mentioned previously, for those cases with very large errors (inf) or not-a-number warnings (NaN), the algorithm proceeded to bisecting those domains in order to continue with the search. The relative error used to determine solutions was of $1 * 10^{-6}$, which is indicated in the graphs with a flat line.

The Bisection method converges linearly in a semilog plot toward the solutions which are found at the end of the process, as it is expected. However, the IN/GB finds solutions prior to the end of the process and with fewer subdivisions of the original interval.

## CONCLUSIONS

This paper has introduced the use of INTLAB in computations for the determination of several solutions without the need for initial guesses and with a single run of the algorithm. The IN/GB method was implemented in INTLAB with four examples. The first two examples were algebraic equations that serve as a reference to verify the effectiveness of the algorithm and describe some important findings of the IN/GB method. The third and fourth cases are related to the computation of multiple solutions that represent properties of substances and fluids. Convergence to a solution is faster with the IN/GB method in general, but there are exceptions to this rule particularly when a singular Jacobian matrix may occur.

## ACKNOWLEDGMENT

The author wishes to acknowledge the helpful comments and review provided by Professor Emmanuel Haven at the University of Leicester, UK.

## REFERENCES

1.   R. B. Kearfott, Rigorous Global Search: Continuous Problem, Kluwer Academic Publishers, Dordrecht, The Netherlands, 1996.

2.   B. Stradi and E. Haven, "Optimal investment strategy via interval arithmetic," International Journal of Theoretical and Applied Finance, vol. 8, no. 2, pp. 185–206, 2005.

3.   B. A. Stradi-Granados and E. Haven, "The use of interval arithmetic in solving a non-linear rational expectation based multiperiod output-inflation process model: the case of the IN/GB method," European Journal of Operational Research, vol. 203, no. 1, pp. 222–229, 2010.

4.   B. A. Stradi, J. F. Brennecke, P. Kohn, and M. A. Stadtherr, "Reliable computation of mixture critical points," AIChE Journal, vol. 47, no. 1, pp. 212–221, 2001.

5.   H. Gecegormez and Y. Demirel, "Phase stability analysis using interval Newton method with NRTL model," Fluid Phase Equilibria, vol. 237, no. 1-2, pp. 48–58, 2005.

6.   Z. Galias, "Proving the existence of long periodic orbits in 1D maps using interval Newton method and backward shooting," Topology and its Applications, vol. 124, no. 1, pp. 25–37, 2002.

7.   M. L. Michelsen and R. A. Heidemann, "Calculation of critical points from cubic two-constant equations of state," AIChE Journal, vol. 27, no. 2, pp. 521–523, 1981.

8.   H. Hoteit, E. Santiso, and A. Firoozabadi, "An efficient and robust algorithm for the calculation of gas-liquid critical point of multicomponent petroleum fluids," Fluid Phase Equilibria, vol. 241, no. 1-2, pp. 186–195, 2006.

9.   D. V. Nichita and S. Gomez, "Efficient and reliable mixture critical points calculation by global optimization," Fluid Phase Equilibria, vol. 291, no. 2, pp. 125–140, 2010.

10.  M. L. Michelsen, "Calculation of phase envelopes and critical points for multicomponent mixtures," Fluid Phase Equilibria, vol. 4, no. 1-2, pp. 1–10, 1980.

11.  M. L. Michelsen, "Phase equilibrium calculations. What is easy and what is difficult?" Computers and Chemical Engineering, vol. 17, no. 5-6, pp. 431–439, 1993.

12.  M. L. Michelsen and H. Kistenmacher, "On composition-dependent interaction coefficeints," Fluid Phase Equilibria, vol. 58, no. 1-2, pp. 229–230, 1990.

13.  M. L. Michelsen and R. A. Heidemann, "Calculation of tri-critical points," Fluid Phase Equilibria, vol. 39, no. 1, pp. 53–74, 1988.

14.  M. L. Michelsen, "Calculation of critical points and phase boundaries in the critical region," Fluid Phase Equilibria, vol. 16, no. 1, pp. 57–76, 1984.

15.  J. Cai, H. Liu, Y. Hu, and J. M. Prausnitz, "Critical properties of polydisperse fluid mixtures from an equation of state," Fluid Phase Equilibria, vol. 168, no. 1, pp. 91–106, 2000.

16.  T. Lindvig, L. L. Hestkjar, A. F. Hansen, M. L. Michelsen, and G. M. Kontogeorgis, "Phase equilibria for complex polymer solutions," Fluid

Phase Equilibria, vol. 194–197, pp. 663–673, 2002.

17. M. Cismondi and M. L. Michelsen, "Global phase equilibrium calculations: critical lines, critical end points and liquid-liquid-vapour equilibrium in binary mixtures," Journal of Supercritical Fluids, vol. 39, no. 3, pp. 287–295, 2007.

18. C. Carstensen and M. S. Petković, "On iteration methods without derivatives for the simultaneous determination of polynomial zeros," Journal of Computational and Applied Mathematics, vol. 45, no. 3, pp. 251–266, 1993.

19. C. D. Maranas and C. A. Floudas, "Finding all solutions of nonlinearly constrained systems of equations,"Journal of Global Optimization, vol. 7, no. 2, pp. 143–182, 1995.

20. R. B. Kearfott and M. Novoa III, "Algorithm 681 INTBIS, a portable interval Newton/bisection package,"ACM Transactions on Mathematical Software, vol. 16, no. 2, pp. 152–157, 1990.

21. S. M. Rump, "INTLAB: INTerval LABoratory," in Developments in Reliable Computing, T. Csendes, Ed., pp. 77–104, Kluwer Academic Publishers, Dodrecht, The Netherlands, 1999.

22. "Matlab R2011b (64 bit)," Mathworks, Natick, Mass, USA, 2011.

23. S. C. Chapra and R. P. Canale, Numerical Methods for Engineers, McGraw-Hill, New York, NY, USA, 6th edition, 2010.

24. S. Nakamura, Numerical Analysis and Graphical Visualization, Prentice-Hall, Upper Saddle River, NJ, USA, 2nd edition, 2001.

25. S. Attaway, MAtlab: A Practical Introduction to Programming and Problem Solving, Butterworth-Heinemann, New York, NY, USA, 2nd edition, 2011.

26. K. Ozaki, T. Ogita, S. M. Rump, and S. Oishi, "Fast algorithms for floating-point interval matrix multiplication," Journal of Computational and Applied Mathematics, vol. 236, no. 7, pp. 1795–1814, 2012

27. S. M. Rump and T. Ogita, "Super-fast validated solution of linear systems," Journal of Computational and Applied Mathematics, vol. 199, no. 2, pp. 199–206, 2007.

28. R. C. Reid, J. M. Prausnitz, and B. E. Poling, The Properties of Gases and Liquids, McGraw-Hill, New York, NY, USA, 4th edition, 1987.

29. A. Neumaier, Interval Methods for Systems of Equations, Cambridge University Press, Cambridge, UK, 1990.

30. R. Moore, Interval Analysis, Prentice-Hall, Upper Saddle River, NJ, USA, 1966.

# CITATION

## CHAPTER 1

Viliam Fedák, Tibor Balogh and Pavel Záskalický (2012). Dynamic Simulation of Electrical Machines and Drive Systems Using MATLAB GUI, MATLAB - A Fundamental Tool for Scientific Computing and Engineering Applications - Volume 1, Prof. Vasilios Katsikis (Ed.), ISBN: 978-953-51-0750-7, InTech, DOI: 10.5772/48519.

## CHAPTER 2

A. Ramsaroop and K. Kanny, "Using MATLAB to Design and Analyse Composite Lami-nates," Engineering, Vol. 2 No. 11, 2010, pp. 904-916. doi: 10.4236/eng.2010.211114.

## CHAPTER 3

Daniele Borio and Eduardo Cano (2012). Semi-Analytic Techniques for Fast MATLAB Simulations, MATLAB - A Fundamental Tool for Scientific Computing and Engineering Applications - Volume 2, Prof. Vasilios Katsikis (Ed.), ISBN: 978-953-51-0751-4, InTech, DOI: 10.5772/46470.

## CHAPTER 4

K. Wang and J. Zhou, "Kinematical Analysis and Simulation of High-Speed Plate Carrying Manipulator Based on Matlab," Engineering, Vol. 4 No. 12, 2012, pp. 850-856. doi: 10.4236/eng.2012.412108.

## CHAPTER 5

Hasan Ozturk, Zeki Kiral, Binnur Goren Kiral, Dynamic Analysis of Elastically Supported Cracked Beam Subjected to a Concentrated Moving Load, doi.org/10.1590/1679-78252195

## CHAPTER 6

M. Salah, R. Amer and M. Matbuly, "The Differential Quadrature Solution of Reaction-Diffusion Equation Using Explicit and Implicit Numerical Schemes," Applied Mathematics, Vol. 5 No. 3, 2014, pp. 327-336. doi: 10.4236/am.2014.53033.

## CHAPTER 7

Monika Žecová and Ján Terpák, "Fractional Heat Conduction Models and Thermal Diffusivity Determination," Mathematical Problems in Engineering, vol. 2015, Article ID 753936, 9 pages, 2015. doi:10.1155/2015/753936

## CHAPTER 8

Lin, C. (2014) Optimization of Bearing Locations for Maximizing First Mode Natural Frequency of Motorized Spindle-Bearing Systems Using a Genetic Algorithm. Applied Mathematics, 5, 2137-2152. doi: 10.4236/am.2014.514208.

## CHAPTER 9

M. Salah, R. Amer and M. Matbuly, "An Efficient Method to Solve Thermal Wave Equation," Applied Mathematics, Vol. 5 No. 3, 2014, pp. 542-552. doi: 10.4236/am.2014.53052.

## CHAPTER 10

Jianbin Hao and Banqiao Wang, "Parameter Sensitivity Analysis on Deformation of Composite Soil-Nailed Wall Using Artificial Neural Networks and Orthogonal Experiment," Mathematical Problems in Engineering, vol. 2014, Article ID 502362, 8 pages, 2014. doi:10.1155/2014/502362

## CHAPTER 11

Wen Zhang, Mounir Ghogho, and Baolun Yuan, "Mathematical Model and Matlab Simulation of Strapdown Inertial Navigation System," Modelling and Simulation in Engineering, vol. 2012, Article ID 264537, 25 pages, 2012. doi:10.1155/2012/264537

## CHAPTER 12

Benito A. Stradi-Granados, "Interval Arithmetic for Nonlinear Problem Solving," International Journal of Engineering Mathematics, vol. 2013, Article ID 768474, 11 pages, 2013. doi:10.1155/2013/768474

# INDEX